Topology in Collective Magnetization Dynamics

Online at: https://doi.org/10.1088/978-0-7503-5082-2

Topology in Collective Magnetization Dynamics

Zhixiong Li, Yunshan Cao and Peng Yan
School of Physics and State Key Laboratory of Electronic Thin Films and Integrated Devices, University of Electronic Science and Technology of China, Chengdu, China

IOP Publishing, Bristol, UK

ISBN 978-0-7503-5082-2 (ebook)
ISBN 978-0-7503-5080-8 (print)
ISBN 978-0-7503-5083-9 (myPrint)
ISBN 978-0-7503-5081-5 (mobi)

DOI 10.1088/978-0-7503-5082-2

Version: 20230501

IOP ebooks

British Library Cataloguing-in-Publication Data: A catalogue record for this book is available from the British Library.

Published by IOP Publishing, wholly owned by The Institute of Physics, London

IOP Publishing, No.2 The Distillery, Glassfields, Avon Street, Bristol, BS2 0GR, UK

US Office: IOP Publishing, Inc., 190 North Independence Mall West, Suite 601, Philadelphia, PA 19106, USA

Contents

Preface

Landau's theory of phase transition has shaped people's understanding of the phases of matter for a long time, and it was believed that the mechanism of phase transition was due to spontaneous symmetry breaking. Remarkably, the discovery of the quantum Hall effect in the 1980s overturned this understanding and enabled people to realize the existence of a new phase, i.e. the so-called topological phase. Topological phases can support peculiar chiral edge/surface states that have great potential for applications in quantum science and information industry. Pursuing topological phases in various materials and structures has therefore become one of the central topics in the current physical sciences and engineering.

In modern magnetism and spintronics, magnons (the quanta of spin waves) and magnetic solitons (such as the vortex, skyrmion, and domain wall) are two important excitations. In recent years, the topological insulators and semimetals emerging in magnonic and magnetic solitonic systems have attracted much attention due to their interesting physics and potential applications for designing novel spintronic devices. The aim of this book is to provide an introduction to the research status of topological phases in classical magnetism. In chapter 1 we introduce the general concept of topological insulators and semimetals. In chapter 2 we elaborate on the typical topological invariants and pedagogical models that are essential to understand the topological properties of magnons and magnetic solitons. In chapter 3 we introduce basics of magnon excitation, propagation, manipulation, and detection. In chapter 4 we focus on the topological phases of magnons, for instance, the magnon Hall effect, topological magnon insulators, Dirac (Weyl) magnon semimetals, topological magnon polarons, and higher-order topological magnons. Appealing proposals of novel topological magnonic devices are also highlighted. In chapter 5 we analyze the intrinsic dynamics and topological phases of twisted magnons. Various magnetic solitons including the domain wall, vortex, bubble, and skyrmion are discussed in chapter 6. We generalize the collective-coordinate approach to describe the dynamics of interacting magnetic solitons. In chapter 7 we focus on the topological phases of magnetic soliton crystals, by investigating theoretically the first-order topological insulators in low-dimensional systems and higher-order topological insulators in breathing lattices. The magnetic soliton topological semimetals are also introduced. In the last chapter (chapter 8) we discuss the experimental fabrication of magnetic soliton crystals and the detection of the edge modes in both magnonic and solitonic lattices. Challenges and prospects are also provided.

We hope this book is useful to researchers who want to learn about topology in magnetic systems. However, the field of topological magnetism is developing very fast, and a lot of new results are emerging. Owing to the limited knowledge of the authors, it is inevitable that there exist some shortcomings in the book. Any feedback and criticism from our readers are welcome.

Zhixiong Li, Yunshan Cao, and Peng Yan
Chengdu, China
January 2023

Acknowledgments

The authors would like to thank their students, postdocs, and colleagues Beining Zhang, Zhenyu Wang, Huanhuan Yang, Chen Wang, Lingling Song, Xiaofan Wang, Wenrui Yang, Weiwei Bao, Tianlin Yu, Yuanyuan Jiang, Huaiyang Yuan, Zhaozhuo Zeng, Xianglong Yao, Zhejunyu Jin, Xiansi Wang, Ke Xia, and Xiangrong Wang for stimulating discussions and collaborations in past years. This work was supported by the National Key Research Development Program under Contract No. 2022YFA1402802 and the National Natural Science Foundation of China (NSFC) (Grants No. 12074057, No. 11604041, and No. 11704060). Zhixiong Li acknowledges financial support from the China Postdoctoral Science Foundation (Grant No. 2019M663461) and the NSFC (Grant No. 11904048).

Author biographies

Zhixiong Li

Dr Zhixiong Li received his PhD degree in condensed matter physics from the Central South University (CSU), Changsha, China, in 2017. After postdoc training from 2018 to 2022 in the theoretical spintronics group in the University of Electronic Science and Technology of China (UESTC), Chengdu, China, he is now an assistant professor in CSU. His main research interests include the collective dynamics of magnetic solitons, topological insulators, and semimetals in magnetic systems.

Yunshan Cao

Dr Yunshan Cao obtained her PhD degree in condensed matter physics from Peking University in 2012. She is now an associate professor at the University of Electronic Science and Technology of China (UESTC). Her current interests include quantum transport in mesoscopic devices and cavity spintronics.

Peng Yan

Professor Peng Yan obtained his PhD degree in condensed matter physics from the Hong Kong University of Science and Technology (HKUST) in 2011. After postdoc training at Delft University of Technology (TUD) in the Netherlands and Mainz University in Germany, he is now a full professor at the University of Electronic Science and Technology of China (UESTC). His current interests are focused on both the topological and quantum properties of magnons and magnetic textures.

Symbols

Abbreviations

AFM	antiferromagnetic
AIAO	all-in-all-out
BLS	Brillouin light scattering
BZ	Brillouin zone
BVMSW	backward volume magnetostatic spin wave
CCW	counterclockwise
CMOS	complementary metal-oxide semiconductor
CPW	coplanar waveguide
CW	clockwise
DDI	dipole–dipole interaction
DM	Dzyakoshinskii–Moriya
DW	domain wall
FDM	frequency-division multiplexing
FFT	fast Fourier transformation
FOTI	first-order topological insulator
FVMSW	forward volume magnetostatic spin wave
HOTI	higher-order topological insulator
ISHE	inverse spin Hall effect
LA	longitudinal acoustic
LLG	Landau–Lifshitz–Gilbert
MC	magnonic crystal
MHE	magnon Hall effect
MOKE	magneto-optical Kerr effect
MSSW	magnetostatic surface spin wave
NN	nearest neighbor
NNN	next-nearest neighbor
OAM	orbital angular momentum
Py	permalloy
RKKY	Ruderman–Kittel–Kasuya–Yosida
SAM	spin angular momentum
SDM	spatial-division multiplexing
SEM	scanning electron microscopy
SHE	spin Hall effect
SOC	spin–orbit coupling
SPE	spin pumping effect
SSH	Su–Schrieffer–Heeger
STNO	spin torque nano-oscillator
STT	spin transfer torque
STXM	scanning transmission x-ray microscopy
SW	spin wave
TA	transverse acoustic
TEM	transmission electron microscope
TI	topological insulator
TKNN	Thouless–Kohmoto–Nightingale–Nijs
TMFC	twisted magnon frequency comb

TMHE	topological magnon Hall effect
TM	twisted magnon
TNLSM	topological nodal-line semimetal
TNPSM	topological nodal-point semimetal
TNSSM	topological nodal-surface semimetal
TSM	topological semimetal
WN	Weyl node
WSM	Weyl semimetal
YIG	yttrium iron garnet
1D	one-dimensional
2D	two-dimensional
3D	three-dimensional

IOP Publishing

Topology in Collective Magnetization Dynamics

Zhixiong Li, Yunshan Cao and Peng Yan

Chapter 1

Introduction

In this chapter, we give a brief introduction to topological insulators and topological semimetals. We also discuss the collective excitations in magnetic system, i.e. spin waves and magnetic solitons.

1.1 Topological insulators

In 1879 the American physicist Edwin Hall [1] found that when a longitudinal charged current (I_x) passes through a two-dimensional metal sample subject to a perpendicular magnetic field, the electrons will deflect their motion because of the Lorentz force and accumulate at the boundary of the sample, leading to the emergence of a transverse voltage, which is called the Hall effect, as shown in figure 1.1(a). About 100 years later, in 1980, Klitzing *et al* [2] discovered the quantum mechanical version of the Hall effect, i.e. the so-called integer quantum Hall effect. It was found that for a two-dimensional electron gas system at very low temperature, when the applied magnetic field is strong enough, the Hall conductance will exhibit a quantum plateau

$$\sigma_{xy} = \nu e^2/h, \tag{1.1}$$

where e is the electron charge, h is the Planck constant, and ν is an integer known as the filling factor. This effect can be understood simply as the electron circular motion around the magnetic field (see figure 1.1(b)) [3]. The radius of the cycle can be expressed as $R_n = \sqrt{\frac{\hbar}{eB}(2n + 1)}$, with $\hbar$ the reduced Plank constant, B the magnetic field strength, and n a quantum number. Because the Landau levels are discrete, the quantized Hall conductance is therefore determined by the number of the filled Landau level. When electrons are in the bulk region, they are localized or pinned in a very small area due to the strong Lorentz force. However, when electrons are close to the boundary, the situation is totally different. They bounce back from the boundary and skip forward, leading to a conducting channel along the edge, which

doi:10.1088/978-0-7503-5082-2ch1

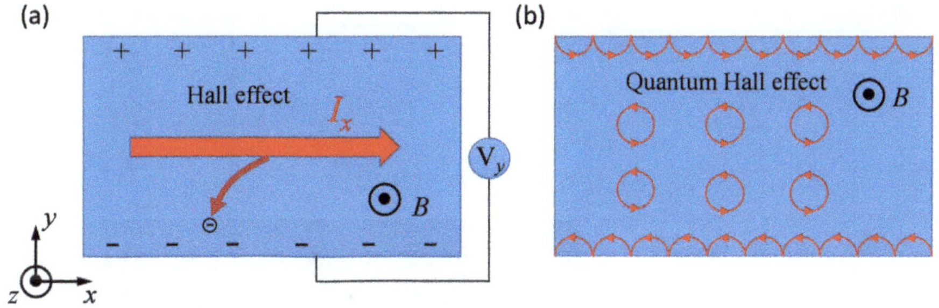

Figure 1.1. Illustrations of the (a) Hall and (b) quantum Hall effect. The magnetic fields are perpendicular to the film surface.

forms an edge state, as shown in figure 1.1(b). Each conducting channel will generate a quantum conductance e^2/h. In 1982 Thouless *et al* [4] further proved that the filling factor ν in equation (1.1) is actually a topological invariant called the Chern number. In the same year, the fractional quantum Hall effect (filling factor ν is a rational fraction) was discovered by Tsui *et al* [5], in which both the electron–electron interaction and Landau quantization play a key role.

The quantum Hall effect emerging in electron gas systems [2, 5–7] made people realize the existence of novel topological phases. These topological phases exhibit gapped band structures and can support chiral edge/surface modes which are absent in ordinary insulators and are called topological insulators (TIs). The topologically protected edge/surface modes enable the system to be robust against moderate disorder and defects, which can be used to define a resistance that only depends on fundamental physical constants. Therefore, it provides an accurate and standardized definition of the ohm. Over the last forty years, owing to the exotic properties and the potential application in topological devices, research on the TIs in various materials have become one of the central topics for physical science and engineering [8–12]. Although the TIs were initially investigated in electronic systems [8, 9, 13–15], the concept has been extended to various fields, for example, photonics [16–22], acoustics [23–29], mechanics [29–35], electric circuits [36–43], and spintronics [44–50]. Figure 1.2 shows the topological edge states reported in these systems.

In TIs there exists an important principle, i.e. the so-called bulk-boundary correspondence, which indicates that the bulk band can determine the properties of edge/surface states [8, 9, 51, 52]. Generally speaking, an ordinary n-dimensional TI can exhibit $(n - 1)$-dimensional robust edge/surface states, and these are called first-order TIs (FOTIs), as plotted in figure 1.3(a). Very recently, the concept of TIs has been extended to higher-order cases, i.e. the so-called higher-order topological insulators (HOTIs) [53–59]. A kth-order TI can support $(n - k)$-dimensional topological edge states $(2 \leqslant k \leqslant n)$, for instance, hinge states and corner states [60]. Schematic diagrams summarizing the current classification of HOTIs are provided in figures 1.3(b) and (c). Because of the exotic features, in the last few years HOTIs have been studied widely across various academic disciplines, such as photonics [61–72], acoustics [73–87], mechanics [88–90], electronics [91–99], and magnetics [100–105]. To realize HOTIs, one of the most popular methods is to construct a so-called 'breathing' lattice. Figure 1.4

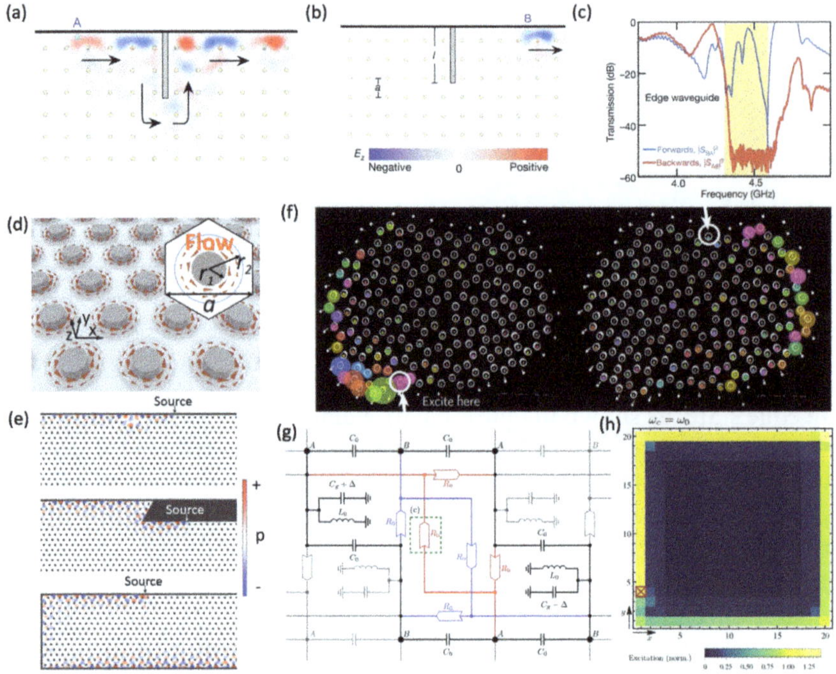

Figure 1.2. (a) and (b) The chiral edge state in a photonic lattice (theoretical calculations). The red and blue regions denote the out-of-plane component of the electric field. The gray region represents a large obstacle which acts as a defect. The black arrows show the direction of the power flow. A and B are the source and feed antennae, respectively. (c) The experimental detection of the one-way topological edge state shown in (a) and (b). The red and blue lines plot the transmission from antenna A to antenna B and antenna B to antenna A, respectively. The yellow region denotes the frequency range of the bandgap. (d) An illustration of a two-dimensional triangular acoustic topological insulator, with a the lattice constant, and r_1 and r_2 the geometric parameters. (e) Demonstration of the robustness of acoustic one-way edge states against disorder based on the model in (d). (f) An edge state propagates clockwise in an amorphous experimental gyroscopic network with white arrows representing the positions of the excitation. (g) A topolectrical Chern circuit with various circuit elements. (h) The chiral voltage propagation for the edge state in topolectrical circuits. ((a), (b), (c) Reproduced with permission from [16]. Copyright 2009 Macmillan. (d), (e) Reproduced with permission from [24]. Copyright 2015 The American Physical Society. (f) Reproduced with permission from [33]. Copyright 2018 the Authors. (g), (h) Reproduced with permission from [39]. Copyright 2019 The American Physical Society.)

shows models of two-dimensional HOTIs in different systems, in which the corresponding modes with oscillations highly confined to corners are also presented. Noticeably, the conventional bulk-boundary correspondence is not applicable to HOTIs. Therefore, new topological invariants, such as bulk polarization (Wannier center) [55, 75, 106, 107], Green's function zeros [108], and the Z_N Berry phase [109–114], should be taken into account to describe the topological properties of HOTIs. The appearance of HOTIs has thus broaden the category of TIs. More importantly, the higher-order topological edge modes (such as corner states) emerging in HOTIs have various potential applications in spintronic imaging [82, 103], topologically protected nano-oscillators, etc. We will discuss this point in following chapters.

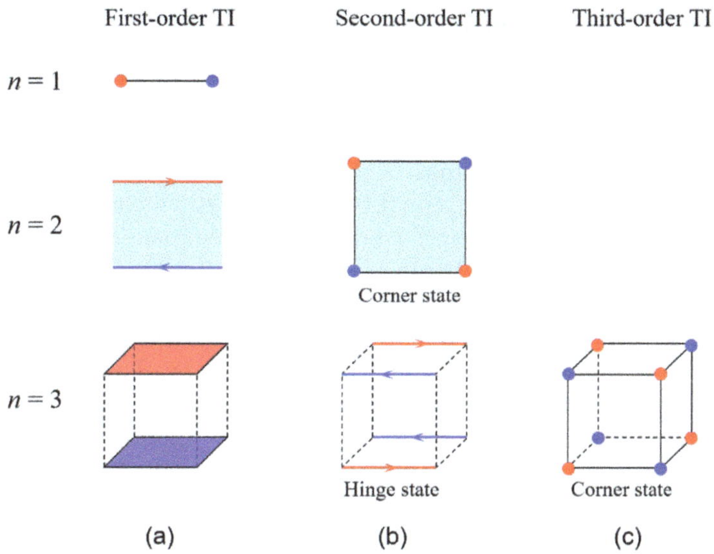

Figure 1.3. Schematic plots for different TIs. (a) The first-order TIs with $(n - 1)$-dimensional edge/surface states. (b) The second-order TIs with $(n - 2)$-dimensional edge states (corner and hinge states). (c) The third-order TI with $(n - 3)$-dimensional edge states (corner states). (Reproduced with permission from [60]. Copyright 2021 Elsevier.)

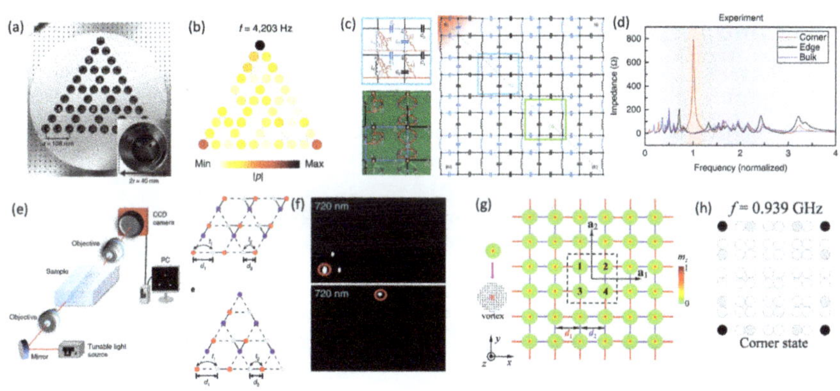

Figure 1.4. (a) Photograph of the fabricated breathing triangle-shaped finite acoustic structure. (b) Measured acoustic pressure distribution in the non-trivial phase at a typical frequency for the corner state. (c) Topolectrical circuits for a breathing square lattice. Here the blue and black circuit elements correspond to weak and strong bonds, respectively. (d) Frequency scan of the impedance between two nearest-neighbor sites at the corner, at the edge and in the bulk. (e) Experimental set-up for detecting a photonic topological corner state. Lattice structures with rhombic and triangular geometry are also shown. t_1 and t_2 denote alternating bonds. (f) Light output from a rhombic (top) and triangular (bottom) lattice with the system in the HOTI phase. (g) A sketch map of a two-dimensional breathing square lattice based on magnetic vortices. (h) The oscillation distribution of vortex gyrotropic motion for the topological corner state. ((a), (b) Reproduced with permission from [75]. Copyright 2019 the Authors. (c), (d) Reproduced with permission from [92]. Copyright 2018 the Authors. (e), (f) Reproduced with permission from [63]. Copyright 2019 the Authors. (g), (h) Reproduced with permission from [103]. Copyright 2020 The American Physical Society.)

1.2 Topological semimetals

In addition to the TIs with gapped band structures, the topological semimetals (TSMs) [115] represent another important topological non-trivial phase with gapless dispersion relations. The TSMs can be divided into three groups according to the dimensionality of the band crossings.

The first group has zero-dimensional band crossings which manifest as nodes or nodal points. The typical topological nodal-point semimetals (TNPSMs) include Weyl semimetals [116–118] and Dirac semimetals [119–122]. The Weyl semimetals are described by twofold degenerate points called Weyl points or nodes in band structure, where the degenerate points originate from the linear crossings of the bands. The Weyl points (or nodes) are characterized by the monopoles of Berry curvature in momentum space and they must emerge in pairs with opposite topological charge because of the no-go theory. Due to the existence of band inversion between two paired Weyl nodes, topologically robust Fermi-arc-like surface states therefore form. Noticeably, only when the time-reversal symmetry or inversion symmetry or both of them are broken, can the Weyl semimetal phase appear. In comparison, the Dirac semimetals are described by the Dirac points with fourfold degenerate crossings. Moreover, the Dirac semimetals hold both the inversion and time-reversal symmetries. However, their topological stability needs to resort to extra crystalline symmetries, such as rotation symmetry [121, 122]. Generally speaking, the Dirac and Weyl semimetals respect the Lorentz invariance and the Dirac and Weyl cones are isotropic with linear band dispersion in all momentum directions. However, if the Hamiltonian of system includes a tilting term, the Weyl (or Dirac) cone will become tilted and the Lorentz invariance is violated. Depending on the strength of the tilting term, the Weyl (Dirac) semimetals can be further divided into type-I, type-II, and type-III. The simplest 2×2 Weyl Hamiltonian reads [123]

$$H = \pm\nu_0\left(\sigma_x k_x + \sigma_y k_y + \sigma_z k_z\right) + tk_z I_0, \tag{1.2}$$

where ν_0 is the Fermi velocity, k_j ($j = x, y, z$) is the j-component of wave vector, σ_x, σ_y, and σ_z are the 2×2 Pauli matrices and $tk_z I_0$ is the tilting term with I_0 the identity matrix and t the strength of the tilting. Solving equation (1.2) we obtain the energy spectrum

$$E_\pm = \pm\nu_0(k_x + k_y) \pm \nu_0\left(1 \pm \frac{t}{\nu_0}\right)k_z. \tag{1.3}$$

If $t < \nu_0$ the two linearly Weyl bands always have opposite Fermi velocities along any k-direction [124]. In this case, although the cone is tilted, the physical properties are still similar to the standard Weyl cone and the corresponding Fermi surface reduces to a point, which is the typical feature of type-I Weyl semimetals (see figures 1.5(a) and (d)). However, if $t > \nu_0$, the two linear Weyl bands have the same sign of Fermi velocities along all k-directions. The Weyl cone is seriously tilted and the Fermi surface becomes open electron and hole pockets, and these are called

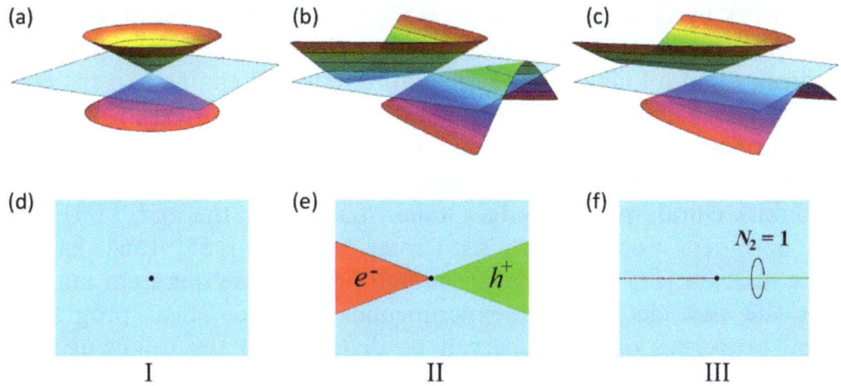

Figure 1.5. Schematic plots of the band structures of (a) type-I, (b) type-II, and (c) type-III Weyl (Dirac) semimetals. (d)–(f) The corresponding Fermi surfaces of three types of Weyl (Dirac) semimetals. (Reproduced with permission from [124]. Copyright 2018 the American Physical Society.)

type-II Weyl semimetals (see figures 1.5(b) and (e)). When $t = v_0$ one of the two bands would have a zero Fermi velocity along the k_z-direction. The Fermi surface becomes a line. This phase is called a type-III Weyl semimetal (see figures 1.5(c) and (f)). For Dirac semimetals, we can obtain similar results. Very recently, it has been proved that a Weyl semimetal can support higher-order topological edge modes (hinge states), which is called a higher-order Weyl semimetal [125–128].

The second group of TSMs has a one-dimensional band-crossing line in the momentum space, and is generally called a topological nodal-line semimetal (TNLSM) [129–132]. Similar to TNPSMs, the TNLSMs also exhibit surface states, i.e. the so-called drumhead surface states. However, because there are no symmetries protecting the nodal lines, the drumhead surface states are not topologically protected and can be destroyed by small perturbations. Lastly, the third group of TSM has a two-dimensional band-crossing surface, and is called a topological nodal-surface semimetal (TNSSM) [133–135]. To date, various TSMs have been extensively investigated in different physical systems due to their exotic characteristics and potential applications [136–142]. The current book is thus necessary to cover the status of this exciting field.

1.3 Collective excitations in magnets

In magnetic systems, the spin wave (SW) [143–146] characterizes the collective oscillation of magnetic moments, which represents one of the most important elementary excitations. The quantized quasiparticle of SW is called magnon. Remarkably, magnons can carry, transmit, and process information [147–152], which is similar to charged currents but without Joule heating. Utilizing magnons, one can design various spintronic devices. More importantly, the magnonic devices have many exciting advantages compared to ordinary electronic devices, which will be discussed in detail in chapter 3. Although conventional magnons have been studied extensively, the topological phases of magnons were rarely reported until the pioneering work by Onose *et al* in 2010 [153]—the experimental observation of the magnon Hall effect (MHE) in insulating pyrochlore ferromagnet $Lu_2V_2O_7$. The MHE indicates that a longitudinal temperature gradient can lead to the emergence of transverse thermal currents. To explain the origin of the MHE, Matsumoto *et al* [49, 154] proposed a model of magnon edge currents. Then the concept of 'topological magnon insulator' was used [44, 45, 47] to characterize the system which exhibits chiral magnon edge states. Later, the thermal MHE was also observed in other systems, such as garnet magnets [155, 156], kagome and pyrochlore magnets [157, 158], and frustrated pryocholore quantum magnets [159] etc. Over the past decade, the experimental and theoretical progress on the topological properties of magnons can be divided into the following directions: (i) topological magnon insulators in different lattices, for example, kagome (or pyrochlore) [46, 160–167] and honeycomb [168–173] lattices; (ii) topological magnons originating from different mechanisms, for instance, the Dzyaloshinskii–Moriya (DM) interaction [174–178] because of the inversion symmetry broken [179, 180], magnetostatic interaction [181–183], pseudodipolar exchange interaction [184, 185] and magnetic spin texture [186–191] etc; (iii) Dirac and Weyl magnons [169, 172, 192–202]; (iv) topological magnon polarons due to the coupling between spin and lattice [203–207]; (v) higher-order topological magnon insulators [100, 208–210]; (vi) topological magnon insulators in antiferromagnets and ferrimagnets [211–216]; (vii) topological magnons from the Bogoliubov–de Gennes perspective [217, 218]; and (viii) exotic topological magnonic devices [185, 219], for example, magnon diodes, magnon beam splitters, magnon interferometers, and magnon logic gates etc. Notably, magnons carrying orbital angular momentum, i.e. the twisted magnons (TMs), have attracted growing attention recently [220–222]. The topological features of TMs will be presented in chapter 5.

In magnetic materials, in addition to the magnon, the magnetic soliton is another important excitation. Magnetic solitons [223] are shape-preserving and self-localized spin textures with typical examples including the magnetic vortex [224, 225], bubble [226–228], skyrmion [229–231], and domain wall [31, 56, 232] etc. These magnetic solitons have potential applications for designing new spintronic devices [233–240] owing to their outstanding characteristics, such as small size, easy manipulation, and high stability. In recent years, magnetic solitons have become one of the central topics in condensed matter physics. The collective oscillation of magnetic soliton arrays shows the feature of waves [241–255], which is similar to other (quasi-) particles. Remarkably, by mapping the massless Thiele's equation into the Haldane model [256], Kim *et al* [257] showed that magnetic vortices or bubbles arranged as a two-dimensional honeycomb lattice can exhibit topological edge modes, which was subsequently verified by micromagnetic simulations [258]. In addition, Li *et al* [259] and Go *et al* [260] investigated the Su–Schrieffer–Heeger (SSH) model [261] based on one-dimensional magnetic soliton arrays and found localized edge modes with topological origins. Very recently, Li *et al* predicted second-order TIs in two-dimensional breathing kagome [101], honeycomb [102], and square [103] lattices and the third-order TIs in a three-dimensional breathing square lattice [104] consisting of magnetic vortices. In addition, it has been confirmed that the emerging corner states are topologically robust and can be immune to moderate disorder and defects due to the generalized chiral symmetry. Topological Weyl semimetal phases of magnetic soliton lattices have been reported as well [262]. The collective dynamics of a magnetic soliton lattice is dominated by the generalized Thiele's equation, which leads to a wavelike equation and it is different from the corresponding wave equations for electronic, photonic, and acoustic systems in the following respects: (i) the nonzero topological charge of a magnetic soliton generates a gyration term which models an effective magnetic field acting on a quasiparticle. The time-reversal symmetry is therefore broken. (ii) The inertial effect of a magnetic soliton is described by a second-order mass term. The high-frequency motion of a magnetic soliton is characterized by a third-order non-Newtonian gyration term. (iii) The coupling between magnetic solitons is anisotropic. (iv) The ordinary chiral symmetry based on a bipartite lattice should be replaced by generalized chiral symmetry.

In this book, we will give a thorough introduction to both the insulating and metallic states of topological magnons and magnetic solitons. The typical topological invariants and pedagogical topological models are discussed in chapter 2; chapter 3 focuses on the main content of magnonics which include the generation, propagation, manipulation, and detection of magnons. A summary of topological magnons is given in chapter 4. Twisted magnons are discussed in chapter 5. Chapter 6 gives a review of various magnetic solitons and properties of collective dynamics of a magnetic soliton lattice. Novel topological phases emerging in magnetic soliton lattices are introduced in chapter 7. In chapter 8 we present the future prospects and challenges for topology in dynamic magnetism.

References

[1] Hall E H 1879 *Am. J. Math.* **2** 287–92
[2] Klitzing K V, Dorda G and Pepper M 1980 *Phys. Rev. Lett.* **45** 494–7
[3] Laughlin R B 1981 *Phys. Rev.* B **23** 5632(R)
[4] Thouless D J *et al* 1982 *Phys. Rev. Lett.* **49** 405
[5] Tsui D C, Stormer H L and Gossard A C 1982 *Phys. Rev. Lett.* **48** 1559
[6] Klitzing K V 1986 *Rev. Mod. Phys.* **58** 519
[7] Stormer H L, Tsui D C and Gossard A C 1999 *Rev. Mod. Phys.* **71** S298
[8] Hasan M Z and Kane C L 2010 *Rev. Mod. Phys.* **82** 3045
[9] Qi X L and Zhang S C 2011 *Rev. Mod. Phys.* **83** 1057
[10] Xiao D, Chang M C and Niu Q 2010 *Rev. Mod. Phys.* **82** 1959
[11] Fu L, Kane C L and Mele E J 2007 *Phys. Rev. Lett.* **98** 106803
[12] Hsieh D *et al* 2009 *Nature* **460** 1101–5
[13] Moore J E 2010 *Nature* **464** 194–8
[14] Kane C L and Mele E J 2005 *Phys. Rev. Lett.* **95** 226801
[15] Kane C L and Mele E J 2005 *Phys. Rev. Lett.* **95** 146802
[16] Wang Z *et al* 2009 *Nature* **461** 772–5
[17] Khanikaev A B *et al* 2013 *Nat. Mater.* **12** 233–9
[18] Ozawa T *et al* 2019 *Rev. Mod. Phys.* **91** 015006
[19] Lu L, Joannopoulos J D and Soljačić M 2014 *Nat. Photon.* **8** 821–9
[20] Khanikaev A B and Shvets G 2017 *Nat. Photon.* **11** 763–73

[21] Bandres M A *et al* 2018 *Science* **359** eaar4005
[22] Harari G *et al* 2018 *Science* **359** eaar4003
[23] Zhang X *et al* 2018 *Commun. Phys.* **1** 97
[24] Yang Z *et al* 2015 *Phys. Rev. Lett.* **114** 114301
[25] He C *et al* 2016 *Nat. Phys.* **12** 1124–9
[26] Fleury R, Khanikaev A B and Alù A 2016 *Nat. Commun.* **7** 11744
[27] Chen Z G and Wu Y 2016 *Phys. Rev. Appl.* **5** 054021
[28] He C *et al* 2019 *Phys. Rev. Lett.* **123** 195503
[29] Ma G, Xiao M and Chan C T 2019 *Nat. Rev. Phys.* **1** 281–94
[30] Huber S D 2016 *Nat. Phys.* **12** 621–3
[31] Paulose J, Chen B G and Vitelli V 2015 *Nat. Phys.* **11** 153–6
[32] Nash L M *et al* 2015 *Proc. Natl. Acad. Sci. USA* **112** 14495–500
[33] Mitchell N P *et al* 2018 *Nat. Phys.* **14** 380–5
[34] Patil V P *et al* 2020 *Science* **367** 71–5
[35] Chen B G *et al* 2016 *Phys. Rev. Lett.* **116** 135501
[36] Lee C H *et al* 2018 *Commun. Phys.* **1** 39
[37] Helbig T *et al* 2020 *Nat. Phys.* **16** 747–50
[38] Zhang Z Q *et al* 2019 *Phys. Rev.* B **100** 184202
[39] Hofmann T *et al* 2019 *Phys. Rev. Lett.* **122** 247702
[40] Liu S *et al* 2020 *Phys. Rev. Appl.* **13** 014047
[41] Hofmann T *et al* 2020 *Phys. Rev. Res.* **2** 023265
[42] Wang Y *et al* 2020 *Nat. Commun.* **11** 2356
[43] Yu R, Zhao Y X and Schnyder A P 2020 *Nat. Sci. Rev.* **7** 1288–95
[44] Zhang L *et al* 2013 *Phys. Rev.* B **87** 144101
[45] Mook A, Henk J and Mertig I 2014 *Phys. Rev.* B **90** 024412
[46] Chisnell R *et al* 2015 *Phys. Rev. Lett.* **115** 147201
[47] Rückriegel A, Brataas A and Duine R A 2018 *Phys. Rev.* B **97** 081106(R)
[48] Chernyshev A L and Maksimov P A 2016 *Phys. Rev. Lett.* **117** 187203
[49] Matsumoto R and Murakami S 2011 *Phys. Rev. Lett.* **106** 197202
[50] Malz D, Knolle J and Nunnenkamp A 2019 *Nat. Commun.* **10** 3937
[51] Hatsugai Y 1993 *Phys. Rev. Lett.* **71** 3697
[52] Hatsugai Y 1993 *Phys. Rev.* B **48** 11851
[53] Benalcazar W A, Bernevig B A and Hughes T L 2017 *Science* **357** 61–6
[54] Benalcazar W A, Bernevig B A and Hughes T L 2017 *Phys. Rev.* B **96** 245115
[55] Ezawa M 2018 *Phys. Rev. Lett.* **120** 026801
[56] Song Z, Fang Z and Fang C 2017 *Phys. Rev. Lett.* **119** 246402
[57] Langbehn J *et al* 2017 *Phys. Rev. Lett.* **119** 246401
[58] Schindler F *et al* 2018 *Sci. Adv.* **4** eaat0346
[59] Queiroz R and Stern A 2019 *Phys. Rev. Lett.* **123** 036802
[60] Li Z X, Cao Y S and Yan P 2021 *Phys. Rep.* **915** 1–64
[61] Xie B Y *et al* 2018 *Phys. Rev.* B **98** 205147
[62] Noh J *et al* 2018 *Nat. Photon.* **12** 408–15
[63] Hassan A E *et al* 2019 *Nat. Photon.* **13** 697–700
[64] Mittal S *et al* 2019 *Nat. Photon.* **13** 692–6
[65] Chen X D *et al* 2019 *Phys. Rev. Lett.* **122** 233902
[66] Xie B Y *et al* 2019 *Phys. Rev. Lett.* **122** 233903

[67] Ota Y *et al* 2019 *Optica* **6** 786–9
[68] Li M *et al* 2020 *Nat. Photon.* **14** 89–94
[69] Kim M, Jacob Z and Rho J 2020 *Light. Sci. Appl.* **9** 130
[70] He L *et al* 2020 *Nat. Commun.* **11** 3119
[71] Zhang L *et al* 2020 *Adv. Sci.* **7** 1902724
[72] Kim M and Rho J 2020 *Nanophotonics* **9** 3227–34
[73] Serra-Garcia M *et al* 2018 *Nature* **555** 342–5
[74] Liu Z 2019 *Nat. Mater.* **18** 98–9
[75] Xue H *et al* 2019 *Nat. Mater.* **18** 108–12
[76] Ni X *et al* 2019 *Nat. Mater.* **18** 113–20
[77] Xue H *et al* 2019 *Phys. Rev. Lett.* **122** 244301
[78] Zhang X *et al* 2019 *Nat. Phys.* **15** 582–8
[79] Zhang X *et al* 2019 *Nat. Commun.* **10** 5331
[80] Chen Z G *et al* 2019 *Phys. Rev.* B **100** 075120
[81] Zhang Z *et al* 2019 *Phys. Rev. Lett.* **122** 195501
[82] Zhang Z *et al* 2019 *Adv. Mater.* **31** 1904682
[83] López M R *et al* 2019 *Commun. Phys.* **2** 132
[84] Qi Y *et al* 2020 *Phys. Rev. Lett.* **124** 206601
[85] Xue H *et al* 2020 *Nat. Commun.* **11** 2442
[86] Ni X *et al* 2020 *Nat. Commun.* **11** 2108
[87] Weiner M *et al* 2020 *Sci. Adv.* **6** eaay4166
[88] Fan H *et al* 2019 *Phys. Rev. Lett.* **122** 204301
[89] Wakao H *et al* 2020 *Phys. Rev.* B **101** 094107
[90] Wu Q *et al* 2020 *Phys. Rev. Appl.* **14** 014084
[91] Peterson C W *et al* 2018 *Nature* **555** 346–50
[92] Imhof S *et al* 2018 *Nat. Phys.* **14** 925–9
[93] Lu L 2018 *Nat. Phys.* **14** 875–7
[94] Ezawa M 2018 *Phys. Rev.* B **98** 201402(R)
[95] Serra-Garcia M, Süsstrunk R and Huber S D 2019 *Phys. Rev.* B **99** 020304(R)
[96] Bao J *et al* 2019 *Phys. Rev.* B **100** 201406(R)
[97] Yang H *et al* 2020 *Phys. Rev. Res.* **2** 022028(R)
[98] Song L *et al* 2020 *Nano Lett.* **20** 7566–71
[99] Ni X *et al* 2020 *Phys. Rev. Appl.* **13** 064031
[100] Sil A and Ghosh A K 2020 *J. Phys.: Condens. Matter* **32** 205601
[101] Li Z X *et al* 2019 *npj Comput. Mater.* **5** 107
[102] Li Z X *et al* 2020 *Phys. Rev. Appl.* **13** 064058
[103] Li Z X *et al* 2020 *Phys. Rev.* B **101** 184404
[104] Li Z X *et al* 2021 *Phys. Rev.* B **103** 214442
[105] Li Z X, Cao Y S and Yan P 2021 *Topics in Applied Physics* vol 138 (Cham: Springer) pp 405–40
[106] King-Smith R D and Vanderbilt D 1993 *Phys. Rev.* B **47** 1651(R)
[107] Vanderbilt D and King-Smith R D 1993 *Phys. Rev.* B **48** 4442
[108] Slager R J *et al* 2015 *Phys. Rev.* B **92** 085126
[109] Zak J 1989 *Phys. Rev. Lett.* **62** 2747
[110] Kariyado T, Morimoto T and Hatsugai Y 2018 *Phys. Rev. Lett.* **120** 247202
[111] Hatsugai Y and Maruyama I 2011 *Europhys. Lett.* **95** 20003

[112] Mizoguchi T, Araki H and Hatsugai Y 2019 *J. Phys. Soc. Jpn.* **88** 104703
[113] Araki H, Mizoguchi T and Hatsugai Y 2020 *Phys. Rev. Res.* **2** 012009(R)
[114] Kudo K, Yoshida T and Hatsugai Y 2019 *Phys. Rev. Lett.* **123** 196402
[115] Armitage N P, Mele E J and Vishwanath A 2018 *Rev. Mod. Phys.* **90** 015001
[116] Wan X *et al* 2011 *Phys. Rev.* B **83** 205101
[117] Lv B Q *et al* 2015 *Phys. Rev.* X **5** 031013
[118] Liu D F *et al* 2019 *Science* **365** 1282–5
[119] Young S M *et al* 2012 *Phys. Rev. Lett.* **108** 140405
[120] Borisenko S *et al* 2014 *Phys. Rev. Lett.* **113** 027603
[121] Liu Z K *et al* 2014 *Science* **343** 864–7
[122] Liu Z K *et al* 2014 *Nat. Mater.* **13** 677–81
[123] Lv B Q, Qian T and Ding H 2021 *Rev. Mod. Phys.* **93** 025002
[124] Huang H, Jin K H and Liu F 2018 *Phys. Rev.* B **98** 121110(R)
[125] Wang H X *et al* 2020 *Phys. Rev. Lett.* **125** 146401
[126] Ghorashi S A A, Li T and Hughes T L 2020 *Phys. Rev. Lett.* **125** 266804
[127] Wei Q *et al* 2021 *Nat. Mater.* **20** 812–7
[128] Song L *et al* 2022 *Nat. Commun.* **13** 5601
[129] Burkov A A, Hook M D and Balents L 2011 *Phys. Rev.* B **84** 235126
[130] Chiu C K and Schnyder A P 2014 *Phys. Rev.* B **90** 205136
[131] Fang C *et al* 2015 *Phys. Rev.* B **92** 081201(R)
[132] Weng H *et al* 2015 *Phys. Rev.* B **92** 045108
[133] Liang Q F *et al* 2016 *Phys. Rev.* B **93** 085427
[134] Zhong C *et al* 2016 *Nanoscale* **8** 7232–9
[135] Fu B B *et al* 2019 *Sci. Adv.* **5** eaau6459
[136] Gao H *et al* 2018 *Phys. Rev. Lett.* **121** 106404
[137] Burkov A A and Balents L 2011 *Phys. Rev. Lett.* **107** 127205
[138] Xu S Y *et al* 2015 *Science* **349** 613–7
[139] Huang S M *et al* 2015 *Nat. Commun.* **6** 7373
[140] Deng K *et al* 2016 *Nat. Phys.* **12** 1105–10
[141] Xiong J *et al* 2015 *Science* **350** 413–6
[142] Wieder B J *et al* 2016 *Phys. Rev. Lett.* **116** 186402
[143] Bloch F 1930 *Z. Phys.* **61** 206–19
[144] Kubo R 1952 *Phys. Rev.* **87** 568
[145] Dyson F J 1956 *Phys. Rev.* **102** 1217
[146] Prabhakar A and Stancil D D 2009 *Spin Waves Theory and Applications* (Berlin: Springer) https://doi.org/10.1007/978-0-387-77865-5
[147] Khitun A, Bao M and Wang K L 2010 *J. Phys. D: Appl. Phys.* **43** 264005
[148] Chumak A V, Serga A A and Hillebrands B 2014 *Nat. Commun.* **5** 4700
[149] Kruglyak V V, Demokritov S O and Grundler D 2010 *J. Phys. D: Appl. Phys.* **43** 264001
[150] Chumak A V *et al* 2015 *Nat. Phys.* **11** 453–61
[151] Lenk B *et al* 2011 *Phys. Rep.* **507** 107–36
[152] Serga A A, Chumak A V and Hillebrands B 2010 *J. Phys. D: Appl. Phys.* **43** 264002
[153] Onose Y *et al* 2010 *Science* **329** 297–9
[154] Matsumoto R and Murakami S 2011 *Phys. Rev.* B **84** 184406
[155] Madon B *et al* 2014 arXiv: 1412.3723
[156] Tanabe K *et al* 2016 *Phys. Status Solidi* B **253** 783–7

[157] Hirschberger M *et al* 2015 *Phys. Rev. Lett.* **115** 106603
[158] Ideue T *et al* 2012 *Phys. Rev.* B **85** 134411
[159] Hirschberger M *et al* 2015 *Science* **348** 106–9
[160] Mook A, Henk J and Mertig I 2014 *Phys. Rev.* B **89** 134409
[161] Mook A, Henk J and Mertig I 2015 *Phys. Rev.* B **91** 174409
[162] Mook A, Henk J and Mertig I 2015 *Phys. Rev.* B **91** 224411
[163] Pereiro M *et al* 2014 *Nat. Commun.* **5** 4815
[164] Laurell P and Fiete G A 2017 *Phys. Rev. Lett.* **118** 177201
[165] Chisnell R *et al* 2016 *Phys. Rev.* B **93** 214403
[166] Seshadri R and Sen D 2018 *Phys. Rev.* B **97** 134411
[167] Owerre S A 2018 *J. Phys.: Condens. Matter* **30** 245803
[168] Kim S K *et al* 2016 *Phys. Rev. Lett.* **117** 227201
[169] Chen L *et al* 2018 *Phys. Rev.* X **8** 041028
[170] McClarty P A *et al* 2018 *Phys. Rev.* B **98** 060404(R)
[171] Pantaleón P A, Carrillo-Bastos R and Xian Y 2019 *J. Phys.: Condens. Matter* **31** 085802
[172] Owerre S A 2016 *J. Phys.: Condens. Matter* **28** 386001
[173] Owerre S A 2016 *J. Appl. Phys.* **120** 043903
[174] Li B and Kovalev A A 2018 *Phys. Rev.* B **97** 174413
[175] Moulsdale C *et al* 2019 *Phys. Rev.* B **99** 214424
[176] Malki M and Uhrig G S 2019 *Phys. Rev.* B **99** 174412
[177] Iacocca E and Heinonen O 2017 *Phys. Rev. Appl.* **8** 034015
[178] Owerre S A 2017 *J. Phys.: Condens. Matter* **29** 185801
[179] Dzyaloshinsky I 1958 *J. Phys. Chem. Solids* **4** 241–55
[180] Moriya T 1960 *Phys. Rev. Lett.* **4** 228
[181] Shindou R *et al* 2013 *Phys. Rev.* B **87** 174402
[182] Lisenkov I *et al* 2014 *Phys. Rev.* B **90** 104417
[183] Pirmoradian F *et al* 2018 *Phys. Rev.* B **98** 224409
[184] Wang X S, Su Y and Wang X R 2017 *Phys. Rev.* B **95** 014435
[185] Wang X S, Zhang H W and Wang X R 2018 *Phys. Rev. Appl.* **9** 024029
[186] Hoogdalem K A V, Tserkovnyak Y and Loss D 2013 *Phys. Rev.* B **87** 024402
[187] Mochizuki M *et al* 2014 *Nat. Mater.* **13** 241–6
[188] Roldán-Molina A, Nunez A S and Fernández-Rossier J 2016 *New J. Phys.* **18** 045015
[189] Mook A *et al* 2017 *Phys. Rev.* B **95** 020401(R)
[190] Garst M, Waizner J and Grundler D 2017 *J. Phys. D: Appl. Phys.* **50** 293002
[191] Díaz S A *et al* 2020 *Phys. Rev. Res.* **2** 013231
[192] Fransson J, Black-Schaffer A M and Balatsky A V 2016 *Phys. Rev.* B **94** 075401
[193] Boyko D, Balatsky A V and Haraldsen J T 2018 *Phys. Rev.* B **97** 014433
[194] Pershoguba S S *et al* 2018 *Phys. Rev.* X **8** 011010
[195] Costa A T *et al* 2020 *2D Mater.* **7** 045031
[196] Yuan B *et al* 2020 *Phys. Rev.* X **10** 011062
[197] Mook A, Henk J and Mertig I 2016 *Phys. Rev. Lett.* **117** 157204
[198] Su Y, Wang X S and Wang X R 2017 *Phys. Rev.* B **95** 224403
[199] Su Y and Wang X R 2017 *Phys. Rev.* B **96** 104437
[200] Zyuzin V A and Kovalev A A 2018 *Phys. Rev.* B **97** 174407
[201] Owerre S A 2018 *Sci. Rep.* **8** 10098
[202] Li F Y *et al* 2016 *Nat. Commun.* **7** 12691

[203] Takahashi R and Nagaosa N 2016 *Phys. Rev. Lett.* **117** 217205
[204] Zhang X *et al* 2019 *Phys. Rev. Lett.* **123** 167202
[205] Park S and Yang B J 2019 *Phys. Rev.* B **99** 174435
[206] Go G, Kim S K and Lee K J 2019 *Phys. Rev. Lett.* **123** 237207
[207] Zhang S *et al* 2020 *Phys. Rev. Lett.* **124** 147204
[208] Hirosawa T *et al* 2020 *Phys. Rev. Lett.* **125** 207204
[209] Mook A *et al* 2021 *Phys. Rev.* B **104** 024406
[210] Park M J, Lee S B and Kim Y B 2021 *Phys. Rev.* B **104** L060401
[211] Owerre S A 2017 *J. Appl. Phys.* **121** 223904
[212] Nakata K *et al* 2017 *Phys. Rev.* B **96** 224414
[213] Yao W *et al* 2018 *Nat. Phys.* **14** 1011–5
[214] Kim K S *et al* 2019 *Phys. Rev.* B **100** 064412
[215] Kawano M and Hotta C 2019 *Phys. Rev.* **B 99** 054422
[216] Li Y M, Xiao J and Chang K 2018 *Nano Lett.* **18** 3032–7
[217] Akagi Y 2020 *J. Phys. Soc. Jpn.* **89** 123601
[218] Kondo H, Akagi Y and Katsura H 2020 *Prog. Theor. Exp. Phys.* **2020** 12A104
[219] Shindou R *et al* 2013 *Phys. Rev.* B **87** 174427
[220] Jiang Y Y *et al* 2020 *Phys. Rev. Lett.* **124** 217204
[221] Jia C *et al* 2019 *Nat. Commun.* **10** 2077
[222] Li Z X *et al* 2022 *Phys. Rev.* B **105** 174433
[223] Kosevich A M, Ivanov B A and Kovalev A S 1990 *Phys. Rep.* **194** 117–238
[224] Wachowiak A *et al* 2002 *Science* **298** 577–80
[225] Waeyenberge B V *et al* 2006 *Nature* **444** 461–4
[226] Makhfudz I, Krüger B and Tchernyshyov O 2012 *Phys. Rev. Lett.* **109** 217201
[227] Moon K W *et al* 2014 *Phys. Rev.* B **89** 064413
[228] Petit D *et al* 2015 *Appl. Phys. Lett.* **106** 022402
[229] Rößler U K, Bogdanov A N and Pfleiderer C 2006 *Nature* **442** 797–801
[230] Mühlbauer S *et al* 2009 *Science* **323** 915–9
[231] Jiang W *et al* 2015 *Science* **349** 283–6
[232] Catalan G *et al* 2012 *Rev. Mod. Phys.* **84** 119
[233] Parkin S S P, Hayashi M and Thomas L 2008 *Science* **320** 190–4
[234] Jonietz F *et al* 2010 *Science* **330** 1648–51
[235] Parkin S and Yang S H 2015 *Nat. Nanotechnol.* **10** 195–8
[236] Yu X Z *et al* 2012 *Nat. Commun.* **3** 988
[237] Pribiag V S *et al* 2007 *Nat. Phys.* **3** 498–503
[238] Zhang S *et al* 2015 *New J. Phys.* **17** 023061
[239] Hrkac G *et al* 2015 *J. Phys. D: Appl. Phys.* **48** 453001
[240] Fert A, Cros V and Sampaio J 2013 *Nat. Nanotechnol.* **8** 152–6
[241] Shibata J and Otani Y 2004 *Phys. Rev.* B **70** 012404
[242] Galkin A Y, Ivanov B A and Zaspel C E 2006 *Phys. Rev.* B **74** 144419
[243] Vogel A *et al* 2010 *Phys. Rev. Lett.* **105** 037201
[244] Jung H *et al* 2011 *Sci. Rep.* **1** 59
[245] Vogel A *et al* 2014 *Phys. Rev.* B **89** 104403
[246] Hänze M *et al* 2014 *Appl. Phys. Lett.* **104** 182405
[247] Behncke C *et al* 2015 *Phys. Rev.* B **91** 224417
[248] Hänze M *et al* 2015 *Phys. Rev.* B **91** 104428

[249] Hänze M *et al* 2016 *Sci. Rep.* **6** 22402
[250] Behncke C *et al* 2018 *Sci. Rep.* **8** 186
[251] Gareeva Z V and Guslienko K Y 2018 *J. Phys. Commun.* **2** 035009
[252] Han D S *et al* 2013 *Sci. Rep.* **3** 2262
[253] Han D S, Jeong H B and Kim S K 2013 *Appl. Phys. Lett.* **103** 112406
[254] Kim J *et al* 2017 *Sci. Rep.* **7** 45185
[255] Mruczkiewicz M *et al* 2016 *Phys. Rev.* B **93** 174429
[256] Haldane F D M 1988 *Phys. Rev. Lett.* **61** 2015
[257] Kim S K and Tserkovnyak Y 2017 *Phys. Rev. Lett.* **119** 077204
[258] Li Z X *et al* 2018 *Phys. Rev.* B **98** 180407(R)
[259] Li Z X *et al* 2021 *Phys. Rev.* B **103** 054438
[260] Go G *et al* 2020 *Phys. Rev.* B **101** 134423
[261] Su W P, Schrieffer J R and Heeger A J 1979 *Phys. Rev. Lett.* **42** 1698
[262] Li Z X *et al* 2022 *Phys. Rev. Appl.* **17** 024054

IOP Publishing

Topology in Collective Magnetization Dynamics

Zhixiong Li, Yunshan Cao and Peng Yan

Chapter 2

Topological band theory

This chapter discusses topological band theory. First we introduce several typical topological invariants which are useful for describing topological phases with different orders. Then we focus on two pedagogical topological models, i.e. the Su–Schrieffer–Heeger model and the Haldane model.

2.1 Topological invariants

The phase transition theory proposed by Landau [1] has dominated the principles of the classification of matter for a long time. According to this theory, different matter phases can be described by different local order parameters and the phase transition between different phases originates from spontaneous symmetry breaking. However, the emergence of the quantum Hall effect [2, 3] overturned this understanding and enabled people to realize the existence of a new phase of matter, i.e. the topological phase, which goes beyond the Landau–Ginzburg–Wilson paradigm [4, 5] and is characterized by a few quantized parameters—the so-called topological invariants. Quantized parameters do not change when the system undergoes adiabatic deformations as long as the bulk gap remains open. Very recently, several new topological invariants have also been proposed to describe the higher-order topological phases. In this section we give a detailed introduction to typical topological invariants [6–8].

To illustrate the general theory, we consider a single-particle Hamiltonian which respects the translation invariance with a Bravais lattice $H(\mathbf{r}) = H(\mathbf{r} + \mathbf{b})$, where $\mathbf{r}$ denotes the displacement vector and $\mathbf{b}$ represents the crystal lattice vector. Based on Bloch's theory, the eigenfunction of the system has the following form

$$|\psi_{n,\mathbf{k}}\rangle = e^{i\mathbf{k}\cdot\mathbf{r}}|u_{n,\mathbf{k}}\rangle. \tag{2.1}$$

Noticeably, here $u_{n,\mathbf{k}}(\mathbf{r})$ has a period of $\mathbf{b}$, i.e. $u_{n,\mathbf{k}}(\mathbf{r}) = u_{n,\mathbf{k}}(\mathbf{r} + \mathbf{b})$ with $\mathbf{k}$ being the wave vector. It is worth noting that the quantities $|u_{n,\mathbf{k}}\rangle$ and $u_{n,\mathbf{k}}(\mathbf{r})$ are different and they satisfy the relation $u_{n,\mathbf{k}}(\mathbf{r}) = \langle \mathbf{r}|u_{n,\mathbf{k}}\rangle$. The Bloch Hamiltonian can therefore be obtained by a unitary transformation

$$H(\mathbf{k}) = \mathrm{e}^{-\mathrm{i}\mathbf{k}\cdot\mathbf{r}}H(\mathbf{r})\mathrm{e}^{\mathrm{i}\mathbf{k}\cdot\mathbf{r}}. \tag{2.2}$$

Afterwards, the Hamiltonian can be diagonalized as

$$H(\mathbf{k})|u_{n,\,\mathbf{k}}\rangle = E_{n,\,\mathbf{k}}|u_{n,\,\mathbf{k}}\rangle, \tag{2.3}$$

where the eigenvalue $E_n(\mathbf{k})$ has the period of the reciprocal lattice vector $\mathbf{K}$, i.e. $E_n(\mathbf{k}) = E_n(\mathbf{k} + \mathbf{K})$, and n denotes the band index.

It is well-known that a gauge is defined by the phase choice in the Brillouin zone, and all physical observables should be gauge invariant. The Berry phase represents one of the most important topological invariants to describe the topological phase. Next, we derive the gauge independent form of the Berry phase. First, we assume that the Hamiltonian changes with 'time' via a few parameters (for example the electric field, magnetic field, etc) which can be denoted by vector $\mathbf{R} = (R_1, R_2, R_3, \ldots)$ with $R_i = R_i(t)$. Then we consider the system under adiabatic evolution, i.e. the parameter $\mathbf{R}(t)$ varies very slowly along the path C. Introducing the instantaneous orthonormal basis from the instantaneous eigenstates of $H(\mathbf{R}(t))$ for every $\mathbf{R}(t)$, we have the following equation:

$$H(\mathbf{R}(t))|v_n(\mathbf{R}(t))\rangle = E_n(\mathbf{R}(t))|v_n(\mathbf{R}(t))\rangle. \tag{2.4}$$

Noticeably, the eigenstate $|v_n(\mathbf{R}(t))\rangle$ in equation (2.4) is not unique. That is to say that there exists an uncertain phase $f(t)$

$$|\Phi(t)\rangle = \mathrm{e}^{-\mathrm{i}f(t)}|v_n(\mathbf{R}(t))\rangle \tag{2.5}$$

that keeps equation (2.4) invariant. Due to the existence of this uncertain phase, the basis function of $|v(\mathbf{R}(t))\rangle$ in equation (2.4) cannot be determined completely. To eliminate this uncertainty, we can make a phase choice. Naturally, we can assume the phase of the basis functions is smooth and single valued along the closed route C. Therefore, the evolution of the system with time is described by the time-dependent Schrödinger equation [9]

$$\mathrm{i}\hbar\partial_t|\Phi(t)\rangle = H(\mathbf{R}(t))|\Phi(t)\rangle. \tag{2.6}$$

Under the adiabatic approximation, the system will stay in one of the instantaneous eigenstates. Here we suppose the eigen-energies are non-degenerate, which is reasonable for a gapped system. Substituting equations (2.4) and (2.5) into equation (2.6), we have

$$E_n(\mathbf{R}(t))|v_n(\mathbf{R}(t))\rangle = \hbar\frac{\mathrm{d}f(t)}{\mathrm{d}t}|v_n(\mathbf{R}(t))\rangle + \mathrm{i}\hbar\frac{\mathrm{d}}{\mathrm{d}t}|v_n(\mathbf{R}(t))\rangle. \tag{2.7}$$

By taking a scalar product with $\langle v_n(\mathbf{R}(t))|$ at both sides of equation (2.7) and supposing that the state is normalized, i.e. $\langle v_n(\mathbf{R}(t))|v_n(\mathbf{R}(t))\rangle = 1$, we obtain

$$E_n(\mathbf{R}(t)) - \mathrm{i}\hbar\langle v_n(\mathbf{R}(t))|\frac{\mathrm{d}}{\mathrm{d}t}|v_n(\mathbf{R}(t))\rangle = \hbar\frac{\mathrm{d}f(t)}{\mathrm{d}t}. \tag{2.8}$$

We thus can derive the phase $f(t)$ as

$$f(t) = \frac{1}{\hbar}\int_0^t E_n(\mathbf{R}(t'))\mathrm{d}t' - \mathrm{i}\int_0^t \langle v_n(\mathbf{R}(t'))|\frac{\mathrm{d}}{\mathrm{d}t'}|v_n(\mathbf{R}(t'))\rangle \mathrm{d}t'. \tag{2.9}$$

Here the first term denotes the ordinary dynamical phase, while the second term is the so-called Berry phase. The eigenstate $|\Phi(t)\rangle$ therefore can be written as

$$|\Phi(t)\rangle = \mathrm{e}^{\mathrm{i}\gamma_c}\exp\left[-\frac{\mathrm{i}}{\hbar}\int_0^t E_n(\mathbf{R}(t'))\mathrm{d}t'\right]|v_n(\mathbf{R}(t))\rangle, \tag{2.10}$$

where γ_c is the Berry phase

$$\gamma_c = \mathrm{i}\int_0^t \langle v_n(\mathbf{R}(t'))|\frac{\mathrm{d}}{\mathrm{d}t'}|v_n(\mathbf{R}(t'))\rangle \mathrm{d}t' = \mathrm{i}\int_0^t \langle v_n(\mathbf{R}(t'))|\nabla_{\mathbf{R}}|v_n(\mathbf{R}(t'))\rangle \frac{\mathrm{d}\mathbf{R}}{\mathrm{d}t'}\mathrm{d}t'. \tag{2.11}$$

Furthermore, the Berry phase can be expressed as

$$\gamma_c = \int_C \mathrm{d}\mathbf{R}\cdot\mathbf{A}_n(\mathbf{R}), \tag{2.12}$$

where the vector

$$\mathbf{A}_n(\mathbf{R}) = \mathrm{i}\langle v_n(\mathbf{R})|\nabla_{\mathbf{R}}|v_n(\mathbf{R})\rangle \tag{2.13}$$

is called the Berry connection. In general, $\mathbf{A}_n(\mathbf{R})$ is gauge dependent. If we perform a gauge transformation $|v_n(\mathbf{R}(t))\rangle \to \mathrm{e}^{\mathrm{i}\chi(\mathbf{R})}|v_n(\mathbf{R}(t))\rangle$, the Berry connection transforms in the general way: $\mathbf{A}_n(\mathbf{R}) \to \mathbf{A}_n(\mathbf{R}) - \frac{\partial\chi(\mathbf{R})}{\partial\mathbf{R}}$, then the phase factor γ_c will has a variation $-\int_C \frac{\partial\chi(\mathbf{R})}{\partial\mathbf{R}}\cdot\mathrm{d}\mathbf{R} = \chi(\mathbf{R}(t=0)) - \chi(\mathbf{R}(t=T))$. Considering the system evolution along a closed route C, according to the wave function single-valued condition, we have $\chi(\mathbf{R}(t=T)) - \chi(\mathbf{R}(t=0)) = 2m\pi$, where m is an integer. Therefore, the γ_c is gauge invariant (mod by 2π) [10] for a closed route C

$$\gamma_c = \oint_C \mathrm{d}\mathbf{R}\cdot\mathbf{A}_n(\mathbf{R}) = \int_S \mathrm{d}\mathbf{S}\cdot\Omega_n(\mathbf{R}) \ (\mathrm{mod}\ 2\pi), \tag{2.14}$$

where the Stokes' theorem is used and $\partial S = C$. In addition, we suppose the Berry connection is analytic over S. Then we consider the three-dimensional parameter space (R_x, R_y, R_z). The Berry curvature $\Omega_n(\mathbf{R})$ can be expressed as the curl of the Berry connection

$$\Omega_n(\mathbf{R}) = \nabla_{\mathbf{R}}\times\mathbf{A}_n(\mathbf{R}). \tag{2.15}$$

In addition, by utilizing the identity

$$\langle v_m(\mathbf{R})|\nabla_{\mathbf{R}}|v_n(\mathbf{R})\rangle = \frac{\langle v_m(\mathbf{R})|\nabla_{\mathbf{R}}H(\mathbf{R})|v_n(\mathbf{R})\rangle}{E_n(\mathbf{R}) - E_m(\mathbf{R})}, \quad m \neq n, \tag{2.16}$$

we can obtain the alternative expression of Berry curvature

$$\Omega_n(\mathbf{R}) = \mathrm{i}\sum_{m\neq n}\frac{\langle v_m(\mathbf{R})|\nabla_{\mathbf{R}}H(\mathbf{R})|v_n(\mathbf{R})\rangle\times\langle v_n(\mathbf{R})|\nabla_{\mathbf{R}}H(\mathbf{R})|v_m(\mathbf{R})\rangle}{[E_m(\mathbf{R}) - E_n(\mathbf{R})]^2}. \tag{2.17}$$

Noticeably, the derivation of the Berry phase above is applicable for all systems characterized by a parameter-dependent Hamiltonian. For Bloch bands, the corresponding Berry connection, Berry curvature, and Berry phase have the definitions as $\mathbf{A}_n(\mathbf{k}) = \mathrm{i}\langle v_n(\mathbf{k})|\nabla_{\mathbf{k}}|v_n(\mathbf{k})\rangle$, $\Omega_n(\mathbf{k}) = \nabla_{\mathbf{k}} \times \mathrm{i}\langle v_n(\mathbf{k})|\nabla_{\mathbf{k}}|v_n(\mathbf{k})\rangle$, and $\gamma_c = \oint_C \mathrm{d}\mathbf{k} \cdot \mathbf{A}_n(\mathbf{k}) \pmod{2\pi}$ [11], respectively.

The Berry phase is an important topological invariant to judge whether the system is a first-order topological insulator (FOTI) phase. In particular, when the Berry phase is determined along a non-contractible loop L in the Brillouin zone, it is called the Zak phase [12]:

$$Z = \int_L \mathrm{d}\mathbf{k} \cdot \mathbf{A}_n(\mathbf{k}) \pmod{2\pi}. \tag{2.18}$$

The Zak phase is particularly powerful to identify the topological insulating phase for a 1D lattice, in which the integral range $-\pi/a$ to π/a (a denotes the lattice constant) is usually adopted. Considering a one-dimensional system respecting inversion symmetry, the system is in the FOTI state if the Zak phase is not zero. However, in numerical calculations, the Bloch wave function $|v_n(\mathbf{k})\rangle$ usually has an uncertain phase factor. Therefore, there will exist some problems in calculating the Zak phase directly through equation (2.18). Recently, Wang *et al* [13] showed that the Wilson-loop approach can avoid this difficulty of the gauge choice. The Zak phase is the integral of the Berry connection over the Brillouin zone. In the numerical calculation one can divide the Brillouin zone into many small segments and the integral can be treated as the summation of the contributions from each small segment. For example, we can divide the Brillouin zone into N segments (see figure 2.1(a)). For every small segment k_i ($1 \leqslant i \leqslant N+1$ and $k_{N+1} = k_1$), the Berry phase Z_i can be expressed as

$$Z_i = \mathrm{i}\langle v_n(k_i)|\frac{|v_n(k_{i+1})\rangle - |v_n(k_i)\rangle}{\mathrm{d}k}\mathrm{d}k = \mathrm{i}\langle v_n(k_i)|v_n(k_{i+1})\rangle - \mathrm{i}\langle v_n(k_i)|v_n(k_i)\rangle. \tag{2.19}$$

By using the normalized condition $\langle v_n(k_i)|v_n(k_i)\rangle = 1$ and the approximation $\mathrm{e}^{-\mathrm{i}Z_i} \approx 1 - \mathrm{i}Z_i$ (consider a sufficiently small segment), equation (2.19) can be simplified as

$$\mathrm{e}^{-\mathrm{i}Z_i} = \langle v_n(k_i)|v_n(k_{i+1})\rangle. \tag{2.20}$$

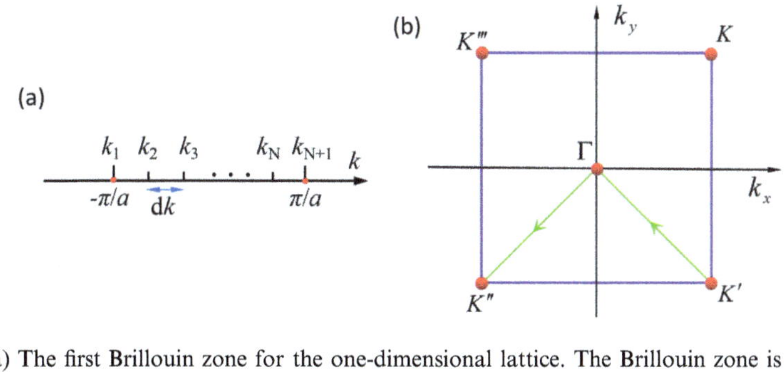

Figure 2.1. (a) The first Brillouin zone for the one-dimensional lattice. The Brillouin zone is divided into N segments. (b) The first Brillouin zone for the two-dimensional square lattice. The corresponding high-symmetry points are marked by red dots.

Then, the total Berry phase is the summation of the Berry phase in each small segment,

$$\mathrm{e}^{-\mathrm{i}Z} = \prod_{i=1}^{N} \mathrm{e}^{-\mathrm{i}Z_i} = \prod_{i=1}^{N} \langle v_n(k_i)|v_n(k_{i+1})\rangle. \tag{2.21}$$

It is worth noting that the above formalism is gauge invariant, because each Bloch state $|v_n(\mathbf{k})\rangle$ for k_i appears twice in the above product and one is the complex conjugate to the other, therefore, the arbitrary phase factors will be canceled.

In addition to the Berry phase, the Chern number is another important topological invariant used to characterize an FOTI. For a two-dimensional system, the Chern number can be expressed as the integration of the Berry curvature over the Brillouin zone [14]:

$$C_n = \frac{1}{2\pi}\int_{\mathrm{BZ}} \mathrm{d}^2k\Omega_{n,z}(\mathbf{k}). \tag{2.22}$$

Generally speaking, the Chern number is an integer [15, 16] and the value is invariant only if the band gap experiences a closing and reopening. The system is in an FOTI (trivial) phase if C_n is finite (zero). Noticeably, the Chern number always vanishes when the system holds the time-reversal symmetry. Therefore, to obtain the topological Chern insulator [17], we must break the time-reversal symmetry of the system. In general, the Berry phase is used to describe the FOTI in a one-dimensional system, while the Chern number is the tool to identify the FOTI in a two-dimensional system. In addition, there exist several other topological invariants to describe an FOTI, for example, the winding number [14] and Z_2 invariant [18, 19] etc. Different topological invariants are applicable to systems with different symmetries.

Different to the FOTIs, new topological invariants should be considered to describe the HOTIs. The bulk polarization and Z_N Berry phase represent two typical topological invariants for characterizing HOTIs. The bulk polarization is calculated as the integration of the Berry connection [20, 21]

$$P_j = \frac{1}{S}\iint_{\mathrm{BZ}} A_j \mathrm{d}^2k, \tag{2.23}$$

where S denotes the area of the first Brillouin zone and Berry connection $A_j = -\mathrm{i}\langle\psi|\partial k_j|\psi\rangle$ with $j = x, y$. The bulk polarization (P_x, P_y) is also known as the Wannier center. Nonzero P_x and P_y imply that the system is in HOTI phase. The Z_N Berry phase [22–26] is evaluated as

$$\theta = \int_{L_1} \mathrm{Tr}[\mathbf{A}(\mathbf{k})]\cdot \mathrm{d}\mathbf{k} \pmod{2\pi}, \tag{2.24}$$

where the wave function in the Berry connection is a matrix consisting of the eigenvectors for bands below the gap. L_1 denotes an integral path in Brillouin zone. Considering the system with N-fold rotational symmetry (C_N symmetry), there exist other $N - 1$ equivalent integral paths which lead to the same θ. In addition, if we consider the integral along the sum of all N paths, we obtain $\theta = 0$. Therefore, the Z_N Berry phase is quantized with $\theta = \frac{2n\pi}{N}$ $(n = 0, 1, 2, \ldots, N - 1)$. Here we take a two-dimensional square lattice as an example to illustrate this point, with the corresponding first Brillouin zone shown in figure 2.1(b). We choose the green line segment as the integral path (L_1: $K' \to \Gamma \to K''$) to calculate the Z_N Berry phase. The high-symmetry points K, K', K'', and K''' are equivalent due to the C_4 symmetry. Therefore, there exist three other equivalent integral paths (L_2: $K'' \to \Gamma \to K'''$, L_3: $K''' \to \Gamma \to K$, L_4: $K \to \Gamma \to K'$) leading to the same θ. In addition, one can easily find that the integral along the path $L_1 + L_2 + L_3 + L_4$ is zero. The Z_4 Berry phase thus must be quantized with $\theta = n\pi/2$ ($n = 0, 1, 2, 3$).

Noticeably, the Chern number discussed above is based on fermionic system. As we know, bosons and fermions follow different statistics, therefore the Chern number also has different definition in these two systems. In this book, we introduce the magnons and magnetic solitons which belong to the bosonic excitations. Next, we present the definition of Chern number in bosonic system [27, 28]. Considering a generic bosonic Hamiltonian of the quadratic form

$$\hat{H} = \frac{1}{2}\sum_{\mathbf{k}}[\beta_{\mathbf{k}}^{\dagger}\ \beta_{-\mathbf{k}}] \cdot H_{\mathbf{k}} \cdot \begin{bmatrix} \beta_{\mathbf{k}} \\ \beta_{-\mathbf{k}}^{\dagger} \end{bmatrix}, \tag{2.25}$$

where $\beta_{\mathbf{k}}^{\dagger} = [\beta_{1,\mathbf{k}}^{\dagger}, \dots, \beta_{N,\mathbf{k}}^{\dagger}]$ represent the boson creation operators, $\mathbf{k} = (k_x, k_y)$ is wave vector, and N is the number of internal degrees of freedom in unit cell. The Hamiltonian $H_{\mathbf{k}}$ includes N by N normal parts ($\mathbf{a}_{\mathbf{k}}$ and $\mathbf{a}_{-\mathbf{k}}^{*}$) and N by N anomalous parts ($\mathbf{b}_{\mathbf{k}}$ and $\mathbf{b}_{-\mathbf{k}}^{*}$),

$$H_{\mathbf{k}} = \begin{bmatrix} \mathbf{a}_{\mathbf{k}} & \mathbf{b}_{\mathbf{k}} \\ \mathbf{b}_{-\mathbf{k}}^{*} & \mathbf{a}_{-\mathbf{k}}^{*} \end{bmatrix}. \tag{2.26}$$

Different from the fermionic case, this Hamiltonian is diagonalized with paraunitary matrix $T_{\mathbf{k}}$ instead of unitary matrix [29],

$$T_{\mathbf{k}}^{\dagger} H_{\mathbf{k}} T_{\mathbf{k}} = \begin{bmatrix} E_{\mathbf{k}} & \\ & E_{-\mathbf{k}} \end{bmatrix}, \tag{2.27}$$

where $E_{\mathbf{k}}$ is a diagonal matrix and the diagonal elements characterize the bulk band structure. A projection operator can be defined as

$$Q_n = T_{\mathbf{k}} \Gamma_n \sigma_3 T_{\mathbf{k}}^{\dagger} \sigma_3, \tag{2.28}$$

where Γ_n is a diagonal matrix which takes +1 for the nth diagonal element and zero otherwise, and a diagonal matrix $\boldsymbol{\sigma}_3$ takes +1 in the particle space while −1 in the hole space. The Berry curvature therefore can be expressed as

$$B_n = \mathrm{i}\epsilon_{\mu\nu}\ \mathrm{Tr}[Q_n(\partial_{k_\mu} Q_n)(\partial_{k_\nu} Q_n)]. \tag{2.29}$$

At last, the Chern number in a bosonic system has the following form

$$C_n = \frac{1}{2\pi}\int_{\mathrm{BZ}} \mathrm{d}\mathbf{k} B_n(\mathbf{k}), \tag{2.30}$$

where the integration is calculated over the first Brillouin zone.

Generally speaking, to obtain the complete phase diagram of the system, one should determine the topological invariants for characterizing both FOTIs and HOTIs simultaneously. In addition, by mapping the Hamiltonian into the well-known topological models, we can have a comprehensive understanding of the topological properties. The pedagogical topological models including the Su–Schrieffer–Heeger model [30] and the Haldane model [17] will be introduced in the following sections.

2.2 The Su–Schrieffer–Heeger model

To characterize the dynamics of one-dimensional polyacetylene, Su, Schrieffer, and Heeger [30] proposed a simple tight-binding model based on spontaneous dimerization, which is called the Su–Schrieffer–Heeger (SSH) model. Over the past decades, the SSH model has been realized in many different systems [15, 31–33] and attracted a lot of attention for demonstrating the fundamental topological physics. Figure 2.2(a) shows

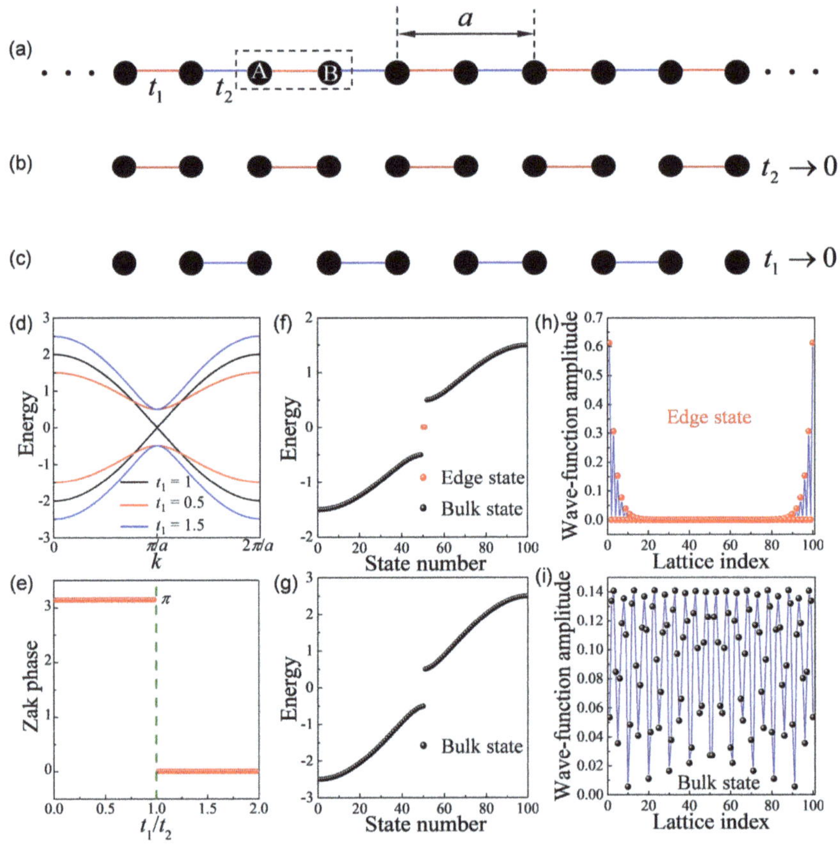

Figure 2.2. (a) A sketch of the SSH model. The configurations of finite systems under two limits: (b) $t_2 \to 0$ and (c) $t_1 \to 0$. (d) The dispersion relation of a system with different t_1, where t_2 is fixed to 1. (e) The calculated Zak phase for different ratios t_1/t_2. The eigenvalues of a finite system considering 100 unit cells with (f) $t_1/t_2 = 0.5$ and (g) $t_1/t_2 = 1.5$. The intensity distribution of the wave function for the (h) edge state and (i) bulk state. (Reproduced with permission from [7]. Copyright 2021 Elsevier.)

the sketch map of SSH model, where A and B denote two sublattices, and t_1 and t_2 are the alternate coupling parameters. For simplicity, we assume the coupling parameters are real and nonnegative, i.e. $t_1,\ t_2 \geqslant 0$. The Hamiltonian can be expressed as

$$H = \sum_n (t_1 c_{A,\,n}^\dagger c_{B,n} + t_2 c_{A,\,n+1}^\dagger c_{B,n}) + \text{h.c.}, \tag{2.31}$$

with $c_{A,n}$ (or $c_{B,n}$) the annihilation operators for the A (or B) sublattice of the nth cell. After the Fourier transformation, we obtain the Hamiltonian in a matrix form

$$H = \begin{pmatrix} 0 & t_1 + t_2 \mathrm{e}^{-ika} \\ t_1 + t_2 \mathrm{e}^{ika} & 0 \end{pmatrix} = [t_1 + t_2 \cos(ka)]\sigma_x + t_2 \sin(ka)\sigma_y, \tag{2.32}$$

where $\sigma_x = \begin{pmatrix} 0 & 1 \\ 1 & 0 \end{pmatrix}$ and $\sigma_y = \begin{pmatrix} 0 & -\mathrm{i} \\ \mathrm{i} & 0 \end{pmatrix}$ are the Pauli matrices, and k and a are wave vector and lattice constant, respectively. The dispersion relation of the system can therefore be solved as

$$E = \pm\sqrt{t_1^2 + t_2^2 + 2t_1t_2 \cos ka}\,. \tag{2.33}$$

When $t_1 = t_2$ the dispersion relation is gapless. However, a gap will appear at $k = \pi/a$ if $t_1 \neq t_2$, which leads to an insulating phase. The corresponding band structures for different parameters are plotted in figure 2.2(d). By using the Zak phase (equation (2.18)), one can judge whether these emerging insulating phases are topological. Figure 2.2(e) shows the Zak phase Z varying with the ratios t_1/t_2. We can clearly see that Z is quantized to π (0) for $t_1/t_2 < 1$ ($t_1/t_2 > 1$), which implies that the system supports two topologically different phases for $t_1/t_2 < 1$ and $t_1/t_2 > 1$.

Considering a system in the topological phase, the bulk-boundary correspondence implies that the robust edge states will appear in a finite system. As an example we choose $t_1/t_2 = 0.5$, the corresponding spectrum of a finite system containing 100 unit cells is plotted in figure 2.2(f). We can see that the system exhibits a degenerate 'zero mode', as marked by red dots. In addition, the wave function of this degenerate state is highly confined at both ends of the lattice (see figure 2.2(h)), showing the typical characteristic for the edge state. In addition, the bulk states are also identified (marked with black dots), with the wave function spreading over the whole lattice (see figure 2.2(i)). The spectrum for the trivial phase is also shown in figure 2.2(g), where we set $t_1/t_2 = 1.5$. In this situation the system can only support the bulk states. One can intuitively understand these results as follows. On the one hand, there are no uncoupled unit cells in the limit of $t_2 \to 0$ (see figure 2.2(b)). Therefore, the system is in trivial phase and can only support bulk modes. On the other hand, we find two isolated lattices appearing at the end of the lattice in the limit of $t_1 \to 0$ (see figure 2.2(c)). Edge states thus emerge and the system is in the topological phase.

Remarkably, the 'zero-energy' edge states emerging in the SSH model are topologically protected by the chiral symmetry. Generally speaking, the bipartite-lattice system holds the chiral symmetry if the Hamiltonian fulfills the relation

$$\Gamma_2 H \Gamma_2^\dagger = -H, \tag{2.34}$$

where Γ_2 is the chiral operator and satisfies the following conditions [34]. (i) Γ_2 must be Hermitian and unitary, i.e. $\Gamma_2^\dagger = \Gamma_2$. For simplicity, this condition can also be expressed as $\Gamma_2^\dagger\Gamma_2 = \Gamma_2^2 = I$, where I denotes a 2×2 identity matrix. (ii) Γ_2 must be local. In other words, the matrix elements between sites from different unit cells in Γ_2 should vanish. Considering a finite system, the chiral operator takes the following form [35]:

$$\Gamma_2 = \sum_n c_{A,n}^\dagger c_{A,n} - \sum_n c_{B,n}^\dagger c_{B,n}. \tag{2.35}$$

By utilizing (2.35) one can confirm that the Hamiltonian (2.31) for describing a finite system respects the chiral symmetry. As long as the chiral symmetry is not broken, the finite lattice can always exhibit 'zero-energy' edge states in the topological phase [36–38]. However, the 'zero-energy' edge states become fragile once the chiral symmetry is broken [39, 40]. Therefore, we conclude that the 'zero-energy' edge states emerging in the SSH model are protected by the chiral symmetry. With the help of both theoretical analysis and numerical calculation, we can verify this point. Let's assume that the finite system has a eigenstate $|\psi\rangle$ with eigenvalue ϵ. Due to the chiral symmetry, the system must have another eigenstate $\Gamma_2|\psi\rangle$ with eigenvalue $-\epsilon$. If $\epsilon = 0$ the two eigenstates $|\psi\rangle$ and $\Gamma_2|\psi\rangle$ are degenerate. Then we can reshuffle these two eigenstates as $\Psi_\pm = (|\psi\rangle \pm \Gamma_2|\psi\rangle)/\sqrt{2}$. The degenerate states $\Psi_\pm$ appear only in one sublattice (A or B) for the reason that $(I \pm \Gamma_2)/\sqrt{2}$ projects states into only one sublattice (A or B). Furthermore, if the eigenstate appear only in sublattice A or sublattice B, the eigenvalue must be zero, i.e. $\epsilon = 0$, because $\Gamma_2|\psi\rangle = \pm|\psi\rangle \propto |\psi\rangle$. We thus obtain $\epsilon = -\epsilon$ [35]. Naturally, the state with zero eigenvalue must be the 'zero-energy' edge mode. We consider two situations for numerical calculations: (i) the disorder is introduced into the coupling terms (off-diagonal elements) of (2.31), which does not break the chiral symmetry, the edge states are therefore stable (see figure 2.3(a)); (ii) the disorder is introduced to the on-site terms (diagonal elements) of (2.31), then the chiral symmetry is broken and the edge states are fragile (see figure 2.3(b)).

2.3 The Haldane model

In 1980 Klitzing *et al* [2] observed that the Hall conductance of two-dimensional electron gas is always an integral multiple of a fixed value under the condition of low temperature and strong magnetic field, known as the quantum Hall effect. To explain the emerging quantum Hall effect, Thouless *et al* [41] proved theoretically that the quantized conductance can be described by the topological structure of the momentum space, which is called the Thouless–Kohmoto–Nightingale–Nijs (TKNN) principle. The realization of the quantum Hall effect usually needs a very strong magnetic field [2, 3] (tens of teslas). In 1988 theoretical physicist Haldane proposed the first model of a quantum Hall system without an external magnetic field based on a honeycomb lattice, i.e. graphene. Graphene is a two-dimensional structure composed of carbons, of which the corresponding conduction band and valence band touch each other at specific points in momentum space, i.e. the so-

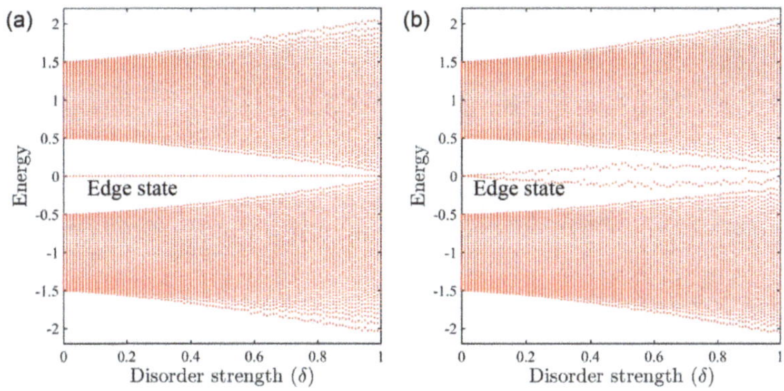

Figure 2.3. The eigenvalues of the finite system containing 100 unit cells with different strengths of disorder, where we choose $t_1/t_2 = 0.5$ and the disorder term is added into the (a) off-diagonal and (b) diagonal elements of the Hamiltonian (2.31). We assume that the coupling coefficients t_1 and t_2 have a variation $t_1 \to t_1(1 + \delta W)$ and $t_2 \to t_2(1 + \delta W)$, which act as the disorder in (a), while the disorder is added in (b) by considering that the on-site terms suffer a random fluctuation, i.e. δW. Here δ denotes the strength of disorder and W represents a random number uniformly distributed between −1 and 1. (Reproduced with permission from [7]. Copyright 2021 Elsevier.)

called Dirac points [42]. Near these points the bands exhibit linear crossings. In this section we introduce this important topological model.

In the graphene structure, the two-dimensional tight-binding model reads

$$H = \sum_{\langle ij \rangle} t_1 c_i^{\dagger} c_j + \text{h.c.}. \tag{2.36}$$

After Fourier transformation, Hamiltonian (2.36) becomes

$$H = \sum_{\mathbf{k}} t_1 (e^{i\mathbf{a}_1 \cdot \mathbf{k}} + e^{i\mathbf{a}_2 \cdot \mathbf{k}} + e^{i\mathbf{a}_3 \cdot \mathbf{k}}) a_{A,\mathbf{k}}^{\dagger} a_{B,\mathbf{k}} + \text{h.c.}, \tag{2.37}$$

where $\mathbf{a}_1 = (\frac{\sqrt{3}}{2}a, \frac{1}{2}a)$, $\mathbf{a}_2 = (-\frac{\sqrt{3}}{2}a, \frac{1}{2}a)$, and $\mathbf{a}_3 = (0, -a)$ are the displacements from B site to its three nearest neighbor A sites (see figure 2.4(a)) with a the lattice constant, and $\mathbf{k} = (k_x, k_y)$ the wave vector. Near the high-symmetry points $\pm K = (\pm\frac{4\pi}{3\sqrt{3}a}, 0)$, by using the Taylor expansion for the wave vector as $\mathbf{k} = \mathbf{K} + \mathbf{p}$ with $|\mathbf{p}| \ll |\mathbf{K}|$, we can obtain the effective Hamiltonian

$$H = \sum_{\mathbf{k}} -\frac{3t_1}{2}(p_x a + i p_y a) a_{A,\mathbf{k}}^{\dagger} a_{B,\mathbf{k}} + \text{h.c.}, \tag{2.38}$$

In terms of the basis $\{\phi^A(\mathbf{p}), \phi^B(\mathbf{p})\}$, the Hamiltonian reads $h(\mathbf{p}) = -\frac{3t_1}{2}(\pm p_x \sigma_x - p_y \sigma_y)$, where σ_x and σ_y are Pauli matrices. One can clearly see that the dispersion relation of graphene is gapless and the Dirac points exhibit the characteristics of linear crossing.

At the Dirac points, due to the existence of both inversion (P) and time-reversal (T) symmetries, the energy spectrum is degenerate. When at least one of these

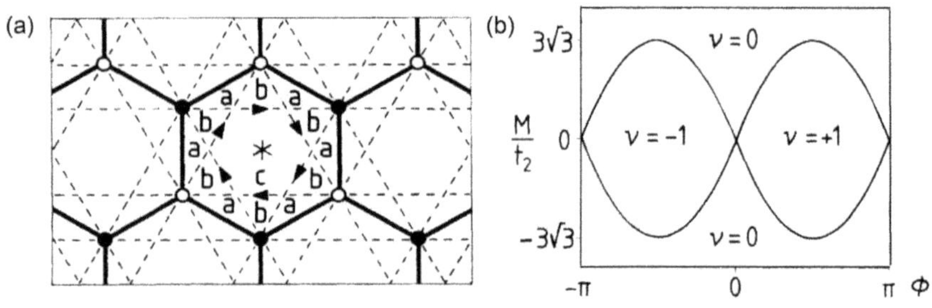

Figure 2.4. (a) The sketch of the honeycomb model, where the solid and dashed line segments denote the nearest neighbor (NN) and next nearest neighbor (NNN) couplings, respectively. The white and black balls represent A and B sublattices. The black arrows on NNN bonds mark the directions of positive phase hopping. (b) The phase diagram of system. The zero-field quantum Hall effect ($\sigma^{xy} = \nu e^2/h$, where σ^{xy} denotes the transverse conductivity and $\nu = \pm 1$) appears when $|M/t_2| < 3\sqrt{3}|\sin\phi|$. (Reproduced with permission from [17]. Copyright 1988 The American Physical Society.)

symmetries is broken the degeneracy can be relieved and gap will appear at Dirac points, then the topological phase emerges. In 1988 Haldane [17] proved that the T symmetry can be broken via a specific magnetic field. The strength of this magnetic field is zero averaged over the whole unit cell, while it brings a periodic magnetic-flux density $B(\mathbf{r})$ along the direction perpendicular to the lattice plane. The Haldane model can be expressed by adding the next nearest neighbor (NNN) coupling

$$H = \sum_{\langle ij \rangle} t_1 c_i^\dagger c_j + \sum_{\langle\langle ij \rangle\rangle} t_2 c_i^\dagger c_j. \tag{2.39}$$

We can determine the displacement of the nearest A–A and B–B coupling with relations: $\mathbf{b}_1 = \mathbf{a}_2 - \mathbf{a}_3$, $\mathbf{b}_2 = \mathbf{a}_3 - \mathbf{a}_1$, and $\mathbf{b}_3 = \mathbf{a}_1 - \mathbf{a}_2$. The closed path of the NNN hopping includes the non-trivial phase $\phi = 2\pi(2\Phi_a + \Phi_b)/\Phi_0$, where $\Phi_0 = |h/e|$ is the flux quantum, Φ_a and Φ_b denote the fluxes within unit cell, as marked with a and b in figure 2.4(a), respectively. The magnetic flux has no effect on parameter t_1, while leads to a change for parameter t_2 as $t_2 \to t_2 \exp(\mathrm{i}\phi)$.

Utilizing the basis $\{\phi^A(\mathbf{k}), \phi^B(\mathbf{k})\}$, we can obtain the Hamiltonian with the following form:

$$\begin{aligned} H(\mathbf{k}) = {} & 2t_2 \cos\phi \sum_i \cos(\mathbf{k}\cdot\mathbf{b}_i)\sigma_0 + \sum_i t_1 \cos(\mathbf{k}\cdot\mathbf{a}_i)\sigma_x + \sum_i t_1 \sin(\mathbf{k}\cdot\mathbf{a}_i)\sigma_y \\ & + [M - 2t_2 \sin\phi \sum_i \sin(\mathbf{k}\cdot\mathbf{b}_i)]\sigma_z. \end{aligned} \tag{2.40}$$

On the one hand, due to the P symmetry, the masses (M) with opposite signs emerge at points $\pm K$. On the other hand, because the T symmetry is broken, a gap appears at points $\pm K$.

To obtain the topological phase, we need $M^2 < (3\sqrt{3}\, t_2 \sin\phi)^2$. Figure 2.4(b) plots the phase diagram of the system. Here we assume $|t_2/t_1| < 1/3$ to guarantee that there exists a finite gap between two bands.

References

[1] Toledano P and Toledano J C 1987 Landau theory of phase transitions, the: application to structural, incommensurate, magnetic and liquid crystal systems (Singapore: World Scientific)
[2] Klitzing K V, Dorda G and Pepper M 1980 *Phys. Rev. Lett.* **45** 494
[3] Tsui D C, Stormer H L and Gossard A C 1982 *Phys. Rev. Lett.* **48** 1559
[4] Landau L D, Lifshitz E M and Pitaevskii L P 1999 *Statistical physics* (New York: Butterworth-Heinemann)
[5] Wilson K G and Kogut J 1974 *Phys. Rep.* **12** 75–199
[6] Cayssol J and Fuchs J N 2021 *J. Phys. Mater.* **4** 034007
[7] Li Z X, Cao Y and Yan P 2021 *Phys. Rep.* **915** 1–64
[8] Xie B *et al* 2021 *Nat. Rev. Phys.* **3** 520–32
[9] Shen S Q 2012 *Topological Insulators* (Berlin: Springer) https://doi.org/10.1007/978-981-10-4606-3
[10] Berry M V 1984 *Proc. R. Soc. Lond. Ser.* A **392** 45–57
[11] Xiao D, Chang M C and Niu Q 2010 *Rev. Mod. Phys.* **82** 1959
[12] Zak J 1989 *Phys. Rev. Lett.* **62** 2747
[13] Wang H X, Guo G Y and Jiang J H 2019 *New J. Phys.* **21** 093029
[14] Chiu C K *et al* 2016 *Rev. Mod. Phys.* **88** 035005
[15] Hasan M Z and Kane C L 2010 *Rev. Mod. Phys.* **82** 3045
[16] Qi X L and Zhang S C 2011 *Rev. Mod. Phys.* **83** 1057
[17] Haldane F D M 1988 *Phys. Rev. Lett.* **61** 2015
[18] Moore J E and Balents L 2007 *Phys. Rev.* B **75** 121306(R)
[19] Qi X L, Hughes T L and Zhang S C 2008 *Phys. Rev.* B **78** 195424
[20] King-Smith R D and Vanderbilt D 1993 *Phys. Rev.* B **47** 1651(R)
[21] Vanderbilt D and King-Smith R D 1993 *Phys. Rev.* B **48** 4442
[22] Kariyado T, Morimoto T and Hatsugai Y 2018 *Phys. Rev. Lett.* **120** 247202
[23] Hatsugai Y and Maruyama I 2011 *Europhys. Lett.* **95** 20003
[24] Mizoguchi T, Araki H and Hatsugai Y 2019 *J. Phys. Soc. Jpn.* **88** 104703
[25] Araki H, Mizoguchi T and Hatsugai Y 2020 *Phys. Rev. Res.* **2** 012009(R)
[26] Kudo K, Yoshida T and Hatsugai Y 2019 *Phys. Rev. Lett.* **123** 196402
[27] Shindou R *et al* 2013 *Phys. Rev.* B **87** 174402
[28] Shindou R *et al* 2013 *Phys. Rev.* B **87** 174427
[29] Colpa J H P 1978 *Physica* A **93** 327–53
[30] Su W P, Schrieffer J R and Heeger A J 1979 *Phys. Rev. Lett.* **42** 1698
[31] Delplace P, Ullmo D and Montambaux G 2011 *Phys. Rev.* B **84** 195452
[32] Li X, Zhao E and Liu W V 2013 *Nat. Commun.* **4** 1523
[33] Ganeshan S, Sun K and Sarma S D 2013 *Phys. Rev. Lett.* **110** 180403
[34] Asbóth J K, Oroszlány L and Pályi A 2016 *A Short Course on Topological Insulators* (Berlin: Springer) https://doi.org/10.1007/978-3-319-25607-8
[35] Chen B H and Chiou D W 2020 *Phys. Lett.* A **384** 126168
[36] Mondragon-Shem I *et al* 2014 *Phys. Rev. Lett.* **113** 046802
[37] Liu T and Guo H 2018 *Phys. Lett.* A **382** 3287–92
[38] Jürß C and Bauer D 2019 *Phys. Rev.* B **99** 195428

[39] Scollon M and Kennett M P 2020 *Phys. Rev.* B **101** 144204
[40] Pérez-González B *et al* 2019 *Phys. Rev.* B **99** 035146
[41] Thouless D J *et al* 1982 *Phys. Rev. Lett.* **49** 405
[42] Neto A H C *et al* 2009 *Rev. Mod. Phys.* **81** 109

IOP Publishing

Topology in Collective Magnetization Dynamics

Zhixiong Li, Yunshan Cao and Peng Yan

Chapter 3

Basics of magnons

This chapter discusses the basics of spin waves (or magnons). First, we derive the magnon dispersion relation in a magnetic ellipsoid, as a pedagogical example, and introduce different types of magnon modes. Then, we focus on the main subjects of magnon spintronics which include the magnon excitation, propagation, manipulation, and detection. Various novel spintronic devices based on magnons are also discussed in the last section.

3.1 Magnon dispersion relation

As early as in 1930, Bloch [1] first predicted that in ferromagnetic materials, because of the short-range exchange and long-range dipolar interactions between magnetic moments, the oscillation of a local magnetic moment can spread over the whole magnet in the form of waves, which are called spin waves (SWs). An illustration of a one-dimensional SW is shown in figure 3.1. A magnon is the quantized quasiparticle of a spin wave. Similar to charged currents, neutral magnon currents can also carry, transmit, and process information [2, 3] in spintronic devices. Compared to ordinary electronic devices, magnonic devices possess a lot of advantages: (i) the propagation of a magnon does not generate electric current, which avoids the extra energy loss and the impact on device performance due to Joule heat [4, 5]; (ii) magnons have many degrees of freedom which can be used conveniently to encode information, for example by utilizing the phase and amplitude of magnons, one can design various logic gates [6–9]; (iii) the magnon diffusion length can reach the order of millimeters or centimeters (much longer than electron spin diffusion length), which is conducive to realize long-distance transmission of information [10–13]; (iv) the wavelength of a magnon can reach a few nanometers or dozens of nanometers, which can greatly improve the information storage density and is beneficial for device miniaturization [3, 14, 15]. Due to these outstanding properties of magnons, a new discipline—magnon spintronics or magnonics [3–5, 16–25]—has emerged in recent years, which focuses on the generation, propagation, manipulation, and detection of magnons. Figure 3.2 shows a schematic of magnon spintronics.

doi:10.1088/978-0-7503-5082-2ch3

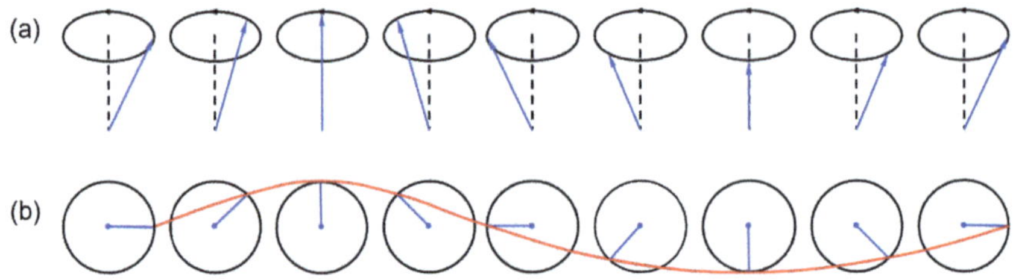

Figure 3.1. The (a) side and (b) top views of the collective dynamic behavior of a one-dimensional spin chain, manifesting as a wave form.

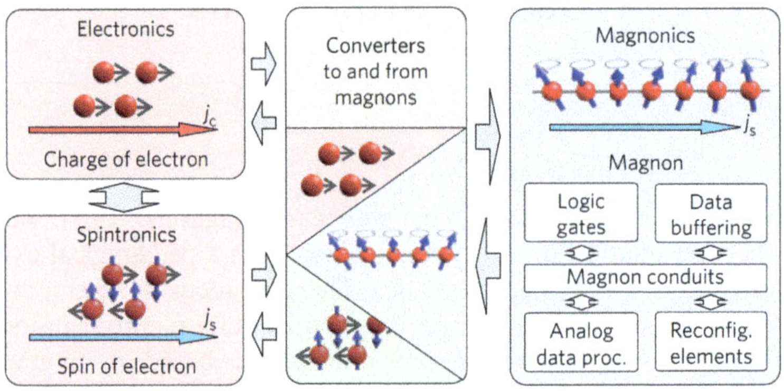

Figure 3.2. The concept of magnon spintronics. The information coded into charge or spin currents is converted into magnon currents, processed within the magnonic system, and converted back. (Reproduced with permission from [5]. Copyright 2015 Nature.)

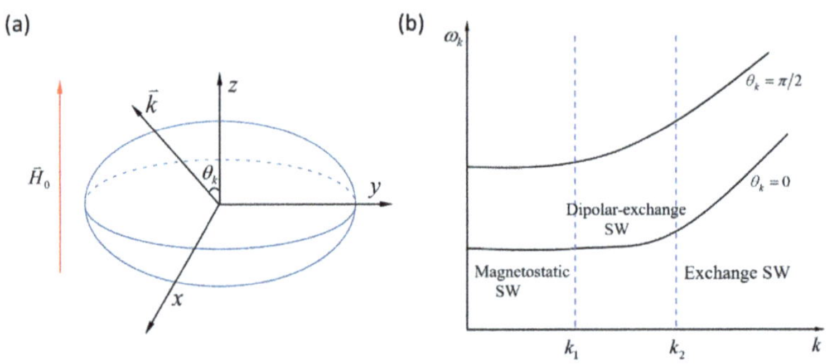

Figure 3.3. (a) An illustration of a magnetic ellipsoid which is magnetized by an external magnetic field $\mathbf{H}_0$ along the z-axis. (b) The SW dispersion relation in a magnetic ellipsoid with different propagation angles θ_k.

Since the dynamic properties of magnons are dominated by the band structure, we first derive the dispersion relation of the SW. Without loss of generality, we discuss the SW in a magnetic ellipsoid, as shown in figure 3.3(a). The magnetic moments are saturated by a static external magnetic $\mathbf{H}_0$ along the z-axis. The magnetization dynamics of the ellipsoid is dominated by the Landau–Lifshitz–Gilbert (LLG) equation [26]

$$\frac{\partial \mathbf{M}}{\partial t} = -\gamma \mathbf{M} \times \mathbf{H}_{\text{eff}} + \frac{\alpha}{M_s} \mathbf{M} \times \frac{\partial \mathbf{M}}{\partial t}, \tag{3.1}$$

where γ is the gyromagnetic ratio, α is the Gilbert damping constant, $\mathbf{M}$ is local magnetization, M_s is the saturation magnetization, and

$$\mathbf{H}_{\text{eff}} = \mathbf{H}_0 + \mathbf{H}_d + \mathbf{H}_{\text{an}} + \mathbf{H}_{\text{ex}} + \mathbf{H}_{\text{dip}} \tag{3.2}$$

is the effective field. Considering the spin wave excitation, i.e. the magnetic moment precession around the equilibrium position with a very small angle, we have

$$\mathbf{M}(\mathbf{r}, t) = M_s\hat{z} + \mathbf{m}(\mathbf{r}, t), \tag{3.3}$$

where $|\mathbf{m}(\mathbf{r}, t)| \ll |\mathbf{M}(\mathbf{r}, t)|$. The component of magnetization therefore can be expressed as $\mathbf{M} = (m_x, m_y, m_z + M_s) \approx (m_x, m_y, M_s)$ (here the linear approximation is considered).

The first term on the right-hand side of equation (3.2) is the external static magnetic field

$$\mathbf{H}_0 = H_0\hat{z}. \tag{3.4}$$

The second term is the static demagnetization field, which comes from the surface magnetic charge of the sample with

$$\mathbf{H}_d = -\mu_0 N_x m_x \hat{x} - \mu_0 N_y m_y \hat{y} - \mu_0 N_z M_s \hat{z} \approx -\mu_0 N_z M_s \hat{z}, \tag{3.5}$$

where N_x, N_y, and N_z are the demagnetizing factors along the x-, y-, and z-axis, respectively. Here, we have used the condition $m_x, m_y \ll M_s$. In addition, the demagnetizing factors satisfy the identity $N_x + N_y + N_z = 1$. The third term is the magnetocrystalline anisotropic field

$$\mathbf{H}_{\text{an}} = \frac{2K}{\mu_0 M_s}\hat{z}. \tag{3.6}$$

Here, we assume the easy-axis along the z-axis. K is the anisotropy constant and μ_0 is the vacuum permeability. The fourth term is the exchange field

$$\mathbf{H}_{\text{ex}} = \frac{2A}{\mu_0 M_s^2}\nabla^2 \mathbf{m}(\mathbf{r}, t), \tag{3.7}$$

with A the exchange stiffness. The last term is the dynamic dipolar field, which is also called the bulk demagnetization field

$$\mathbf{H}_{\text{dip}} = -\frac{\mathbf{k}}{k^2}[\mathbf{k} \cdot \mathbf{m}(\mathbf{r}, t)], \tag{3.8}$$

with $\mathbf{k}$ the wave vector.

Then we consider a plane-wave expansion, $\mathbf{m}(\mathbf{r}, t) = \mathbf{m}e^{i(\omega t - \mathbf{k}\cdot\mathbf{r})}$, with ω the frequency of the SW. In addition, we assume that the SW propagates in the x–z-plane, then the wave vector can be expressed as $\mathbf{k} = (k \sin\theta_k, 0, k\cos\theta_k)$, where θ_k is the angle between wave vector and z-axis (see figure 3.3(a)). Substituting equations

(3.3)–(3.8) into (3.1), by dropping the damping term and high-order quantities, the LLG equation can be simplified to two separate equations,

$$\begin{aligned} &\mathrm{i}\omega m_x + \gamma\left(H_0 - \mu_0 N_z M_s + \frac{2K}{\mu_0 M_s} + \frac{2A}{\mu_0 M_s}k^2\right)m_y = 0, \\ -&\gamma\left(H_0 - \mu_0 N_z M_s + \frac{2K}{\mu_0 M_s} + \frac{2A}{\mu_0 M_s}k^2 + \sin^2\theta_k M_s\right)m_x + \mathrm{i}\omega m_y = 0. \end{aligned} \tag{3.9}$$

Considering that the equation (3.9) has nontrivial solutions, we obtain the SW dispersion relation

$$\omega = \gamma\sqrt{\begin{aligned}&\left(H_0 - \mu_0 N_z M_s + \frac{2K}{\mu_0 M_s} + \frac{2A}{\mu_0 M_s}k^2\right) \\ &\left(H_0 - \mu_0 N_z M_s + \frac{2K}{\mu_0 M_s} + \frac{2A}{\mu_0 M_s}k^2 + \sin^2\theta_k M_s\right)\end{aligned}}. \tag{3.10}$$

Figure 3.3(b) plots the dispersion curves with different θ_k. Depending on their wavelengths, SWs can be divided into three types: (i) the exchange SW, in which the wavelength is very short and the exchange interaction dominates; (ii) the dipolar (or magnetostatic) SW, in which the wavelength is very long and the dipolar interaction dominates; and (iii) the dipolar-exchange SW, in which neither exchange nor dipolar interactions can be ignored.

For practical applications, people are interested in SWs in the thin-film waveguides. In addition, because of the limitations of experimental techniques, the exchange SWs are difficult to excite and most research focuses on dipolar SWs [27–29] with long wavelengths. Depending on the distribution of magnetization and the propagation direction of the SWs, the dipolar SWs can be further divided into three types. The first is the forward volume magnetostatic spin wave (FVMSW), with equilibrium magnetic moments perpendicular to both the film surface and the SW propagating direction. The approximate dispersion relation for the FVMSW is given by [30]

$$f_{\mathrm{FVMSW}} = \gamma\sqrt{H_0\left\{H_0 + 4\pi M_s\left[1 - \frac{1 - \exp(-kd_0)}{kd_0}\right]\right\}}, \tag{3.11}$$

where d_0 is the film thickness. The curve for equation (3.11) is plotted in figure 3.4(a). For in-plane magnetized films, we have the backward volume magnetostatic spin wave (BVMSW) and magnetostatic surface spin wave (MSSW), with the propagation direction of the SW parallel and perpendicular to the applied bias field, respectively. The dispersion relation of the BVMSW is plotted in figure 3.4(b), which can be approximated by the formula [30]

$$f_{\mathrm{BVMSW}} = \gamma\sqrt{H_0\left[H_0 + 4\pi M_s\frac{1 - \exp(-kd_0)}{kd_0}\right]}. \tag{3.12}$$

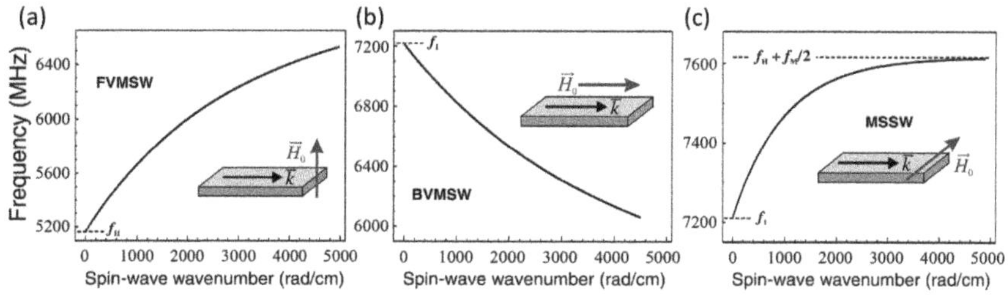

Figure 3.4. The dispersion relation obtained by formulae (3.11), (3.12), and (3.13) for a (a) FVMSW, (b) BVMSW, and (c) MSSW, respectively. The following parameters are used: the effective internal field $H_0 = 1845$ Oe, the saturation magnetization $4\pi M_s = 1750$ G, and the film thickness $d_0 = 5$ μm. (Reproduced with permission from [16]. Copyright 2010 IOP Publishing.)

Interestingly, the BVMSWs shows a negative slope in their dispersion curve, i.e. the group velocity $v_g = \partial f / \partial k < 0$, which indicates that the directions of phase velocity and group velocity of the SW are opposite. On the other hand, the dispersion curve of the MSSW can be calculated by the following formula [30, 31]:

$$f_{\mathrm{MSSW}} = \gamma\sqrt{(H_0 + 2\pi M_s)^2 - (2\pi M_s)^2 \exp(-2kd_0)}, \tag{3.13}$$

as shown in figure 3.4(c). The FVMSWs and BVMSWs are volume modes, which means that the amplitude of the magnetization precession has a sinusoidal distribution across the film thickness, while the MSSWs are surface modes which are localized to one surface of the film, and the distribution of precessional amplitude decays exponentially across the film thickness.

3.2 Magnon excitation

For realizing spintronic devices based on magnons, the first task is to effectively generate magnons. In experiments, the most common method to excite magnons is by means of the microwave antenna technique: first, the microwave antenna is placed on the magnetic film, then a microwave current is sent into the antenna and the induced alternating Oersted field will generate torque and act on the adjacent magnetic moment. The local magnetic moments therefore begin to precess. As a result, SWs are excited and propagate in the ferromagnetic layer [16, 32, 33].

Figure 3.5 shows the illustration of magnon excitation by using microwave antenna for two different types of magnetostatic spin waves. The magnetic moments in yttrium iron garnet (YIG) are saturated by the external magnetic field and the radiofrequency current is applied to the input antenna. On the one hand, for exciting BVMSWs (see figure 3.5(a)), only the vertical component of the induced Oersted field can produce torque to the adjacent magnetic moment, while the horizontal component does not because it is parallel to the magnetic moment. On the other hand, for the excitation of MSSWs (see 3.5(b)), both the horizontal and vertical component of the induced Oersted field can produce torque to the adjacent magnetic moment. Therefore, the excitation efficiency in MSSWs is higher than that of BVMSWs under the same condition. Interestingly, the MSSWs obtained in

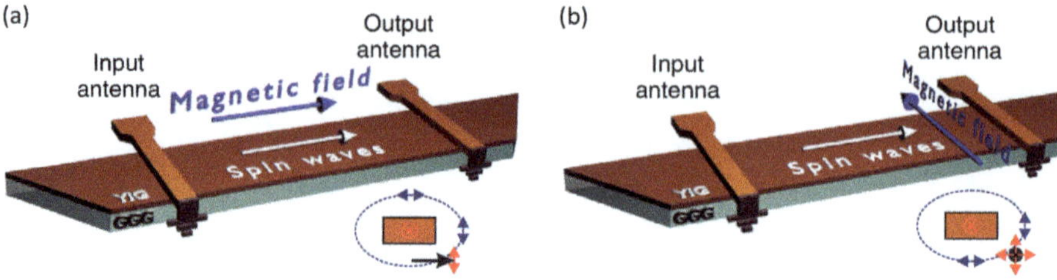

Figure 3.5. Sketches of (a) BVMSW and (b) MSSW experimental set-ups. The insets schematically illustrate the excitation mechanism. (Reproduced with permission from [16]. Copyright 2010 IOP Publishing.)

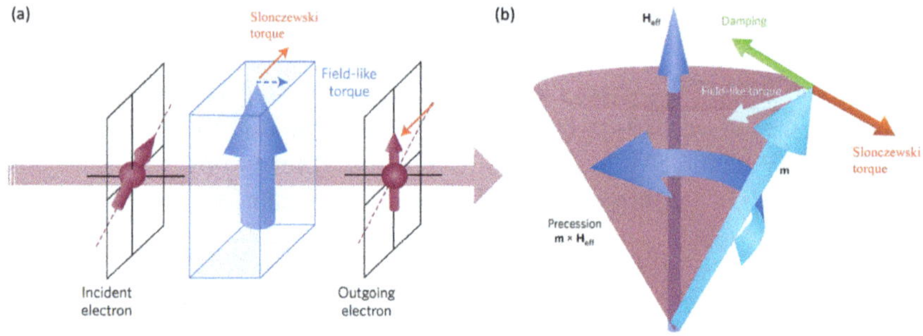

Figure 3.6. (a) An illustration of current-induced torques which originate from the interaction between the spin polarized electron and local magnetization when they are not collinear. (b) A sketch of the magnetization dynamics when STT is included. (Reproduced with permission from [36]. Copyright 2012 Nature.)

figure 3.5(b) are nonreciprocal, which means that the properties of SWs propagating in different directions are distinct [32]. In addition, by tuning the parameters, such as the size of the microwave antenna and the frequency of radiofrequency current, the frequency, wavelength and phase of the generated SWs can be modified easily within a certain range. However, the microwave antenna method only can be used to excite magnetostatic SWs with long wavelengths. It is difficult to excite exchange SWs with short wavelengths, which is mainly because it is difficult to achieve a size of the microwave antenna in the nanometer scale in experiments.

In 1996 Slonczewski [34] and Berger [35] independently predicted that when spin polarized current (the number of electrons with spin-up is different to number of electrons with spin-down) passes through a ferromagnetic layer, the local magnetic moments would suffer from a torque, which is called spin transfer torque (STT). The illustration of current-induced torque is plotted in figure 3.6(a). The STT can be qualitatively understood as follows: when the spin polarized electrons pass through a ferromagnetic layer, if the polarization direction of the incident electron is not collinear with local magnetic moment, the electron would experience a torque that deflects its polarization direction because the polarization direction of the outgoing electron is forced to be along with the direction of the local magnetic moment. According to Newton's third law, the local magnetization will also suffer from an opposite torque, i.e. the STT [36]. Remarkably, by using the STT effect, one can realize a spin torque nano-oscillator (STNO) in a magnetic system and the SWs can be excited in this architecture [37–43]. The principle of the STNO can be explained as follows. At first, when we consider the STT, the LLG equation (3.1) should be modified as [34, 44, 45]

$$\frac{\partial \mathbf{M}}{\partial t} = -\gamma \mathbf{M} \times \mathbf{H}_{\text{eff}} + \frac{\alpha}{M_s}\mathbf{M} \times \frac{\partial \mathbf{M}}{\partial t} + \mathbf{T}_{\text{STT}}. \tag{3.14}$$

Generally speaking, the spin transfer torque $\mathbf{T}_{\text{STT}}$ can be divided into two parts, i.e. the Slonczewski torque $\mathbf{T}_{\text{ST}}$ and a field-like torque $\mathbf{T}_{\text{FLT}}$:

$$\mathbf{T}_{\text{ST}} = -\gamma \frac{a_J}{M_s}\mathbf{M} \times (\mathbf{M} \times \mathbf{m}_p) \tag{3.15}$$

and

$$\mathbf{T}_{\text{FLT}} = -\gamma \xi a_J (\mathbf{M} \times \mathbf{m}_p), \tag{3.16}$$

where $\mathbf{m}_p$ is a unit vector along the magnetization direction of the reference layer, $a_J = \hbar JP/2deM_s$, where $\hbar$ is the reduced Planck constant, d is the thickness of the free layer, J is the current density, $e > 0$ is the charge of the electron, and P is the spin polarization of the current. Then we consider the magnetic moment precessing around the effective field (see figure 3.6(b)). Because the Slonczewski torque $\mathbf{T}_{\text{ST}}$ is parallel to the damping torque, these two torques can be balanced and the magnetization can maintain stable oscillation when we choose a proper polarized current (direction and strength). Here the field-like torque $\mathbf{T}_{\text{FLT}}$ can be regarded as an extra effective field and just changes the frequency of the magnetization's precessional motion. In recent years, magnon excitation by means of STNOs has attracted more and more attention. The main reason is that compared to other methods of magnon excitation, it will be much easier to control the current. In addition, from an experimental point of view, the combination of spintronic devices and integrated circuits is a general trend, which also requires that the generation of SWs can be realized by current.

In 2010, using the micro-focus Brillouin light scattering spectroscopy, Demidov *et al* [41] observed experimentally the propagation of SWs generated by in-plane magnetized nanocontact STNO for the first time. An illustration is shown in figure 3.7(a) where the STNO device (nanocontact) is placed on permalloy (Py) film. The nanocontact contains a nanopatterned polarizing layer ($Co_{70}Fe_{30}$) and a Cu spacer. When a dc current I flows from the polarizer to the Py free layer, because of the STT effect, the local magnetization in the Py film begins to oscillate and the SWs are excited. Two-dimensional maps of the spin wave intensity distributions for an external magnetic field in different directions are shown in figure 3.7(b). One can clearly see that the propagation direction of the SWs is closely related to the direction of the external magnetic field H_0. However, in this case, the propagation distance of the SWs is too short (no more than one micron). Subsequently, with the help of the same technique, Madami *et al* [43] considered the out-of-plane magnetized nanocontact STNO and observed that the propagating SWs are emitted radially from the nanocontact region into the ferromagnetic film up to several micrometers away from the nanocontact. A schematic of the sample is shown in figure 3.7(c). Here a thicker CoFe layer is considered as the fixed magnetic layer

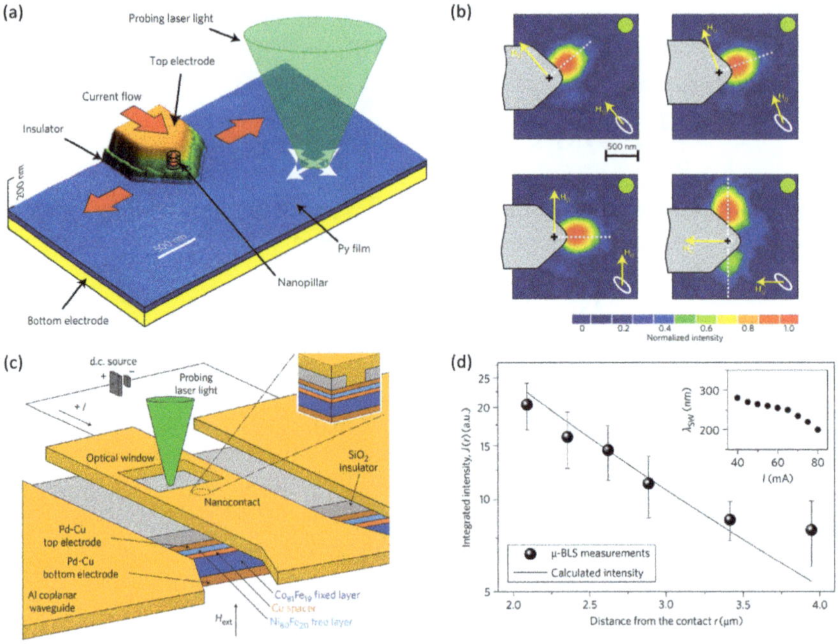

Figure 3.7. A schematic of the magnon excitation through STNO with an (a) in-plane magnetized nanocontact and (c) out-of-plane magnetized nanocontact. Due to the existence of STT, magnons are generated with dc current flowing from the polarizer to the free layer. (b) The normalized intensity maps of spin waves emitted by the STNOs when external magnetic fields are along different directions. (d) The dependence of the magnon intensity on the distance from the STNO. The solid line and dots denote the results from theoretical calculations and experimental measurements, respectively. ((a), (b) Reproduced with permission from [41]. Copyright 2010 Nature. (c), (d) Reproduced with permission from [43]. Copyright 2011 Nature.)

while the thinner NiFe acts as the free magnetic layer. A circular contact is patterned in the middle of the spin valve structure. The perpendicular field H_{ext} is used to control the direction of magnetization in the fixed layer. Figure 3.7(d) plots the SW attenuation as a function of distance from the STNO, from which one can clearly see that the experimental data (black dots) compare well with the theoretical prediction (solid line).

In addition, by utilizing the methods of optical excitation [46, 47] (the energy carried by an ultrashort laser pulse can induce magnetic moments deviating from the equilibrium state, which leads to the oscillation of the magnetic moment, then the SWs are generated) and thermal excitation [48–50] (when a temperature gradient is applied to the magnet, the thermal magnons can be excited because of the spin Seebeck effect), SWs can also be effectively generated in magnetic system.

3.3 Magnon propagation and manipulation

In magnon spintronics, how to effectively control the propagation of the magnon is one of the most important tasks. The band structure of SWs can be significantly tuned by a lot of methods. One appealing method is utilizing magnonic crystals

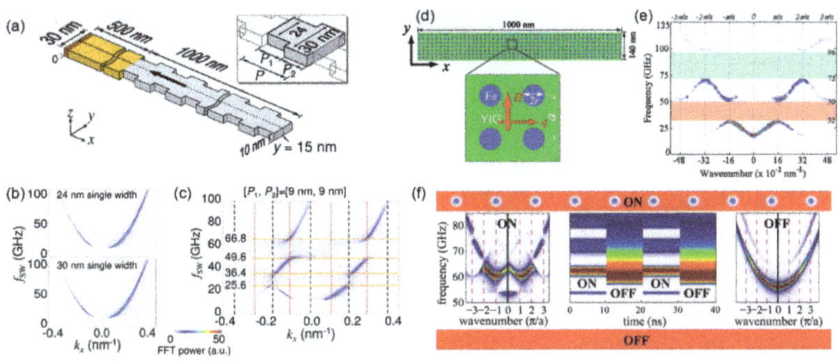

Figure 3.8. (a) A sketch map of MCs in a nanostrip with periodic variation of widths. (b) The SW dispersion relations of single-width nanostrips with 24 and 30 nm. (c) The SW dispersion relations of the MC with $[P_1, P_2] = [9\text{ nm}, 9\text{ nm}]$. (d) A sketch of a two-dimensional MC with a square lattice based on Fe circular nanodots in a YIG film. (e) The SW dispersion curves of the MC described in (d). The dotted lines represent the Brillouin zone boundaries. (f) A schematic diagram and corresponding SW band structures for a uniform magnetized ferromagnetic state and one-dimensional skyrmion lattice. ((a), (b), (c) Reproduced with permission from [53]. Copyright 2009 The American Physical Society. (d), (e) Reproduced with permission from [54]. Copyright 2011 AIP Publishing. (f) Reproduced with permission from [61]. Copyright 2015 American Chemical Society.)

(MCs) [51, 52]. MCs are artificial magnetic structures with periodic variation of geometric or magnetic parameters. Similar to photonic and phononic crystals, the dispersion relations of magnons in MCs exhibit a series of forbidden and allowed bands [53–56]. Generally speaking, MCs can be divided into one-dimensional (1D), two-dimensional (2D), and three-dimensional (3D) configurations.

Lee *et al* [53] investigated 1D MCs based on a nanostrip with periodic variation of width, with the results plotted in figure 3.8(a). They showed that there are no magnonic gaps for a nanostrip with uniform width (see figure 3.8(b)). However, two magnonic band gaps appear (see figure 3.8(c)) when the nanostrip width varies periodically. In addition, they further found that the position, number, and width of the band gaps can be modified by changing the geometric parameters. Similarly, the 2D MCs with periodic variation of magnetic parameters can also show a few allowed and forbidden bands [54], as plotted in figures 3.8(d) and 3.8(e). The corresponding band gaps can be modified by tuning the lattice constant, external magnetic field, etc. The 3D MCs [57, 58] have also been designed to manipulate the magnon band structure. Chumak *et al* [59] and Karenowska *et al* [60] proposed the concept of dynamical MCs, in which the effective field of the magnetic film periodically varies due to the periodic Oersted field induced by the AC current. By controlling the intensity and direction of current, the propagation characteristics of the SWs in the MC can be dynamically controlled. Interestingly, Ma *et al* [61] proposed a dynamical MC based on a magnetic skyrmion lattice. A few band gaps will emerge when magnons propagate in this system, as shown in figure 3.8(f). Moreover, they showed that the width and center frequencies of the band gaps can be tuned by changing the size of the skyrmions via adjusting external magnetic field or the current. In addition, MCs based on other types of magnetic solitons have been designed as well [62–64].

Garcia-Sanchez *et al* [65] reported that the propagation of SWs can be confined within the domain wall. Figure 3.9(a) shows a sketch where the magnetic moments are vertically magnetized and the domain wall structures are formed along the width direction (see the inset). When the SW is excited at one end of the nanostrip, if the exciting frequency is lower than the ferromagnetic resonance frequency of the system, the SW can only propagate in the DW. What is more important, the SW can pass flexibly through the curved magnetic structure by using this method. Subsequently, Wagner *et al* [66] proved that the propagation characteristics of SWs can be tuned by the transverse domain walls in in-plane magnetized magnets, as shown in figure 3.9(b). When the excitation frequency is very low, the SW can only propagate in the domain wall channel. However, if the excitation frequency exceeds a critical value (ferromagnetic resonance frequency), the SW can propagate over the whole nanostrip.

As mentioned above, the magnon excitation can be realized by using the STT effect induced by the out-of-plane spin polarized current. Similarly, it has been proved that the STT can modulate the propagation characteristics of the SW (such as the amplitude, wavelength, and damping). In 2009 Seo *et al* [67] studied in detail the effect of in-plane spin polarized current on SW band structure. The LLG equation including the STT induced by the in-plane spin polarized current can be expressed as [68, 69]

$$\frac{\partial \mathbf{m}}{\partial t} = -\gamma \mathbf{m} \times \mathbf{H}_{\text{eff}} + \alpha \mathbf{m} \times \frac{\partial \mathbf{m}}{\partial t} + (\mathbf{u} \cdot \nabla)\mathbf{m} - \beta \mathbf{m} \times [(\mathbf{u} \cdot \nabla)\mathbf{m}], \tag{3.17}$$

where $\mathbf{u} = u_0\hat{x}$ (here we consider a one-dimensional model, see figure 3.10(a)), $u_0 = \mu_B j_e P/eM_s$ is the magnitude of the adiabatic spin troque, j_e is the current density, μ_B is the Bohr magneton, β is the non-adiabaticity coefficient, and $\mathbf{m}$ is the unit vector of magnetization. Considering a small amplitude SW which decays exponentially in the perturbative limit, we have

$$\mathbf{m} = \hat{x} + \mathbf{m}_0 \exp[\mathrm{i}(\omega t + kx)]\exp[-x/\Lambda], \tag{3.18}$$

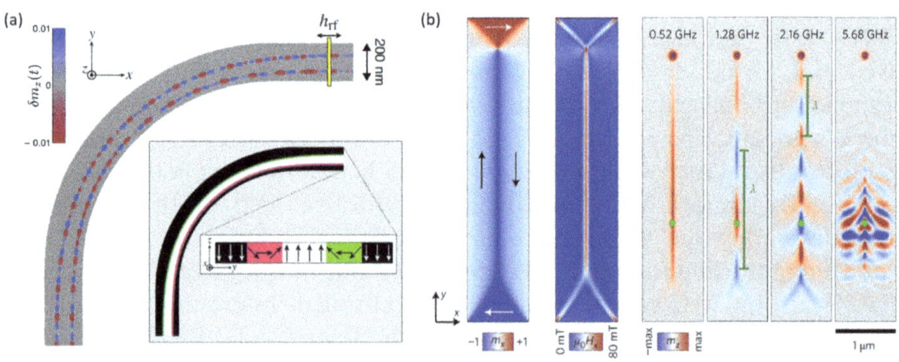

Figure 3.9. (a) The SW channeling around a curved track through two Néel-type domain walls. The microwave antenna with $f = 5$ GHz is located at the top of the track. (b) The domain configuration of an in-plane magnetized Py nanostrip and the amplitude profiles of SW modes with different frequencies. The green dots denote the positions of excitation sources. ((a) Reproduced with permission from [65]. Copyright 2015 The American Physical Society. (b) Reproduced with permission from [66]. Copyright 2016 Nature.)

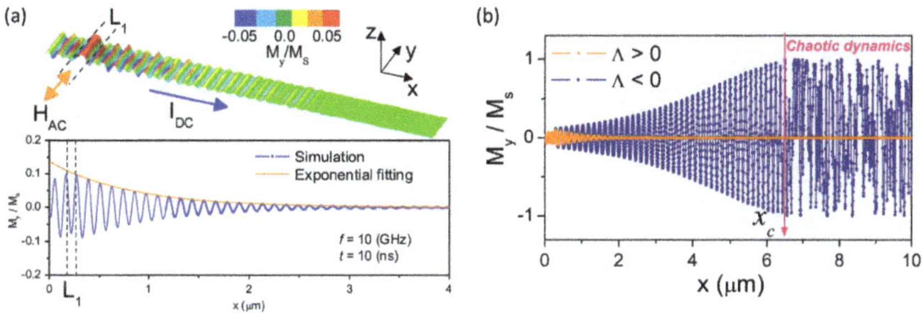

Figure 3.10. (a) Schematics of the propagating SWs along the nanowire. (b) The spatial variation of the normalized M_y component with frequency $f = 10$ GHz. The cases of positive ($\Lambda > 0$) and negative ($\Lambda < 0$) attenuation lengths correspond to the conditions of $\beta = 10\alpha$ and $\beta = 0$, respectively. (Reproduced with permission from [67]. Copyright 2009 The American Physical Society.)

where $|\mathbf{m}_0| \ll 1$ and Λ is the SW attenuation. In addition, the effective field can be expressed as

$$\mathbf{H}_{\text{eff}} = H_k m_x \hat{x} + D\nabla^2 \mathbf{m} - H_d m_z \hat{z}, \tag{3.19}$$

where H_k and H_d are the easy-axis and the hard-axis anisotropy fields, respectively, $D = 2A/M_s$. Substituting equations (3.18) and (3.19) into (3.17), we obtain the approximate expression for the SW attenuation length,

$$\Lambda = \frac{2\gamma Dk}{\alpha\omega - \beta u_0 k}. \tag{3.20}$$

From equation (3.20), one can clearly see that Λ increases or decreases depending on the relative sign of u_0 and k. Specifically, when the direction of the electron motion is consistent with the propagation direction of the SW, the attenuation length will increase. Interestingly, if the current density is large enough (the value $\beta u_0 k/\alpha$ exceeds ω, i.e. $\Lambda < 0$), the SW amplitude grows exponentially rather than attenuates. However, when the direction of the electron motion is opposite to the propagation direction of the SW, the spin wave is obviously suppressed. The greater the current density, the smaller the attenuation length, as shown in figure 3.10(b). Later, the STT effect induced by the out-of-plane spin polarized current has been proved to be able to modulate the SW as well [70]. Similar to the in-plane current, the gain or suppression of SW propagation can also be achieved by adjusting the parameters of out-of-plane spin polarized current. In addition, Zhou *et al* [71] showed that the ferromagnetic system can support left-handed polarized magnons when an appropriate spin polarized current is applied. Interestingly, they further confirmed that the right-handed and left-handed polarized magnons can coexist if the current density exceeds a critical value. These theoretical analyses may provide important references for the future experiments to manipulate the SW handedness through spin polarized current.

In a separate work, Moon *et al* [72] found that the interfacial DM interaction can modify the magnon band structure and lead to the appearance of nonreciprocal magnon propagation. The model is plotted in figure 3.11(a). The dynamics of

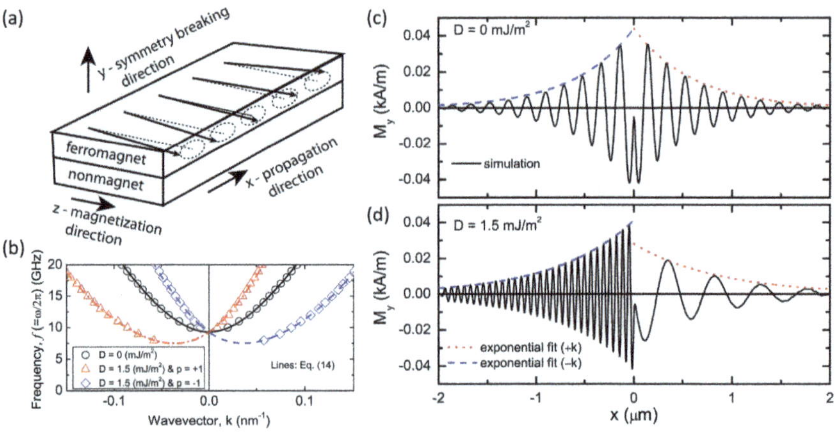

Figure 3.11. (a) A sketch map of the magnetic nanostrip, where the symmetry breaking is along the *y*-axis. (b) The asymmetric SW dispersion curves induced by interfacial DM interaction with the region of the large-*k* limit. Snapshot images of the spin wave (frequency $f = 11$ GHz) distribution along the *x*-axis for (c) $D = 0$ mJ m^{-2} and (d) $D = 1.5$ mJ m^{-2}. (Reproduced with permission from [72]. Copyright 2013 The American Physical Society.)

magnetization in the ferromagnet is given by the LLG equation (equation (3.1)), and the effective field $\mathbf{H}_{\text{eff}}$ can be expressed as

$$\mathbf{H}_{\text{eff}} = H\hat{z} + \frac{2A}{\mu_0 M_s^2}\nabla^2\mathbf{M} - \frac{2D}{\mu_0 M_s^2}\left(\hat{z} \times \frac{\partial \mathbf{M}}{\partial x}\right) + \mathbf{H}_{\text{dipole}}, \tag{3.21}$$

where the first term denotes an external magnetic field with strength *H*, the second and third terms represent the exchange and DM interaction, respectively, and the last term is the dipolar field [28, 73]

$$\mathbf{H}_{\text{dipole}} = -\frac{M_s}{2}(1 - e^{-2|k|d})m_x\hat{x} - M_s(1 - (1 - e^{-2|k|d})/4)m_y\hat{y}. \tag{3.22}$$

Inserting equations (3.21) and (3.22) into equation (3.1), and neglecting small high-order terms, we obtain the magnon dispersion relation as follows

$$\frac{\omega}{\gamma\mu_0} = \sqrt{\left(H + \frac{M_s}{4} + \frac{2A}{\mu_0 M_s}k^2\right)\left(H + \frac{3M_s}{4} + \frac{2A}{\mu_0 M_s}k^2\right) - \frac{e^{-4|k|d}M_s^2}{16}(1 + 2e^{2|k|d})} + \frac{2D}{\mu_0 M_s}k, \tag{3.23}$$

where *d* and *D* are the film thickness and the intensity of DM interaction, respectively. From equation (3.23) we can see that, if $D = 0$, the band structure is symmetrical, i.e. the energy of spin wave is degenerate for $+k$ and $-k$. However, for a nonzero DM interaction, the magnon dipersion becomes asymmetric. In other words, the frequencies of the SW are not equal for $+k$ and $-k$ (see figure 3.11(b)). Micromagnetic simulations verify the theoretical prediction: when $D = 0$, the propagation of SW is symmetrical, while if $D \neq 0$, the propagation characteristics of SW are totally different, as shown in figures 3.11(c) and (d).

It is worth noting that a nonreciprocal SW can also be generated by utilizing a magnetic grating [74, 75], dipolar interaction [76], magnetization gradient [77], magnetoelastic coupling [78], light excitation [79], left-hand microwave [80], etc.

3.4 Magnon detection

The detection of SWs was realized as early as the 1940s. In 1946 Griffiths [81] directly observed a SW with a wave vector $k = 0$ (also called ferromagnetic resonance mode) by using a resonant measurement. Then, the detection of a spin wave with a nonzero wave vector was achieved in 1966 using a Brillouin light scattering technique [82]. Similar to the excitation of SWs, there exist many methods to detect SWs, such as the microwave antenna technique [16] (the inverse of the excitation process, when the propagating spin wave reaches the detection area, it will induce an ac current in the output antenna), optical detection technology including the time-resolved magneto-optical Kerr effect [83–85] and Brillouin light scattering technique [86–88], thermal detection technology (detecting the thermal footprint of magnon motion) [89, 90], electrical detection technology (using spin pumping and inverse spin Hall effects) [91–93], etc.

Among these methods for magnon detection, Brillouin light scattering (BLS) spectroscopy is the most widely used method because of its outstanding advantages [94]. On the one hand, BLS spectroscopy has the ability to map the SW intensity with high spatial resolution and sensitivity. On the other hand, the information on SW frequency and wave vector can be detected simultaneously through BLS spectroscopy. BLS spectroscopy is based on the inelastic scattering of photons by magnons. The scheme of this process is depicted in figure 3.12(a), where a magnon is created (the Stokes process). Comparatively, the annihilation of a magnon upon scattering with a photon is denoted as an anti-Stokes process. The energy and momentum are conserved in both cases, we therefore have the following relations:

$$h\nu_{\text{out}} = h\nu_{\text{in}} \mp h\nu_{\text{sw}} \tag{3.24}$$

and

$$\hbar\mathbf{k}_{\text{out}} = \hbar\mathbf{k}_{\text{in}} \mp \hbar\mathbf{k}_{\text{sw}}, \tag{3.25}$$

where ν and $\mathbf{k}$ are the corresponding frequency and wave vector (see figure 3.12(a)). By analyzing the frequency shift $|\nu_{\text{out}} - \nu_{\text{in}}|$ and wave vector shift $|\mathbf{k}_{\text{out}} - \mathbf{k}_{\text{in}}|$ of the scattered photon, we can obtain the frequency and momentum information of the detected magnons. The energy analysis (equation (3.24)) of the spin wave signal is the basis for BLS spectroscopy. In general, high-contrast tandem Fabry–Pérot interferometers are used to analyze the weak light signals generated by the inelastic scattering of photons and SWs, as shown in figure 3.12(b). A typical result of BLS measurement is shown in figure 3.12(c) in which the spectrum exhibits three peaks. The zero position is the signal of the reference laser, while the peak located at negative (positive) frequency is the Stokes (anti-Stokes) signal. The peak height in the BLS spectrum represents the SW intensity.

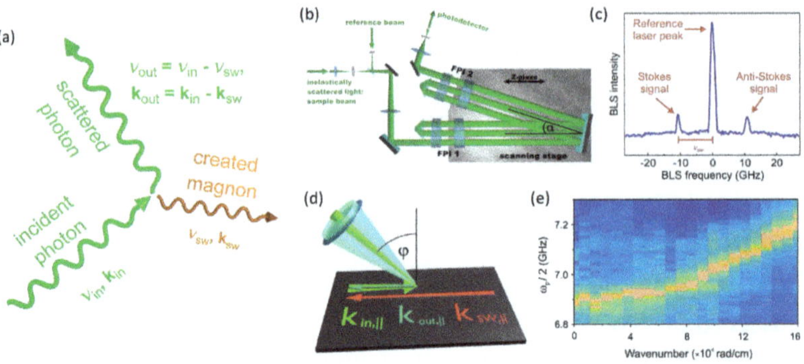

Figure 3.12. (a) A sketch of the BLS process. The incident light with wave vector $\mathbf{k}_{in}$ and frequency ν_{in} generates a spin wave ($\mathbf{k}_{sw}$, ν_{sw}) in the Stokes process. Energy and momentum conservation lead to a reduced wave vector $\mathbf{k}_{out}$ and a smaller frequency ν_{out} of the scattered light. (b) Principle of a tandem Fabry–Pérot interferometer. (c) BLS spectrum showing the central reference laser peak as well as the Stokes and anti-Stokes signals. (d) Schematic of the backscattering geometry. (e) BLS measurement of the spin wave intensity as a function of the wave vector of parametrically excited magnons. (Reproduced with permission from [94]. CC BY 4.0.)

On the other hand, according to equation (3.25) one can realize the wave vector selective detection of SW. Figure 3.12(d) plots the scheme of the backscattering geometry. By tuning the incidence angle φ, one can obtain information on the magnon wave vector through the following relation:

$$k_{\mathrm{sw},\parallel} = 2 \cdot \frac{2\pi}{\lambda_{\mathrm{L}}} \cdot \sin\varphi, \tag{3.26}$$

where λ_{L} is the laser wavelength. Figure 3.12(e) shows a measurement of the SW intensity by utilizing wave vector-resolved BLS spectroscope [95]. Here, every column represents a BLS spectrum detected for a fixed wave vector.

As shown in figure 3.2, in magnon spintronics the spin wave signal ultimately needs to be converted back to the electrical signal, so the electrical detection technique for the magnon becomes particularly important. In addition to the microwave antenna method mentioned above, in recent years people have proposed new electrical detection technologies based on two new spintronic effects, i.e. the spin pumping effect (SPE) [93, 96, 97] and inverse spin Hall effect (ISHE) [91, 92, 98]. Illustrations of the SPE and ISHE are provided in figure 3.13(a). The SPE is the process in which a pure spin current is injected from a ferromagnetic layer to an adjacent heavy metal (Pt for instance) layer, while the ISHE is the inverse process of the spin Hall effect (SHE), i.e. a conversion of spin current to electric current: when the pure spin current $\mathbf{J}_{\mathrm{s}}$ is generated in the Pt through SPE, two electrons with opposite spins travel in the opposite directions along $\mathbf{J}_{\mathrm{s}}$, then the spin–orbit interaction bends these two electrons in the same direction, leading to a charge current $\mathbf{J}_{\mathrm{c}}$ perpendicular to $\mathbf{J}_{\mathrm{s}}$; see figure 3.13(a).

As early as 2002, Tserkovnyak *et al* [99] predicted theoretically that the precession of magnetic moment in the ferromagnetic layer would pump spin current into the adjacent non-magnetic metal layer. In 2006 Costache *et al* [100] proposed an electrical detection method for measuring the spin current induced by spin pumping.

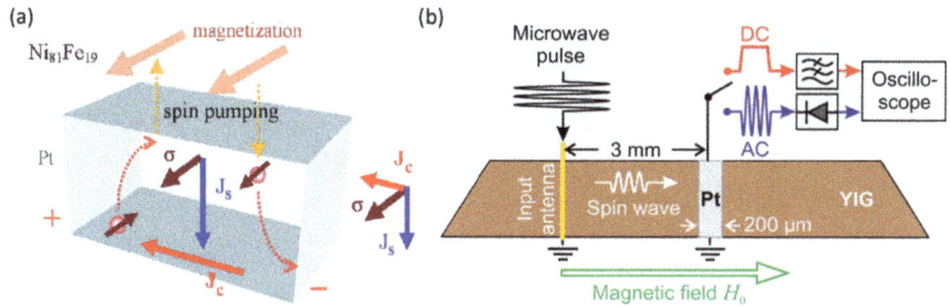

Figure 3.13. (a) A sketch map of the SPE and the ISHE. $\mathbf{J}_S$ and $\mathbf{J}_C$ denote the pure spin current induced by the SPE and an electric current generated by the ISHE, respectively. σ is the spin-polarization vector of the spin current. (b) Schematic illustration of the experimental detection of a magnon by the ISHE. The magnon is excited in the YIG waveguide by utilizing a microstrip antenna and is detected at the 3 mm separated Pt stripe as electromagnetically induced ac and ISHE voltage dc signals. ((a) Reproduced with permission from [92]. Copyright 2006 AIP Publishing. (b) Reproduced with permission from [101]. Copyright 2012 AIP Publishing)

Subsequently, Saitoh *et al* [92] observed the current signal in the heavy metal layer (Pt) induced by the uniform precession of the magnetization in the adjacent ferromagnetic layer through ISHE. In 2012 Chumak *et al* [101] successfully detected the propagating spin wave using SP-ISHE. Figure 3.13(b) shows a schematic illustration: the microwave pulse is applied in the input antenna to excite SWs, then the transmitted SW is detected at the Pt layer. The electrical measurement technology combining SPE and ISHE has become one of the most important methods to detect SWs. Using this method to detect SWs has several advantages over other methods. On the one hand, the electrical measurement technology can detect both the magnetostatic SWs [102, 103] and the exchange SWs with very short wavelengths [93, 104], which is challenging for microwave and optical methods. On the other hand, it is also an indispensable part of magnon spintronics to convert magnon signals into electrical signals by using electrical detection techniques.

3.5 Magnonic devices

As discussed in section 3.1, magnonic devices have many outstanding advantages compared to conventional electronic devices. By utilizing the external forces (magnetic field, current, and light, etc), one can manipulate SWs and realize various logical operations. In 2005 Kostylev *et al* [6] proposed encoding the amplitude of the spin wave into binary and using the current to modulate the phase of the interfering magnon, then to realize a logic gate. Subsequently, Schneider *et al* [7] realized the operation of an exclusive-not-OR (XNOR) logic gate, with an illustration shown in figure 3.14(a). Here magnons are excited at the magnon source and detected at the output, and the current I_1 (I_2) is applied to input 1 (input 2) which serves as the logical input. The logical '0' and '1' are represented by $I = 0$ A and I_π, respectively, where I_π means the current creating a phase shift of π to the magnon. The process of logic operation is as follows: when there are no currents at both inputs, the amplitude of spin wave reaches a relatively large value due to constructive interference, and the output is logic '1'; when a current is applied at one of the

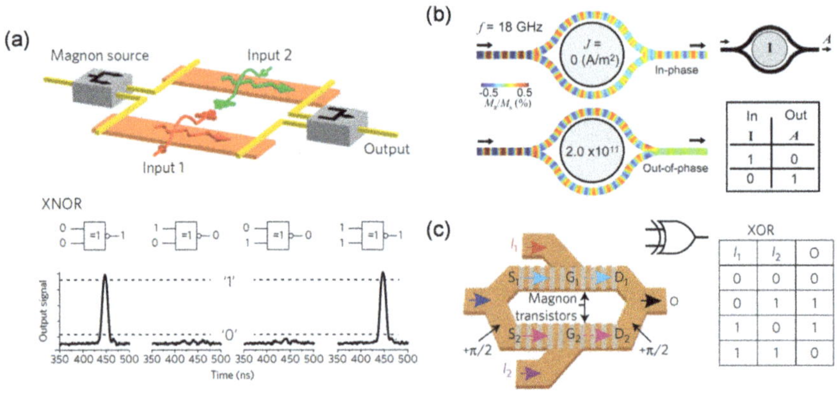

Figure 3.14. (a) A spin wave XNOR logic gate based on a Mach–Zehnder interferometer with electric current-controlled phase shifters. (b) Logical operations of a NOT gate using spin waves. (c) The operational principle of an XOR logic gate based on the magnon transistor. ((a) Reproduced with permission from [7]. Copyright 2008 AIP Publishing. (b) Reproduced with permission from [9]. Copyright 2008 AIP Publishing. (c) Reproduced with permission from [14]. Copyright 2014 the Authors.)

input terminals, due to the modulation of the π phase of the SW by the current, the spin wave experiences interference cancellation and the amplitude becomes relatively small, and the output is logic '0'. In addition, when the currents are applied at both input terminals, the two channel spin waves are both modulated by the π phase, so there still exists constructive interference when they meet, and the output logic is '1'. The truth table and the corresponding output signals are shown at the bottom of figure 3.14(a). Similarly, the NOT logic function can also be realized by current, as shown in figure 3.14(b). In the case of zero current density $J = 0$ (logical input '0'), the two merged SW beams constructively interfere with logical output '1'. In contrast, when a current is applied, the split SWs are merged to destructively interfere and the logical output is '0'. In addition, the XOR logic gate can be achieved through magnon waveguides, as shown in figure 3.14(c). Here the two channel spin waves (blue and magenta arrows) are independently controlled by the input signals I_1 and I_2 (current create $\pi/2$ phase shift). By performing an analysis similar to figures 3.14(a) and (b), we can obtain the corresponding truth table as shown in the figure.

Furthermore, Khitun *et al* [105] showed that the logic operation can also be realized by using the phase of the spin wave to encode information, which has been proved in later experiments [106, 107]. Compared to the logic gate based on the amplitude of a spin wave, it is more convenient to construct the NOT and majority gates utilizing the phase information of the spin wave [108].

Remarkably, in 2014 Chumak *et al* [14] proposed the magnon transistor for all-magnon data processing, as shown in figure 3.15(a). In this three-terminal device, magnons (blue balls) flow from the source to the drain and they are controlled by the extra magnons (red balls) injected into the gate through the nonlinear interaction between magnons (see figure 3.15(b)). The region where magnons interact is

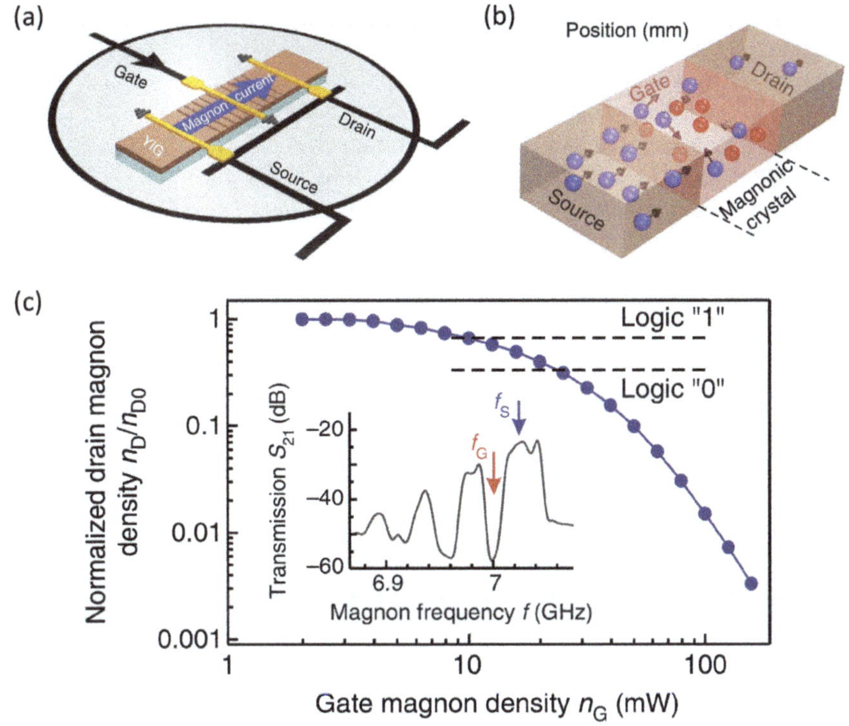

Figure 3.15. (a) Schematics of a magnon transistor. Magnons are injected into the transistor's source and are detected at its drain. Magnons that control the source-to-drain magnon current are injected directly into the magnonic crystal. (b) The transistor's operational principle with a nonlinear four-magnon scattering mechanism. (c) The drain magnon density n_D normalized to the original value n_{D0} is plotted as the function of the magnon density n_G at the gate. (Reproduced with permission from [14]. Copyright 2014 the Authors.)

composed of magnonic crystal which is used to increase the density of the gate magnons. Therefore, the efficiency of the nonlinear four-magnon scattering process is enhanced and the source-to-drain magnon current is suppressed, as shown in figure 3.15(c).

In addition to the magnon logic device and magnon transistor, magnons can also be used to design filters [109], diodes [110, 111], multiplexers [112, 113], etc.

References

[1] Bloch F 1930 *Z. Phys.* **61** 206–19

[2] Khitun A, Bao M and Wang K L 2010 *J. Phys. D: Appl. Phys.* **43** 264005

[3] Lenk B *et al* 2011 *Phys. Rep.* **507** 107–36

[4] Kruglyak V V, Demokritov S O and Grundler D 2010 *J. Phys. D: Appl. Phys.* **43** 264001

[5] Chumak A V *et al* 2015 *Nat. Phys.* **11** 453–61

[6] Kostylev M P *et al* 2005 *Appl. Phys. Lett.* **87** 153501

[7] Schneider T *et al* 2008 *Appl. Phys. Lett.* **92** 022505

[8] Klingler S *et al* 2015 *Appl. Phys. Lett.* **106** 212406

[9] Lee K S and Kim S K 2008 *J. Appl. Phys.* **104** 053909

[10] Cornelissen L J *et al* 2015 *Nat. Phys.* **11** 1022–6
[11] Wesenberg D *et al* 2017 *Nat. Phys.* **13** 987–93
[12] Pirro P *et al* 2014 *Appl. Phys. Lett.* **104** 012402
[13] Liu C *et al* 2018 *Nat. Commun.* **9** 738
[14] Chumak A V, Serga A A and Hillebrands B 2014 *Nat. Commun.* **5** 4700
[15] Dumas R K and Åkerman J 2014 *Nat. Nanotechnol.* **9** 503–4
[16] Serga A A, Chumak A V and Hillebrands B 2010 *J. Phys. D: Appl. Phys.* **43** 264002
[17] Li Z X, Cao Y and Yan P 2021 *Phys. Rep.* **915** 1–64
[18] Wang X S and Wang X R 2021 *J. Appl. Phys.* **129** 151101
[19] Pirro P *et al* 2021 *Nat. Rev. Mater.* **6** 1114–35
[20] Yu H M, Xiao J and Schultheiss H 2021 *Phys. Rep.* **905** 1–59
[21] Yuan H Y *et al* 2022 *Phys. Rep.* **965** 1–74
[22] Baltz V *et al* 2018 *Rev. Mod. Phys.* **90** 015005
[23] Jungwirth T *et al* 2016 *Nat. Nanotechnol.* **11** 231–41
[24] Fukami S, Lorenz V O and Gomonay O 2020 *J. Appl. Phys.* **128** 070401
[25] Šmejkal L *et al* 2018 *Nat. Phys.* **14** 242–51
[26] Gilbert T L 2004 *IEEE Trans. Magn.* **40** 3443–9
[27] Eshbach J R and Damon R W 1960 *Phys. Rev.* **118** 1208
[28] Damon R W and Eshbach J R 1961 *J. Phys. Chem. Solids* **19** 308–20
[29] Damon R W and Vaart H V D 1965 *J. Appl. Phys.* **36** 3453
[30] Stancil D D 1003 *Theory of Magnetostatic Waves* (New York: Springer)
[31] Kalinikos B A and Slavin A N 1986 *J. Phys. C: Solid State Phys.* **19** 7013
[32] Schneider T *et al* 2008 *Phys. Rev.* **B 77** 214411
[33] Demidov V E *et al* 2009 *Appl. Phys. Lett.* **95** 112509
[34] Slonczewski J C 1996 *J. Magn. Magn. Mater.* **159** L1–7
[35] Berger L 1996 *Phys. Rev.* **B 54** 9353
[36] Brataas A, Kent A D and Ohno H 2012 *Nat. Mater.* **11** 372–81
[37] Ji Y, Chien C L and Stiles M D 2003 *Phys. Rev. Lett.* **90** 106601
[38] Katine J A *et al* 2000 *Phys. Rev. Lett.* **84** 3149
[39] Gerhart G *et al* 2007 *Phys. Rev.* **B 76** 024437
[40] Demidov V E *et al* 2011 *Phys. Rev.* **B 83** 060406(R)
[41] Demidov V E, Urazhdin S and Demokritov S O 2010 *Nat. Mater.* **9** 984–8
[42] Bonetti S *et al* 2010 *Phys. Rev. Lett.* **105** 217204
[43] Madami M *et al* 2011 *Nat. Nanotechnol.* **6** 635–8
[44] Boone C T *et al* 2010 *Phys. Rev. Lett.* **104** 097203
[45] Khvalkovskiy A V *et al* 2009 *Phys. Rev. Lett.* **102** 067206
[46] Au Y *et al* 2013 *Phys. Rev. Lett.* **110** 097201
[47] Satoh T *et al* 2012 *Nat. Photon.* **6** 662–6
[48] Ritzmann U, Hinzke D and Nowak U 2014 *Phys. Rev.* **B 89** 024409
[49] Lin S Z *et al* 2014 *Phys. Rev. Lett.* **112** 187203
[50] Bauer G E W, Saitoh E and Wees B J V 2012 *Nat. Mater.* **11** 391–9
[51] Krawczyk M and Grundler D 2014 *J. Phys.: Condens. Matter* **26** 123202
[52] Chumak A V, Serga A A and Hillebrands B 2017 *J. Phys. D: Appl. Phys.* **50** 244001
[53] Lee K S, Han D S and Kim S K 2009 *Phys. Rev. Lett.* **102** 127202
[54] Ma F S *et al* 2011 *Appl. Phys. Lett.* **98** 153107
[55] Ciubotaru F *et al* 2013 *Phys. Rev.* **B 88** 134406

[56] Wang Z K *et al* 2009 *Appl. Phys. Lett.* **94** 083112
[57] Krawczyk M and Puszkarski H 2008 *Phys. Rev.* B **77** 054437
[58] Mamica S *et al* 2012 *Phys. Rev.* B **86** 144402
[59] Chumak A V *et al* 2010 *Nat. Commun.* **1** 141
[60] Karenowska A D *et al* 2012 *Phys. Rev. Lett.* **108** 015505
[61] Ma F S *et al* 2015 *Nano Lett.* **15** 4029–36
[62] Wang X G *et al* 2015 *Europhys. Lett.* **109** 37008
[63] Li Z X *et al* 2015 *J. Magn. Magn. Mater.* **388** 10–5
[64] Wang D W *et al* 2016 *IEEE Trans. Magn.* **53** 1–10
[65] Garcia-Sanchez F *et al* 2015 *Phys. Rev. Lett.* **114** 247206
[66] Wagner K *et al* 2016 *Nat. Nanotechnol.* **11** 432–6
[67] Seo S M *et al* 2009 *Phys. Rev. Lett.* **102** 147202
[68] Zhang S and Li Z 2004 *Phys. Rev. Lett.* **93** 127204
[69] Thiaville A *et al* 2005 *Europhys. Lett.* **69** 990
[70] Xing X J, Yu Y P and Li S W 2009 *Appl. Phys. Lett.* **95** 142508
[71] Zhou Z W *et al* 2019 *Phys. Rev.* B **99** 014420
[72] Moon J H *et al* 2013 *Phys. Rev.* B **88** 184404
[73] Arias R and Mills D L 1999 *Phys. Rev.* B **60** 7395
[74] Chen J *et al* 2019 *Phys. Rev.* B **100** 104427
[75] Yu T *et al* 2019 *Phys. Rev.* B **99** 134424
[76] Gallardo R A *et al* 2019 *Phys. Rev. Appl.* **12** 034012
[77] Borys P *et al* 2021 *Phys. Rev.* B **103** 144411
[78] Tateno S and Nozaki Y 2020 *Phys. Rev. Appl.* **13** 034074
[79] Sadovnikov A V *et al* 2019 *Phys. Rev.* B **99** 054424
[80] Zhang Z Z *et al* 2022 *Phys. Rev.* B **106** 174413
[81] Griffiths J H E 1946 *Nature* **158** 670–1
[82] Fleury P A *et al* 1966 *Phys. Rev. Lett.* **17** 84
[83] Zhang G P *et al* 2009 *Nat. Phys.* **5** 499–502
[84] Bauer H G *et al* 2014 *Appl. Phys. Lett.* **104** 102404
[85] Deb M *et al* 2019 *Phys. Rev. Appl.* **12** 044006
[86] Demokritov S O, Hillebrands B and Slavin A N 2001 *Phys. Rep.* **348** 441–89
[87] Gubbiotti G *et al* 2010 *J. Phys. D: Appl. Phys.* **43** 264003
[88] Stashkevich A A *et al* 2009 *Phys. Rev.* B **80** 144406
[89] An T *et al* 2013 *Nat. Mater.* **12** 549–53
[90] Schultheiss H *et al* 2012 *Phys. Rev. Lett.* **109** 237204
[91] Kelly O D *et al* 2013 *Appl. Phys. Lett.* **103** 082408
[92] Saitoh E, Ueda M and Miyajima H 2006 *Appl. Phys. Lett.* **88** 182509
[93] Sandweg C W *et al* 2011 *Phys. Rev. Lett.* **106** 216601
[94] Sebastian T *et al* 2015 *Front. Phys.* **3** 35
[95] Serga A A *et al* 2012 *Phys. Rev.* B **86** 134403
[96] Tserkovnyak Y, Brataas A and Bauer G E W 2002 *Phys. Rev.* B **66** 224403
[97] Rojas-Sánchez J C *et al* 2014 *Phys. Rev. Lett.* **112** 106602
[98] Miao B F *et al* 2013 *Phys. Rev. Lett.* **111** 066602
[99] Tserkovnyak Y, Brataas A and Bauer G E W 2002 *Phys. Rev. Lett.* **88** 117601
[100] Costache M V *et al* 2006 *Phys. Rev. Lett.* **97** 216603
[101] Chumak A V *et al* 2012 *Appl. Phys. Lett.* **100** 082405

[102] Castel V *et al* 2012 *Appl. Phys. Lett.* **101** 132414
[103] Castel V *et al* 2014 *Phys. Rev.* B **90** 214434
[104] Ando K and Saitoh E 2012 *Phys. Rev. Lett.* **109** 026602
[105] Khitun A and Wang K L 2005 *Superlatt. Microstruct.* **38** 184–200
[106] Khitun A *et al* 2007 *Mater. Res. Soc. Symp. Proc.* **1067** B01–4
[107] Wu Y *et al* 2009 *J. Nanoelectron. Optoelectron.* **4** 394–7
[108] Klingler S *et al* 2014 *Appl. Phys. Lett.* **105** 152410
[109] Kim S K, Lee K S and Han D S 2009 *Appl. Phys. Lett.* **95** 082507
[110] Lan J *et al* 2015 *Phys. Rev.* X **5** 041049
[111] Szulc K *et al* 2020 *Phys. Rev. Appl.* **14** 034063
[112] Vogt K *et al* 2014 *Nat. Commun.* **5** 3727
[113] Wang Q, Chumak A V and Pirro P 2021 *Nat. Commun.* **12** 2636

Chapter 4

Topological magnons

This chapter is devoted to the topological states of magnons. We first introduce the magnon Hall effect, including the experimental observation and theoretical explanation. Then we focus on topological magnon insulators in both collinear and noncollinear magnets. After that, we elaborate on the semimetal phases of magnons, including the Dirac, Weyl, and nodal line states. Peculiar topological magnon polarons and higher-order topological magnons are also highlighted. Finally, we present appealing proposals for topological magnonic devices.

4.1 Magnon Hall effect

After the discovery of the magnon Hall effect (MHE), people began to investigate the topological properties of magnons. It is widely known that when a charge current is subject to a perpendicular magnetic field [1], the Hall effect occurs because of the Lorentz force. However, the magnon belongs to a neutral quasi-particle, the MHE therefore does not require Lorentz force, which is analogous to the anomalous Hall effect emerging in metallic ferromagnets [2].

In 2010 Katsura *et al* [3] predicted the existence of the intrinsic thermal MHE in kagome magnets. Importantly, it provides a feasible method to detect MHE using thermal transport measurements. In the same year, Onose *et al* [4] observed the MHE experimentally in the ferromagnet $Lu_2V_2O_7$ with a pyrochlore lattice. The pyrochlore structure can be regarded as the stacking of alternate kagome and triangular lattices along the [111]-direction. They observed a transverse thermal current after a longitudinal temperature gradient was applied. In this insulating ferromagnet, both the phonons and magnons can carry heat current. On the one hand, they found that the thermal Hall conductivity increases sharply and saturates in the range of a low magnetic field. This result cannot be described by the normal Hall effect with the conductivity proportional to the strength of the magnetic field. On the other hand, they identified that the thermal Hall conductivity decreases in the

doi:10.1088/978-0-7503-5082-2ch4

range of a high magnetic field, which cannot be explained by the phonon mechanism [5–7]. They thus concluded that a transverse heat current originates from the MHE. In a pyrochlore lattice, the midpoint between two apices of the tetrahedron is not a center of inversion symmetry, which leads to a nonzero DM interaction. The DM interaction acts as a vector potential and serves as the 'orbital magnetic field' for the propagation of magnons. The MHE therefore occurs.

Various microscopic theories have been proposed to explain the origin of the MHE. Matsumoto *et al* [8, 9] proved that when a magnon wavepacket suffers from a temperature gradient, an anomalous velocity perpendicular to the gradient will emerge, which leads to the edge magnon currents (see figures 4.1(a) and (b)). By using both the semiclassical theory and linear response formalism, they showed that the thermal Hall conductivity κ^{xy} can be written in terms of the Berry curvature $\Omega_n(\mathbf{k})$ as below

$$\kappa^{xy} = -\frac{k_B^2 T}{\hbar V}\sum_{n,\mathbf{k}} c_2(\rho_n)\Omega_{n,z}(\mathbf{k}), \tag{4.1}$$

where $\rho_n = \rho[\varepsilon_n(\mathbf{k})]$ is the Bose–Einstein distribution function $\rho(\varepsilon) = [e^{(\varepsilon-\mu)/k_B T} - 1]^{-1}$, k_B is the Boltzmann constant, μ is the chemical potential, T is the temperature, and $c_2(\rho) = (1+\rho)(\log\frac{1+\rho}{\rho})^2 - (\log\rho)^2 - 2\text{Li}_2(-\rho)$ with $\text{Li}_2(z)$ the polylogarithm function. Near the band crossing, different bands are close to each other, and the Berry curvature reaches its maximum. As a result, the value of κ^{xy} takes its maximum. We can intuitively understand the MHE with the following processes. On the one hand, when the system is in equilibrium state, the edge magnon currents circulating along the boundary will appear because of the confining potential, as shown in figure 4.1(c). The intensity of the magnon currents at opposite edges are equal which leads to a vanishing

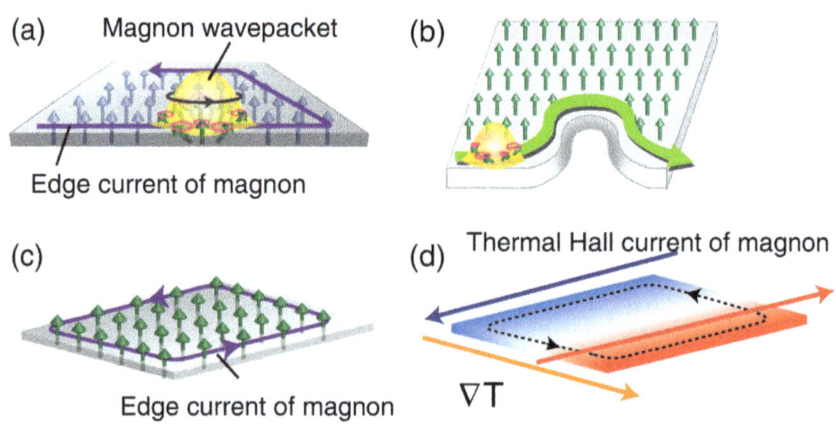

Figure 4.1. (a) A schematic diagram of a magnon wave packet with self-rotation and a magnon edge current. (b) The magnon close to the boundary propagates along the boundary. (c) The edge current of a magnon for an equilibrium system. (d) When the temperature gradient is applied, the transverse heat currents are no longer balanced between two different edges, leading to the emergence of a finite transverse thermal Hall current. (Reproduced with permission from [8]. Copyright 2011 The American Physical Society.)

effective thermal current. On the other hand, when a longitudinal temperature gradient is applied, the thermal magnons will flow from the high-temperature region to the low-temperature region, then the balance of the thermal current at opposite edges is broken, and a finite thermal Hall current therefore appears, as shown in figure 4.1(d). In a follow-up paper, Zhang *et al* [10] proved that the edge magnon currents are actually magnon chiral edge modes originating from the topological magnonic band structure by evaluating the Chern number. In addition, they demonstrated that the chiral edge state mainly localizes around edges and leaks into the bulk with an oscillatory decay. By utilizing the nonequilibrium Green's function method [11], Zhang *et al* [10] demonstrated that the transport of edge states is robust against moderate defects and disorder. In brief, the MHE results from the topological magnonic band structure.

Shortly after its discovery in pyrochlore ferromagnets [4], the MHE has been reported in other magnetic materials. For example, Hirschberger *et al* [12] observed a large thermal Hall conductivity κ_{xy} in a kagome magnet Cu(1,3-bdc), as plotted in figures 4.2(a) and (b). The kagome magnet Cu(1,3-bdc) is comprised of stacked kagome planes and the spin $-\frac{1}{2}$ Cu^{2+} moments interact via an in-plane ferromagnetic exchange coupling [13]. Surprisingly, the measured κ_{xy} exhibits sign reversal with the increase of magnetic field or temperature, which can be explained as the sign switching of the Chern number for magnon bands. In addition, Hirschberger *et al* [14] also observed the MHE in frustrated pryocholore quantum magnet $Tb_2Ti_2O_7$. Moreover, in the unsaturated yttrium iron garnet (YIG) film, Tanabe *et al* [15] observed the magnonic Hall-like effect via sample-edge scattering. Figures 4.3(a)–(c) show measurements for the distribution of temperature when a coplanar waveguide (CPW) is located at different regions with an external magnetic field $B^{\text{ext}} = 109$ mT. One can see that a thermal gradient with ~0.3 K mm^{-1} appears in the YIG, as shown in figure 4.3(a). Further, the sign of thermal gradient reverses when the CPW is placed at the opposite side, as shown in figure 4.3(b). In contrast, no thermal gradient emerges when the CPW is placed at the center of the sample (see figure 4.3(c)). These results suggest that the thermal gradient emerging in figures 4.3(a) and (b) originates from the edge magnons.

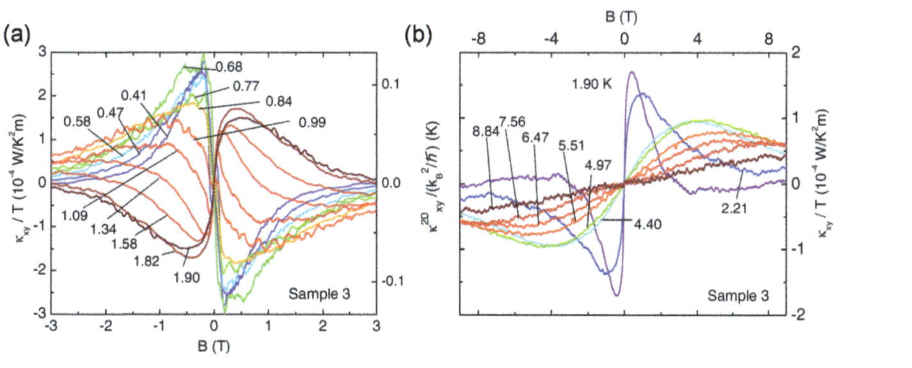

Figure 4.2. (a) and (b) Experimental measurements of thermal Hall conductivity for the kagome magnet Cu (1,3-bdc) with different temperatures. (Reproduced with permission from [12]. Copyright 2015 The American Physical Society.)

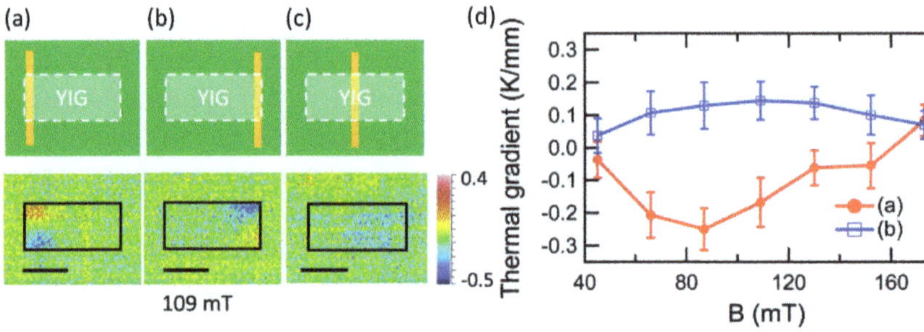

Figure 4.3. (a)–(c) Upper pictures plot the structure of YIG film and the CPW, while the corresponding temperature distribution are shown in the lower panels. (d) The dependence of thermal gradient on the magnetic field. The red dots and blue squares are experimental results for considering the structures in (a) and (b), respectively. (Reproduced with permission from [15]. Copyright 2016 Wiley.)

The magnetic field dependence of the thermal gradient is also shown in figure 4.3(d). One can see that the Hall-like effect is enhanced in the unsaturated regime (between 70 and 150 mT).

Later, Matsumoto *et al* [16] developed a comprehensive theory to describe the MHE in magnets, which is widely applicable to various types of magnets, including ferromagnets with dipole–dipole coupling and antiferromagnets with the DM interaction.

The above discussions about MHE are based on collinear magnets. Remarkably, Hoogdalem *et al* [17] theoretically showed that when magnons move through inhomogeneous magnetizations, the geometrical properties of a smooth noncollinear magnetic spin texture (such as a skyrmion) will induce a fictitious magnetic field and emergent Lorentz force, leading to the magnon thermal Hall effect. This Hall effect comes from the nonzero topological charge of the spin texture, and is thus called the topological magnon Hall effect (TMHE). They further showed that the fictitious magnetic vector potential consists of two comparable contributions which include the contribution from the geometric texture and that from the original spin–orbit term in the free energy. Subsequently, utilizing a Lorentz transmission electron microscope (TEM), Mochizuki *et al* [18] experimentally verified the existence of the TMHE. They observed the unidirectional rotation motion of the skyrmion lattice in Cu_2OSeO_3 and MnSi films, as shown in figure 4.4(a). This result can be understood as follows: the electron-beam irradiation in the Lorentz TEM generates a thermal gradient, leading to the emergence of a thermal magnon current. Then the magnons are deflected by the fictitious magnetic field in the skyrmion lattice. In turn, this results in the rotational motion of the skyrmion lattice via the spin-transfer torque (STT) and/or momentum transfer. In addition, the direction of skyrmion lattice rotation reverses when the sign of the temperature gradient switches, as shown in figure 4.4(b). With the help of the atomistic spin dynamics, Mook *et al* [19] also investigated the MHE induced by the skyrmion lattice and spin spiral. They found that the magnon Hall angle can reach 60%.

The discovery of the MHE helped people understand the topologically nontrivial properties of magnons. The topological states of magnons have promising potential

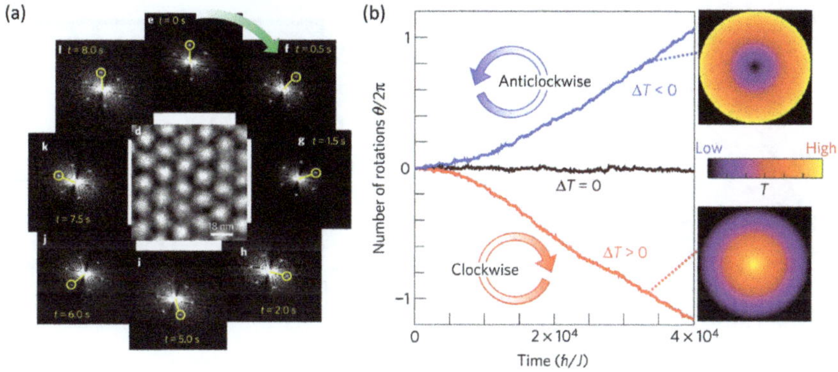

Figure 4.4. (a) The results of Fourier transforms for a skyrmion crystal in MnSi with different times, which shows a clockwise rotation. (b) The number of rotations $\theta/2\pi$ of the skyrmion lattice as a function of time under different cases: equilibrium ($\Delta T = 0$), positive T gradient ($\Delta T > 0$), and negative T gradient ($\Delta T < 0$). (Reproduced with permission from [18]. Copyright 2014 Nature.)

applications for designing various novel robust spintronic devices. Over the past decade, many studies have focused on the magnon topological properties, including magnon topological insulators and semimetals. The concepts and methodology will be elaborated upon in detail in the following sections.

4.2 Topological magnon insulators

In general, there exist two different magnetic configurations that allow topological magnon insulators, i.e. collinear magnets and noncollinear magnetic texture. In what follows, we discuss topological magnon insulators based on a few concrete examples.

4.2.1 Collinear magnets

In 2014 Mook *et al* [20] studied in detail the topological properties of magnons in a kagome lattice (see figure 4.5(a)) based on a quantum-mechanical Heisenberg model [21]:

$$H = -\sum_{n \neq m} J_m^n \hat{\mathbf{s}}_m \cdot \hat{\mathbf{s}}_n + \sum_{n \neq m} \mathbf{D}_m^n (\hat{\mathbf{s}}_m \times \hat{\mathbf{s}}_n) - g\mu_{\mathrm{B}} \sum_n \mathbf{H} \cdot \hat{\mathbf{s}}_n. \tag{4.2}$$

The first term is the exchange interaction, where two spin operators $\hat{\mathbf{s}}_m$ and $\hat{\mathbf{s}}_n$ are coupled via symmetric exchange coefficients $J_m^n = J_n^m$, m and n denote different sites. The second term is the antisymmetric DM interaction, where $\mathbf{D}_m^n$ represents the DM vector between sites m and n ($\mathbf{D}_m^n = -\mathbf{D}_n^m$). The last term describes the Zeeman coupling, with $\mathbf{H}$ the external magnetic field, g the g-factor of the electrons, and μ_{B} being the Bohr magneton.

The so-called 'bulk–boundary correspondence' indicates that the bulk property (Chern number) dominates the surface/edge properties (edge states). Therefore, the Chern number (see chapter 2 for details) is an important quantity to describe the topological properties of a system, which determines both the propagation direction

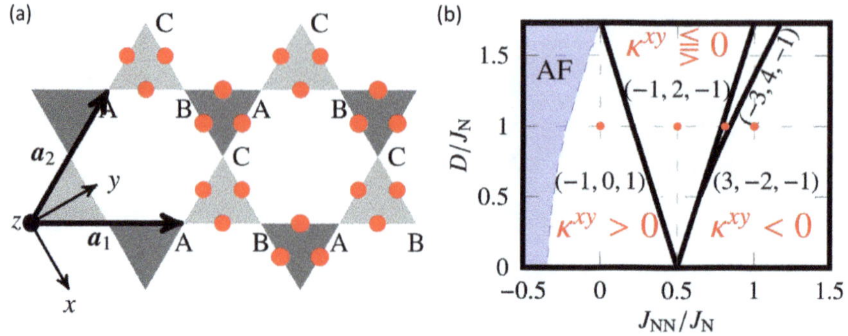

Figure 4.5. (a) Kagome lattice with basis vectors $\mathbf{a}_1$ and $\mathbf{a}_2$. DM vectors are aligned perpendicular to the lattice plane and are represented by red dots. A, B, and C denote identical atoms located at the corners of the triangles. (b) The phase diagram of the system. The Chern numbers of different bands (C_1, C_2, C_3) are also shown. (Reproduced with permission from [20]. Copyright 2014 The American Physical Society.)

and the number of edge states. For the lth band gap, we can define the 'winding number' with the sum of Chern numbers up to the lth band $\nu_l = \sum_{j\leqslant l} C_j$. The value of $|\nu_l|$ denotes the number of topological edge states and $\text{sgn}(\nu_l)$ determines their propagation direction.

By tuning the parameters J_{NN}/J_N and D/J_N, Mook *et al* [22] identified four different topological phases in this ferromagnetic kagome lattice. Here J_N and J_{NN} denote the Heisenberg exchange coefficients describing the nearest and next-nearest neighbor coupling, respectively, D is the DM parameter which only includes the nearest-neighbor interaction. The phase diagram of the system is plotted in figure 4.5(b), and the corresponding transverse thermal conductivity κ^{xy} is also shown. Figures 4.6(a)–(d) plot the band structures of a semi-infinite lattice for different topological phases. We choose the topological phase (3, −2, −1) as an example to illustrate the relation between 'winding number' and edge states (see figure 4.6(d)). First, because $\nu_1 = C_1 = 3$, there exist three topological edge states in the first (lowest) gap with positive group velocity. Second, only one topological edge state with positive group velocity emerges in the second band gap for the reason that $\nu_2 = C_1 + C_2 = 1$. Finally, no topological edge state is allowed above the uppermost band due to the vanishing Chern numbers for the sum of all bands. We can perform a similar analysis for the other cases in figures 4.6(a)–(c). Remarkably, the edge modes emerging in the phases (−1, 2, 1) and (−3, 4, −1) can exhibit bidirectional propagation, which results in the sign reverse of κ^{xy} with increasing temperature. This result can be understood as follows: at low temperatures, the number of edge magnons in the first gap is greater than that in the second gap. Therefore, the sign of κ^{xy} is determined by the edge magnons in the first gap. However, with a temperature increase the number of edge magnons in the second gap increases sharply. When the temperature exceeds the critical value, the heat current is dominated by these edge magnons, the sign of κ^{xy} therefore reverses.

As discussed above, the topological magnons emerging in the ferromagnetic kagome lattice originate from the nonzero DM interaction induced by spin–orbit coupling. In addition, Shindou *et al* [23] theoretically proved that the magnetostatic

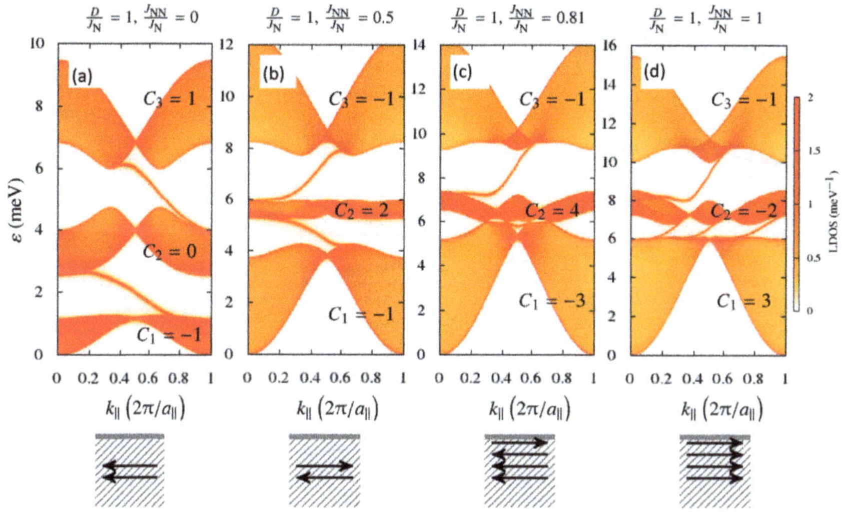

Figure 4.6. (a)–(d) The band structures of semi-infinite lattices allowing different topological phases, as marked by red dots in figure 4.5(b). The topological edge states of the magnons and their propagation direction are also presented in the bottom panel. (Reproduced with permission from [20]. Copyright 2014 The American Physical Society.)

interaction can also endow the magnon band with a non-zero Chern number, and the chirality of edge mode can be tuned via an external magnetic field. The model of decorated square-lattice based on ferromagnetic islands is plotted in figure 4.8(a). The total energy merely contains magnetostatic energy and Zeeman energy

$$E = -\frac{1}{2}(\Delta V)^2 \sum_{i,j}^{i \neq j} M_a(\mathbf{r}_i) f_{ab}(\mathbf{r}_i - \mathbf{r}_j) M_b(\mathbf{r}_j) + H \Delta V \sum_i M_z(\mathbf{r}_j), \tag{4.3}$$

where ΔV is a volume element for each ferromagnetic island, $\mathbf{r}_i$ denotes the spatial location of a ferromagnetic island, $a, b = x, y, z$, M_i $(i = x, y, z)$ is the components of the magnetization, H is the external magnetic, and $f_{ab}(\mathbf{r})$ represents the magnetic dipole–dipole interaction

$$f_{ab}(\mathbf{r}) = \frac{1}{4\pi}\left(\frac{\delta_{a,b}}{|\mathbf{r}|^3} - \frac{3r_a r_b}{|\mathbf{r}|^5}\right). \tag{4.4}$$

Then the Landau–Lifshitz equation is adopted to characterize the magnetic moment dynamics,

$$\partial_t M_a(\mathbf{r}_i) = \epsilon_{abc}\left[H\delta_{b,z} - \Delta V \sum_{j \neq i} f_{bd}(\mathbf{r}_i - \mathbf{r}_j) M_d(\mathbf{r}_j)\right] M_c(\mathbf{r}_i), \tag{4.5}$$

where ϵ_{abc} is the third-order Levi-Civita symbol. By utilizing the tight-binding model and linearizing the above equation, we obtain the magnon Hamiltonian. The magnon dispersion relations for different strength of vertical magnetic field (H)

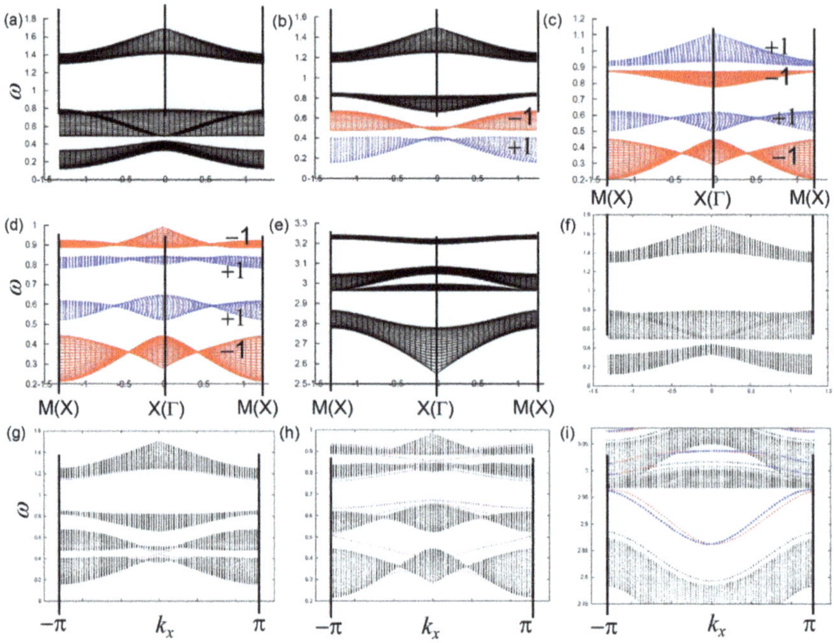

Figure 4.7. The dispersion relations of an infinite system with different strengths of magnetic fields: (a) $H = 0.0$, (b) $H=0.47H_c$, (c) $H=0.76H_c$, (d) $H=0.82H_c$, and (e) $H=2.35H_c$. The dispersion relations of a semi-infinite system under different magnetic fields: (f) $H = 0.0$, (g) $H = 0.47H_c$, (h) $H = 0.82H_c$, and (i) $H = 2.35H_c$. (Reproduced with permission from [23]. Copyright 2013 The American Physical Society.)

are plotted in figure 4.7. When $H = 0$, the magnon Hamiltonian has both mirror and time-reversal symmetries because the out-of-plane component of the magnetic moment vanishes, as shown in figure 4.7(a). In this case, the Chern numbers are always equal to zero for every band and no chiral magnon edge mode emerges (see figure 4.7(f)). As the strength of the magnetic field increases, the lowest and second-lowest bands gradually become closer, and the two bands will touch at the Γ point for $H = 0.24H_c$ and at the X points for $H = 0.67H_c$. Here H_c denotes the saturation field for all magnetic moments being fully magnetized along the direction of the magnetic field. When $0.24 < H/H_c < 0.67$ ($H/H_c > 0.67$), the Chern numbers are +1 and −1 (−1 and +1) for the lowest and second-lowest bands, respectively, as shown in figures 4.7(b)–(d). As a result, the magnon edge state with clockwise (counterclockwise) propagation appears for $0.24 < H/H_c < 0.67$ ($0.67 < H/H_c$), as plotted in figures 4.7(g) and (h). In addition, the three bands touch between the third-lowest and highest bands at the point M for $H = 0.71H_c$, at point X for $H = 0.79H_c$, and at point Γ for $H = 0.85H_c$. When $0.71 < H/H_c < 0.79$ ($0.79 < H/H_c < 0.85$), the Chern numbers are −1 and +1 (+1 and −1) for the third-lowest and the highest bands, respectively, which support the edge state with counterclockwise (clockwise) propagation, as shown in figures 4.7(c), (d), and (h). Considering the limit for a very large field, by using the perturbation analysis, Shindou *et al* [23] proved that the

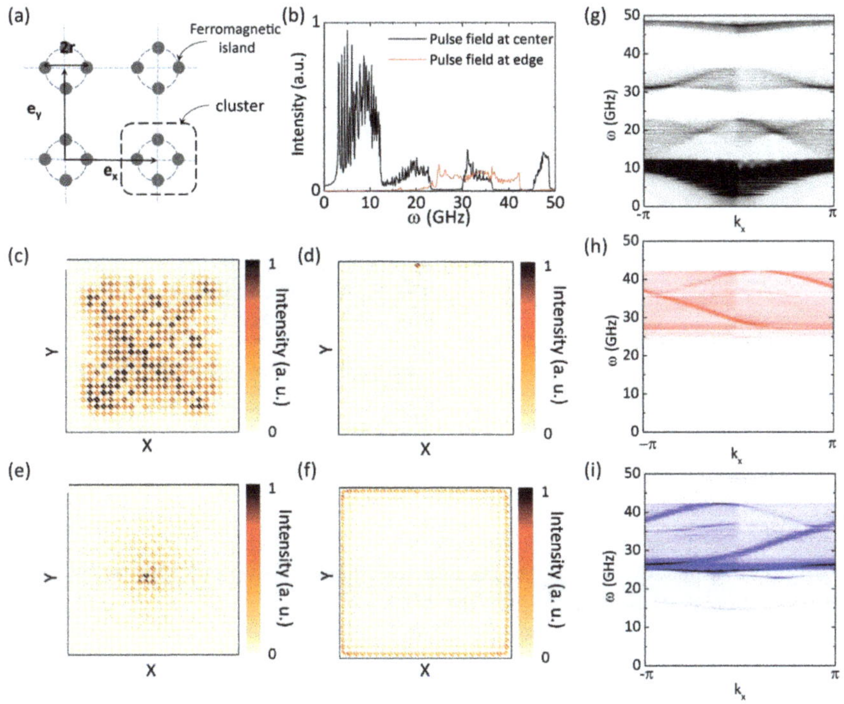

Figure 4.8. (a) A decorated square-lattice composed of ferromagnetic islands. (b) Fourier spectra with the excitation source localized at different positions. The intensity distribution of different modes when the excitation field is confined at the center (c), (e) and the edge (d), (f). The frequency $\omega = 10$ GHz for (c) and (d), $\omega = 29$ GHz for (e) and (f). (g) The band structures with the excitation field at the center. (h), (i) The band structures with the excitation field at the edge. The Fourier analysis is taken only on (h) the upper side ($Y > L/2$) or (i) lower side ($Y < L/2$). (Reproduced with permission from [23]. Copyright 2013 The American Physical Society.)

system respects time-reversal symmetry effectively, and the Chern numbers reduce to zero for all bands and no magnon chiral edge states are allowed, as shown in figures 4.7(e) and (i). However, in this case, the system still supports magnon edge states, which have parabolic band structures and therefore exhibit the feature of bidirectional propagation, as shown in figure 4.7(i). Here the non-chiral spin wave edge modes can be explained by the Tamm–Shockley mechanism [24, 25].

To verify the theoretical prediction for the chiral magnon edge state mentioned above, Shindou *et al* [23] performed micromagnetic simulations based on a square lattice. The magnetic moments are coupled merely through the dipolar interaction. The spins are magnetized in a saturated manner along the z-axis with $H = 1.02H_c$. Figure 4.8(b) plots the magnon spectra for an exciting pulse located at the center and at the edge. One can determine the frequency range supporting the edge mode and bulk mode. The oscillation distributions for different magnon states are plotted in figures 4.8(c)–(f), from which we can easily identify the bulk state for $\omega = 10$ GHz (figures 4.8(c) and (d)) and edge state for $\omega = 29$ GHz (figures 4.8(e) and (f)). In addition, the emerging magnon edge state exhibits the feature of unidirectional propagation, which can be further clarified by the band structure. When the exciting

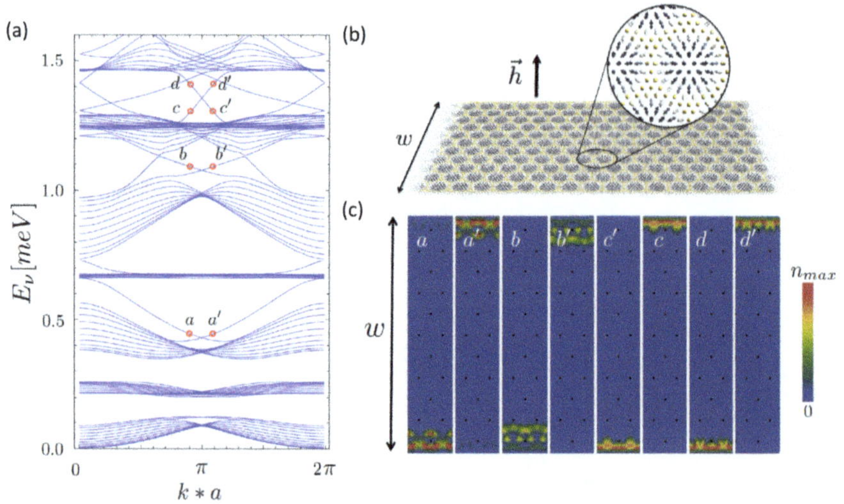

Figure 4.9. (a) The spin wave band structures for a one-dimensional strip. (b) A sketch map of the strip geometry. (c) The spatial distribution of magnon occupation for various edge states with symbols a, a', b, b', c, c', d, and d' marked in (a). The black dots represent the centers of the skyrmions. (Reproduced with permission from [30]. Copyright 2016 IOP Publishing.)

field is located at the center of the system, no magnon edge band is supported in the dispersion curves, as shown in figure 4.8(g). However, when the exciting field is located at the edge, by performing a Fourier analysis over the upper (lower) side of the system, we can clearly see the edge bands allowing magnons to propagate counterclockwise, as shown in figure 4.8(h) (figure 4.8(i)). These results agree well with the theoretical predictions in figure 4.7.

It is noted that the concept of 'magnonic topological insulator' has been extended to antiferromagnets [26, 27] and ferrimagnets [28, 29].

4.2.2 Noncollinear magnetic texture

As discussed in section 4.1, the TMHE can be induced by inhomogeneous magnetic textures, which implies that there exist topological magnons in these systems. In 2016 Roldán-Molina *et al* [30] studied in detail the topological magnons in two-dimensional ferromagnetic skyrmion lattices, as plotted in figure 4.9(b). The corresponding Hamiltonian includes the ferromagnetic exchange interaction, the DM interaction, the Zeeman energy, and the uniaxial anisotropy term:

$$H = -\sum_{\langle i,j \rangle} J_{i,j} \mathbf{S}_i \cdot \mathbf{S}_j + \sum_{\langle i,j \rangle} \mathbf{D}_{i,j} \cdot (\mathbf{S}_i \times \mathbf{S}_j) - \sum_i \mathbf{h} \cdot \mathbf{S}_i - K \sum_i (S_i^z)^2, \tag{4.6}$$

where $\mathbf{S}_i$ and $\mathbf{S}_j$ are the spin operators at sites i and j, $J_{i,j}$ is the exchange stiffness, $\mathbf{D} = D_0(\hat{z} \times \mathbf{n}_{ij})$ with $\mathbf{n}_{ij}$ the unit vector that joins sites i and j, and D_0 the strength of the DM interaction, $\mathbf{h}$ denotes the external magnetic field, and K is the anisotropy constant. The topological magnonic bands can be computed using a quantum mechanical Holstein–Primakoff (HP) boson theory [31]. The spin wave band structure for one-dimensional skyrmion lattice is plotted in figure 4.9(a). One can see that there exist a few bands which support magnon edge modes. Figure 4.9(c) shows the distribution of magnon occupation for various edge states, as marked by red circles in figure 4.9(a), from which we can clearly identify the localized features. These edge modes are chiral and propagate along a specific direction, and the elastic backscattering is forbidden for them because of the topological protection. The physical origin of topological magnons emerging in a skyrmion array can be explained as follows. The magnons in a skyrmion lattice are subject to the Lorentz-type or anomalous force, which is analogous to the quantum anomalous Hall effect introduced in electronic system. The force acting on magnons comes from the fictitious magnetic field induced by the magnetic texture (such as a skyrmion), which leads to the emergence of nonzero Berry curvature. Similar results have also been reported by Díaz *et al* [32]. Figure 4.10(a) shows the corresponding magnon dispersion relation in a one-dimensional skyrmion lattice with red and blue curves denoting the edge magnon bands. It is confirmed that topological magnon bands are robust against disorders (see figure 4.10(b)). In addition, the unit cell of the skyrmion array and the probability density for edge modes are plotted in figures 4.10(c) and (d), respectively. One can clearly observe the localization properties of the edge modes. Hamamoto *et al* [33] also obtained the magnon edge states and reported the quantized topological Hall effect in a skyrmion crystal. Micromagnetic simulations have been performed by Wang *et al* [34] and Xie *et al* [35] to verify the existence of topological magnon edge modes in ferromagnetic skyrmion lattices.

In addition, Díaz *et al* [36] showed that topological magnons can exist in antiferromagnetic skyrmion lattices, with the corresponding magnon dispersion relation being shown in figure 4.11(a). One can clearly identify the magnon band gap, as marked by the area shaded in green. In addition, they also considered a strip with periodic boundary conditions along the x-axis and a finite size along the y-axis. In this situation, the edge magnon bands will appear, as plotted in figure 4.11(b). The magnon edge mode is presented in figure 4.11(c) with its wavefunction being highly localized at the top and bottom of the strip.

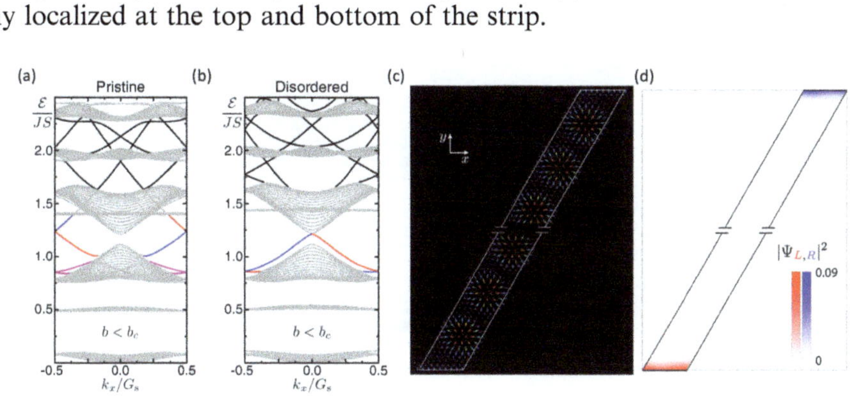

Figure 4.10. The magnon band structures of a one-dimensional infinite strip (a) without and (b) with disorders. (c) The unit cell of the skyrmion lattice for the strip geometry. (d) The spatial distribution of probability density $|\Psi_{L,R}|^2$ for magnonic edge states with left-moving (red) and right-moving (blue) indicated. (Reproduced with permission from [32]. Copyright 2020 The American Physical Society.)

Comparing the topological magnons in ferromagnetic (figures 4.9 and 4.10) and antiferromagnetic (figure 4.11) skyrmion lattices, we can find the differences and similarities between them. On the one side, the antiferromagnetic skyrmion array includes two interpenetrating sublattices, and each of them acts as a ferromagnetic skyrmion lattice. On the other side, the two lowest magnon bands are trivial for the ferromagnetic skyrmion lattice, while it becomes topological for the antiferromagnetic case. For practical applications, this distinction is more beneficial for antiferromagnetic skyrmion lattices because the energy needed to stimulate magnon edge modes is lower than that of ferromagnetic systems (see figures 4.10(a) and 4.11(a)).

In addition to ferromagnetic and antiferromagnetic skyrmion lattices, Kim *et al* [37] reported the magnonic thermal Hall effect in a ferrimagnetic skyrmion crystal in the vicinity of the angular momentum compensation point. They derived the equation of motion for magnons in the background of an arbitrary equilibrium spin texture, which gives rise to the fictitious electromagnetic field for magnons. When the equilibrium spin density $s > 0$ ($s < 0$), the transverse heat current along $+y$ ($-y$) will appear, while at the compensation point, i.e. $s = 0$, the thermal Hall effect is absent, similar to the case in antiferromagnets [26], as shown in figure 4.12(a). This result can be understood qualitatively as follow: when the spin density is negative (positive), there are more positive-chirality (negative-chirality) magnons, causing a negative (positive) thermal Hall conductivity, as shown in figure 4.12(c) (figure 4.12(d)); when the spin density is equal to zero, magnons with opposite chiralities are degenerate and the thermal Hall conductivity vanishes. Therefore, the magnonic thermal Hall conductivity changes its sign when the ferrimagnet is tuned across the compensation point (see figure 4.12(b)).

Although topological magnon band structures in skyrmion lattices have been studied extensively [17, 18, 30, 32, 36, 37], most of the works are theoretical and the experimental evidence is relatively rare. Very recently, with the help of the polarized inelastic neutron scattering technique, Weber *et al* [38] experimentally determined the topological magnon band structure with emergent Landau levels in a skyrmion lattice of MnSi.

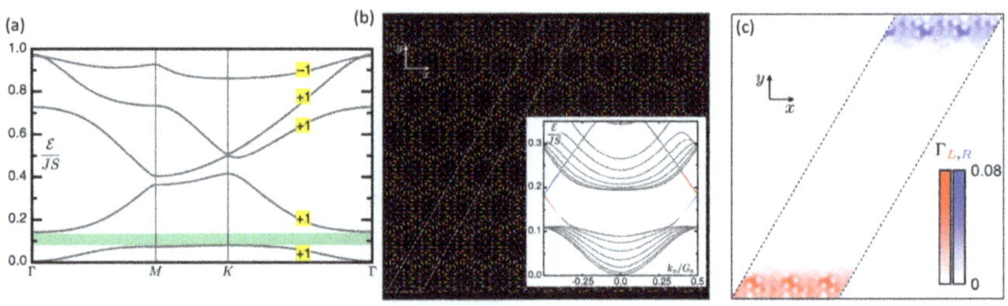

Figure 4.11. (a) The bulk magnon dispersion relation of antiferromagnetic skyrmion crystals. The Chern numbers of different bands are also labeled. The green rectangle marks the magnon gap. (b) A one-dimensional antiferromagnetic skyrmion crystal strip and the corresponding magnon dispersion relation. The white dotted box represents the unit cell. (c) The spatial distribution of the wave function $\Gamma_{L,R}$ with edge states, where the wave functions over 3 times the unit cells are shown. (Reproduced with permission from [36]. Copyright 2019 The American Physical Society.)

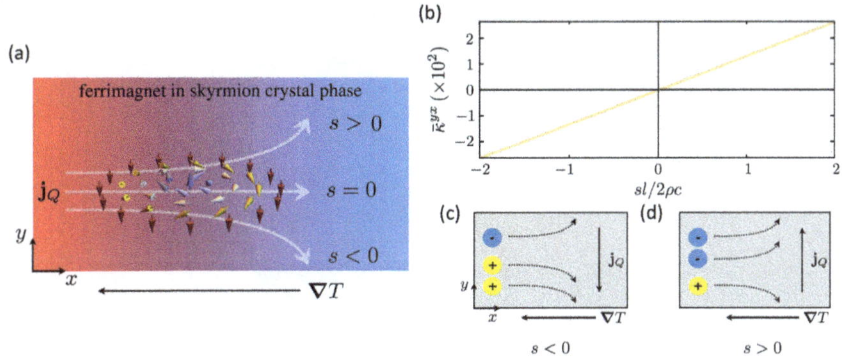

Figure 4.12. (a) Schematic illustration of the magnonic heat flux $\mathbf{j}_Q$ through a ferrimagnet in its skyrmion crystal phase subject to a temperature gradient ∇T. (b) Dependence of the rescaled thermal Hall conductivity on the spin density. (c) and (d) Schematic illustrations of the positive-chirality (+) and the negative-chirality (−) magnon motions subject to a temperature gradient and the resultant transverse energy flux $\mathbf{j}_Q$. (Reproduced with permission from [37]. Copyright 2019 The American Physical Society.)

4.3 Topological magnon semimetals

4.3.1 Dirac magnons

Since the discovery of two-dimensional graphene [39, 40], the Dirac materials exhibiting a Dirac-like band structure have attracted growing interest. Although Dirac quasiparticles were originally investigated in fermionic systems, it has been generalized to a broad field of bosonic systems, for example, photonic crystals [41–44], plasmonic systems [45–47], acoustic metamaterials [48–50], superconducting grains [51], and magnonics [52–57].

In magnetic systems, the simplest two-band structure supporting the Dirac points is the Heisenberg exchange model based on the two-dimensional honeycomb lattice [54]. The corresponding Hamiltonian only contains the exchange interaction

$$H = -\sum_{\langle ij \rangle} J_{ij}\mathbf{S}_i^{(A)} \cdot \mathbf{S}_j^{(B)}, \tag{4.7}$$

where $\mathbf{S}_i^{(A)}$ and $\mathbf{S}_j^{(B)}$ denote the spins located at different sublattices, J_{ij} is the exchange coefficient, and $\langle ij \rangle$ indicates the summation over the nearest neighbors. Considering the uniform ferromagnetic interaction, i.e. $J_{ij} = J > 0$, after the Holstein–Primakoff transformation [31], we obtain the effective Hamiltonian

$$H_{\mathrm{FM}} = \sum_i (\varepsilon_A a_i^\dagger a_i + \varepsilon_B b_i^\dagger b_i) - J\sqrt{S_A S_B}\sum_{\langle ij \rangle}(a_i^\dagger b_j + \mathrm{h.c.}) - 3JNS_AS_B. \tag{4.8}$$

In reciprocal space, letting $a_i = \sum_\mathbf{k} a_\mathbf{k} e^{i\mathbf{k}\cdot\mathbf{r}_i}/\sqrt{N}$ and $b_j = \sum_\mathbf{k} b_\mathbf{k} e^{i\mathbf{k}\cdot\mathbf{r}_j}/\sqrt{N}$, one obtains

$$H_{\mathrm{FM}} = \sum_\mathbf{k} \{\varepsilon_A a_\mathbf{k}^\dagger a_\mathbf{k} + \varepsilon_B b_\mathbf{k}^\dagger b_\mathbf{k} + [\phi(\mathbf{k}) a_\mathbf{k}^\dagger b_\mathbf{k} + \mathrm{h.c.}]\} - 3JNS_AS_B, \tag{4.9}$$

where $\phi(\mathbf{k}) = -J\sqrt{S_A S_B}\sum_i \exp(i\mathbf{k}\cdot\vec{\delta}_i)$ $(i = 1, 2, 3)$ with $\vec{\delta}_i$ being the vectors connecting adjacent spins (see figure 4.13(a)), and $\mathbf{k}$ is the wave vector. The eigenvalues of the system can be determined

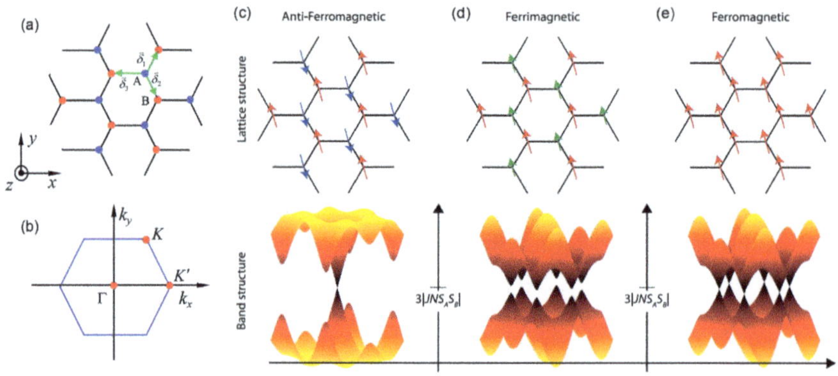

Figure 4.13. (a) Schematic diagram of a honeycomb lattice based on spins, where A and B denote the sublattices. (b) The first Brillouin zone with representative high-symmetry points. The configurations and corresponding magnon dispersion relations for (c) antiferromagnetic, (d) ferrimagnetic, and (e) ferromagnetic states. ((c)–(e) Reproduced with permission from [54]. Copyright 2016 The American Physical Society.)

$$E_{\pm}(\mathbf{k}) = [\varepsilon_A + \varepsilon_B \pm \Omega(\mathbf{k})]/2, \tag{4.10}$$

where $\Omega^2(\mathbf{k}) = \Delta^2 + 4|\phi(\mathbf{k})|^2$ with $\Delta = \varepsilon_A - \varepsilon_B = -3J(S_A - S_B)$.

For a ferromagnetic system, we have $S_{A(B)} = S$ and $\Delta = 0$, the configuration of spins and corresponding magnon dispersion relations are plotted in figure 4.13(e). Near the high-symmetry points $\mathbf{K} = (2\pi/3a,\ 2\sqrt{3}\pi/3a)$ and $\mathbf{K}' = (4\pi/3a,\ 0)$ (see figure 4.13(b)), the energy is degenerate and the bands are linear. For ferrimagnetic structure, we have $S_A \neq S_B$, thus $\Delta \neq 0$, leading to a gap appearing at points $\mathbf{K}$ and $\mathbf{K}'$, and the width of the gap is $3J|S_A - S_B|$ (see figure 4.13(d)). However, if we consider the antiferromagnetic lattice, the energy is degenerate only at the Γ point where the band structure has a linear crossing, as shown in figure 4.13(c). It is noted that the time-reversal symmetry is preserved in this Heisenberg model: under the operation of time reversal ($\mathbf{S}_i^{(A)} \to -\mathbf{S}_i^{(A)}$, $\mathbf{S}_j^{(B)} \to -\mathbf{S}_j^{(B)}$), the Hamiltonian is invariant and thus respects the time-reversal symmetry. Magnons belong to bosons, the time-reversal-operator $\mathcal{T}$ therefore has the definition $\mathcal{T} = \mathrm{i}\sigma_y\mathcal{K}_0$ where $\mathcal{K}_0$ denotes the complex conjugation and $\mathcal{T}$ satisfies the relation $\mathcal{T}^2 = -1$ [53, 58]. The Dirac point can always exist [53] as long as the system respects the pseudo-spin time-reversal symmetry.

In Dirac materials, a gap emerges at the Dirac points when the time-reversal symmetry of the system is broken, which may lead to the TI phase. Similarly, we can obtain a topological magnon insulator based on the same method. For a ferromagnet with a honeycomb lattice, if the Hamiltonian merely includes nearest-neighbor (NN) exchange interaction, the magnon dispersion relation is gapless. Owerre [53] theoretically proved that when a next nearest-neighbor (NNN) DM interaction is considered, the pseudo-spin time-reversal symmetry is broken, and then a gap appears at Dirac points. In addition, Owerre also showed that the NNN exchange interaction can only shift Dirac points. The corresponding Hamiltonian reads

$$H = -J\sum_{\langle lm\rangle}\mathbf{S}_l\cdot\mathbf{S}_m - J'\sum_{\langle\langle lm\rangle\rangle}\mathbf{S}_l\cdot\mathbf{S}_m + \sum_{\langle\langle lm\rangle\rangle}\mathbf{D}_{lm}\cdot\mathbf{S}_l\times\mathbf{S}_m, \tag{4.11}$$

where J and J' denote NN and NNN exchange constants, respectively, and $\mathbf{D}_{lm}$ is the NNN DM interaction with l and m representing the different sites. Noticeably, the time-reversal symmetry here is broken for the reason that under the operation of time reversal ($\mathbf{S}_l \to -\mathbf{S}_l$, $\mathbf{S}_m \to -\mathbf{S}_m$, and $\mathbf{D}_{lm} \to -\mathbf{D}_{lm}$), the Hamiltonian is not invariant. By utilizing the Holstein–Primakoff transformation, we have

$$H = -JS\sum_{\langle lm\rangle}(b_l^\dagger b_m + \text{h. c. }) - \sqrt{(J'S)^2 + (DS)^2}\sum_{\langle\langle lm\rangle\rangle}(\mathrm{e}^{\mathrm{i}\phi_{lm}}b_l^\dagger b_m + \text{h.c.}) + (3JS + 6J'S)\sum_l b_l^\dagger b_l, \tag{4.12}$$

where D is the strength of DM interaction, and $\phi_{lm} = \nu\phi$ is a phase factor representing the magnetic flux induced by DM interaction, with $\nu = \pm 1$ and $\phi = \arctan(D/J^{'})$, which is analogue to the Haldane model in electron system. After some algebra, we obtain the Bogoliubov Hamiltonian

$$H_B(\mathbf{k}) = h_0\sigma_0 + h_x\sigma_x + h_y\sigma_y + h_z\sigma_z, \tag{4.13}$$

where $h_0 = 3JS - 4\sqrt{(J^{'}S)^2 + (DS)^2}\cos\phi\sum_\mu \cos\mathbf{k}\cdot\vec{\rho}_\mu$, $h_x = -JS\sum_\mu \cos\mathbf{k}\cdot\vec{\delta}_\mu$, $h_y = -JS\sum_\mu \sin\mathbf{k}\cdot\vec{\delta}_\mu$, and $h_z = -2\sqrt{(J^{'}S)^2 + (DS)^2}\sin\phi\sum_\mu \sin\mathbf{k}\cdot\vec{\rho}_\mu$, with $\vec{\delta}_\mu$ and $\vec{\rho}_\mu$ ($\mu = 1, 2, 3$) being the NN and NNN vectors, respectively. Assuming $J' = 0$ ($\phi = \pi/2$), in the vicinity of Dirac points, we obtain the expanded form of Hamiltonian

$$H_B(\mathbf{q}) = 3JS\sigma_0 + \frac{1}{2}JS(\sigma_x q_x - \tau\sigma_y q_y) + 3\sqrt{3}DS\tau\sigma_z, \tag{4.14}$$

where $\mathbf{q} = \mathbf{k} - \mathbf{K}_\pm$, $\mathbf{K}_\pm = (2\pi/3, \pm 2\pi/3\sqrt{3})$ are Dirac points, and $\tau = \pm$ characterizes the states at $\mathbf{K}_\pm$. Since the time reversal symmetry is broken due to the DM interaction (the last term in equation (4.14)), the band structure therefore opens a gap at the Dirac points. The system naturally becomes gapless again if we do not consider the DM interaction. Figure 4.14(a) plots the dispersion relation for a semi-infinite system; one can clearly identify the edge magnon bands. The schematic diagram of magnon propagation for edge mode is plotted in figure 4.14(b).

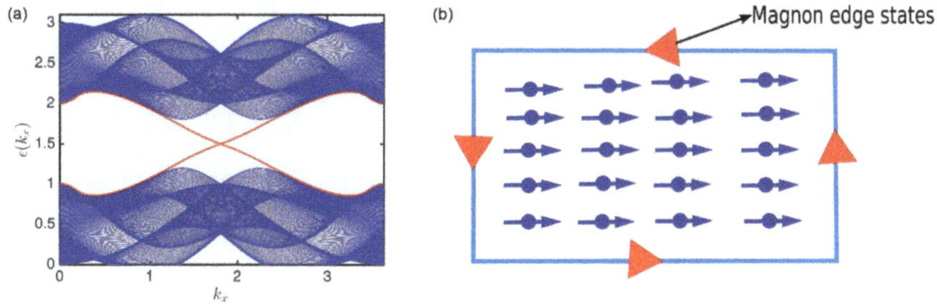

Figure 4.14. (a) The magnon band structure of a one-dimensional strip based on ferromagnetic spins arranged as a honeycomb lattice. Here the NNN DM interaction is considered. (b) A schematic diagram of the magnon topological edge states. (Reproduced with permission from [53]. Copyright 2016 IOP Publishing.)

The chromium trihalides CrX_3 (X = F, Cl, Br, and I) represent a concrete instance of ferromagnets composed of van der Waals-bonded stacks of honeycomb lattices, which exhibit a two-magnon band structure akin to graphene. Pershoguba *et al* [56] investigated the properties of Dirac magnons in CrX_3 theoretically, with the influences of particle statistics and interactions on the stability of Dirac cones also being discussed. They proved that the magnon-magnon interaction can result in a prominent momentum-dependent renormalization of the dispersion relation and strongly momentum-dependent magnon lifetime. There exist obvious differences between Dirac fermions and bosons, as shown in figure 4.15. On the one side, the interacting bosons result in a nondivergent velocity renormalization in the vicinity of Dirac points, while the Coulomb interactions between fermions induce a logarithmic correction of the Dirac velocity. On the other side, the non-Bravais features of the lattice entail multicomponent wave functions for quasiparticles, which provide a route to Dirac quasiparticles and distinct surface states. In addition, Pershoguba *et al* [56] showed that the ferromagnets with honeycomb lattice can exhibit dispersive surface and edge modes.

Soon after the theoretical reports about Dirac magnons in CrX_3 (X = F, Cl, Br, and I), Chen *et al* [52] observed experimentally the band gap at the Dirac point in CrI_3 bulk material. The corresponding crystal structure is plotted in figure 4.16(g).

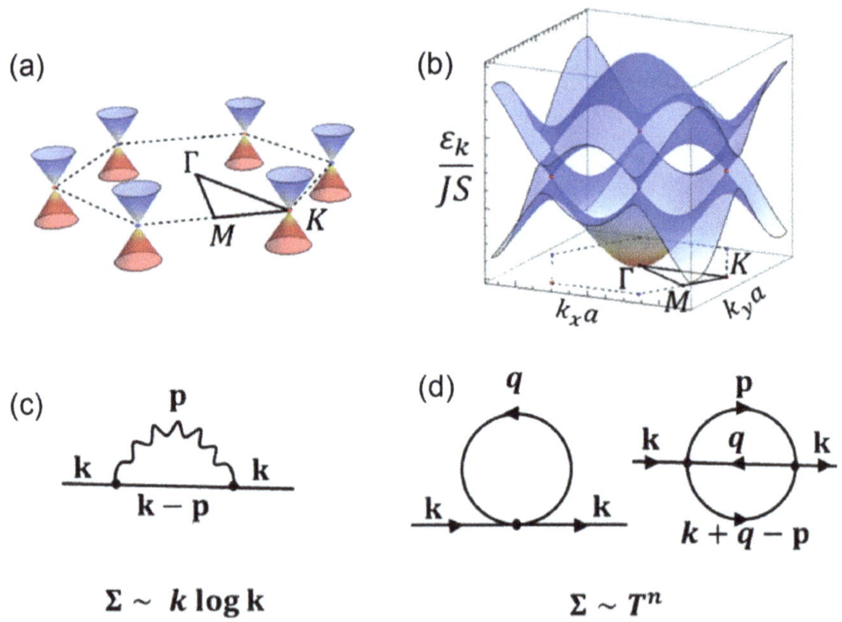

Figure 4.15. The comparison between Dirac fermions and bosons. (a) The dispersion relation of the fermionic system: the properties of the Dirac fermion are dominated by the states at the chemical potential, therefore considering the band structure of the Dirac cones close to the centers is sufficient. (b) The dispersion relation of a bosonic system: because bosons can occupy any state in the Brillouin zone, the information of the whole band structure is necessary. (c) The relevant self-energy diagram of a fermionic system that gives the logarithmic correction to the Dirac fermion velocity. (d) The relevant self-energy diagrams of the bosonic system give a temperature-dependent renormalization and attenuation of the Dirac bosons. (Reproduced with permission from [56]. Copyright 2018 The American Physical Society.)

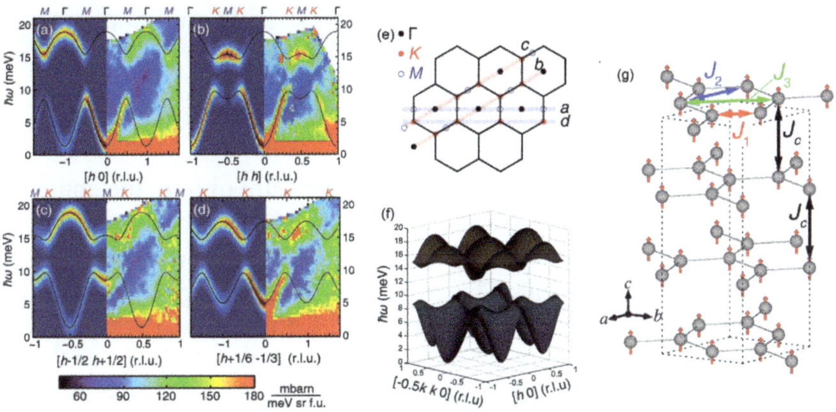

Figure 4.16. (a)–(d) The comparison between experimental measurements of neutron scattering at 5 K and theoretical calculations. (e) The illustration of the displayed cuts within the (h, k)-plane. (f) The magnon band structures of CrI_3. (g) The crystal structure of CrI_3. (Reproduced with permission from [52]. Copyright 2018 The American Physical Society.)

The magnon band structures are obtained with the help of inelastic neutron scattering. The experimentally observed neutron scattering intensities are plotted in figures 4.16(a)–(d) with temperature at 5 K, the determined dispersion curves are also presented for comparison. One can clearly see a large band gap approximately equal to 4 meV at the Dirac points, which most likely originates from the breaking of inversion symmetry induced by the NNN DM interaction. This gap results in the appearance of topological magnon insulator.

In addition to the CrI_3 bulk material, Costa *et al* [57] showed that the CrI_3 monolayer can also exhibit topological magnons. They adopted the itinerant fermion theory to characterize the magnons and observed a gap at the Dirac points. In addition, the topological magnon edge states are explicitly proved in the CrI_3 nanostrip.

In addition to the DM interaction, Wang *et al* [59] proved that the pseudodipolar exchange interaction which originates from the superexchange and atomistic spin–orbit interaction [60, 61] can also lead to a gap emerging at Dirac points and generate topological magnons. The two-dimensional ferromagnet with honeycomb lattice is characterized by the following Hamiltonian

$$H = -\frac{J}{2}\sum_{\langle i,j\rangle}\mathbf{m}_i \cdot \mathbf{m}_j - \frac{1}{2}\sum_{\langle i,j\rangle}F(\mathbf{m}_i, \mathbf{m}_j, \mathbf{e}_{ij}) - \sum_i \frac{K_i}{2}m_{iz}^2 - \mu B\sum_i m_{iz}, \tag{4.15}$$

with $\langle i, j\rangle$ denoting the NN sites. The first term describes the exchange interaction with exchange constant J. The second term is an NN dipole–dipole-like pseudodipolar exchange interaction which arises from the superexchange and atomistic spin–orbit interactions [60, 61]: $F(\mathbf{m}_i, \mathbf{m}_j, \mathbf{e}_{ij}) = F(\mathbf{m}_i \cdot \mathbf{e}_{ij})(\mathbf{m}_j \cdot \mathbf{e}_{ij})$, with $\mathbf{e}_{ij}$ the unit vector connecting sites i and j, and F being the strength of the interaction. The third term describes the magnetocrystalline anisotropy energy with anisotropy coefficient $K_i = K_A$ and K_B for sublattices A and B, respectively, where the easy-axis is assumed along the z-direction. The last term is the Zeeman energy with external magnetic field along the z-axis and B is the strength. Neglecting the damping term, the dynamics of magnetization $\mathbf{m}_i$ can be expressed as

$$\frac{\partial \mathbf{m}_i}{\partial t} = -\mathbf{m}_i \times [J\sum_{j\in\langle i\rangle}\mathbf{m}_j + F\sum_{j\in\langle i\rangle}(\mathbf{m}_j\cdot\mathbf{e}_{ij})\mathbf{e}_{ij} + Km_{iz}\mathbf{e}_z + B\mathbf{e}_z], \tag{4.16}$$

where $K_A = K_B = K$. Solving equation (4.16), we can obtain the magnon band structure for infinite system, as shown in figure 4.17 (b). The width of the band gap is $\Delta_g = \Omega - \sqrt{\Omega^2 - 9F^2/4}$, with $\Omega = 3J + K + B$. Obviously, if $F = 0$, the gap will close and the Dirac cones appear. Therefore, the band gap opening originates from the nonzero pseudodipolar exchange interaction. Considering a semi-infinite nanostrip with zigzag edges (see figure 4.17(a)), the density plot of the wave function for top boundary is plotted in figure 4.17(c). One can easily identify that the propagation direction of magnon edge mode is counterclockwise for the reason that the group velocity of the edge band is negative. Figure 4.17(d) plots the oscillation distribution for the magnon edge states.

Very recently, utilizing an inelastic neutron scattering technique, Yuan *et al* [62] observed the Dirac magnons in a quantum XY magnet $CoTiO_3$ with ilmenite structure. A band gap ~1 meV was determined in the dispersion curves. Here, the gap originates from the bond-anisotropic exchange coupling because of the quantum order by disorder. In addition, they proved that the system with a zigzag edge can support edge states connecting the Dirac points, while the armchair edge cannot. $CoTiO_3$ could be a good candidate for studying the magnon topological phase transitions.

4.3.2 Weyl magnons

In topological magnonic systems, in addition to the Dirac semimetal, another important class of topological phases with gapless band structure is the Weyl semimetals (WSMs) [63–68]. In magnonic WSMs, the band structures have specific

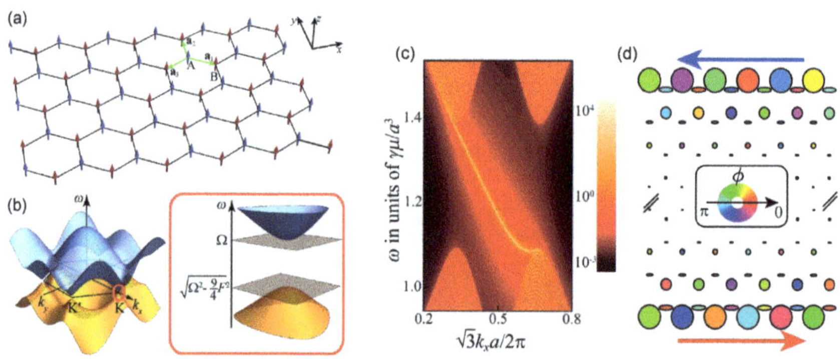

Figure 4.17. (a) An illustration of ferromagnetic spins arranged as a honeycomb lattice. (b) The spin-wave band structure of an infinite system with parameters $J = 0.1$, $F = 0.5$, and $\Omega = 1.3$ (in units of $\mu_0\mu^2/a^3$). (c) The density plot of the wave-function on the top boundary. (d) The oscillation distribution of the magnon edge mode with $\omega = 1.2$ ($\gamma\mu/a^3$). Here the shapes of the symbols describe the trajectories of spin precession, while the sizes of the symbols represent the amplitude of spin precession at each site. (Reproduced with permission from [59]. Copyright 2017 The American Physical Society.)

crossings (called Weyl nodes) in pairs, which is analogous to the WSMs in electronic system [69, 70]. The Weyl nodes (WNs) are monopoles of Berry curvature and are described via different topological charge or chirality. According to the no-go theory, the total topological charges within the whole Brillouin zone must vanish, the WNs therefore emerge in pairs with contrary topological charges [71, 72]. The effective Weyl Hamiltonians can be used to characterize the magnons near the WNs, these special magnons are called Weyl magnons. In WSMs, the topologically protected surface states connect the pairs of WNs and form arcs, and the number of arcs is equal to the number of paired WNs [69, 73, 74].

Mook *et al* [63] and Su *et al* [64] have showed that the pyrochlore ($Lu_2V_2O_7$) ferromagnet is an intrinsic magnon WSM. Here, the material magnetism originates from the vanadium ions. The effective Hamiltonian has the following form which contains exchange interaction, DM interaction, and Zeeman interaction:

$$H = -\sum_{\langle ij\rangle} J_{ij}\mathbf{S}_i \cdot \mathbf{S}_j + \sum_{\langle ij\rangle} \mathbf{D}_{ij} \cdot (\mathbf{S}_i \times \mathbf{S}_j) - g\mu_B \sum_i \mathbf{B} \cdot \mathbf{S}_i, \tag{4.17}$$

where $\langle ij\rangle$ indicates the NN sites, $\mathbf{S}_i$ ($\mathbf{S}_j$) denotes the spin of the V^{4+} ion, J_{ij} is the exchange coefficient, $\mathbf{D}_{ij}$ is the DM vector between sites i and j, and $\mathbf{B}$ represents the external magnetic field. Utilizing Holstein–Primakoff transformation and Bloch theory, the Hamiltonian (4.17) can be simplified as $H = \sum_{\mathbf{k}} b_{\mathbf{k}}^{\dagger} H(\mathbf{k}) b_{\mathbf{k}} + E_0$, where $b_{\mathbf{k}}^{\dagger}$ and $b_{\mathbf{k}}$ are the magnon creation and annihilation operators, respectively, E_0 denotes the energy for zero magnon mode, and $H(\mathbf{k})$ is a matrix as

$$H(\mathbf{k}) = \begin{pmatrix} 3J & JA_1(\mathbf{k}) & JA_2(\mathbf{k}) & JA_3(\mathbf{k}) \\ JA_1(\mathbf{k}) & 3J & J_-A_{12}(\mathbf{k}) & J_+A_{13}(\mathbf{k}) \\ JA_2(\mathbf{k}) & J_+A_{12}(\mathbf{k}) & 3J & J_-A_{23}(\mathbf{k}) \\ JA_3(\mathbf{k}) & J_-A_{13}(\mathbf{k}) & J_+A_{23}(\mathbf{k}) & 3J \end{pmatrix}, \tag{4.18}$$

where $A_\alpha(\mathbf{k}) = -\cos(\mathbf{a}_\alpha \cdot \mathbf{k}/2)$, $A_{\alpha\beta}(\mathbf{k}) = -\cos[(\mathbf{a}_\alpha - \mathbf{a}_\beta) \cdot \mathbf{k}/2]$ with $\alpha, \beta = 1, 2, 3$, $J_\pm = J \pm i\sqrt{2}D/\sqrt{3}$, $J_{ij} = J$ (we consider the uniform exchange interaction), and $D = |\mathbf{D}_{ij}|$ is the DM interaction strength. The basis vectors are $\mathbf{a}_1 = (1, 1, 0)/2$, $\mathbf{a}_2 = (1, 0, 1)/2$, and $\mathbf{a}_3 = (0, 1, 1)/2$.

The dispersion relation of the system can be determined by a diagonalizing Hamiltonian (4.18). Figure 4.18(a) plots the schematic diagram of the Brillouin zone for a pyrochlore lattice. The blue and red dots denote the WNs with opposite topological charges. The hallmark of a magnonic WSM is the magnon arc emerging on the surface of system. For the top surface, the density map of the wavefunction along path $\overline{\Gamma XL\Gamma}$ is plotted in figure 4.18(b), from which one can identify one WN. In addition, we can also observe the topological surface states, as marked in red. Furthermore, if we consider the path $\overline{L\Gamma L}$, we can see one pair of WNs connected via surface states, with the results being plotted in figure 4.18(c). Here, the distance between two WNs is dominated by the DM interaction strength. Considering the given energies with E_c, E_d, and E_e near the WNs, as marked in figure 4.18(b), the corresponding density maps of the wave function are plotted in figures 4.18(d)–(f), in

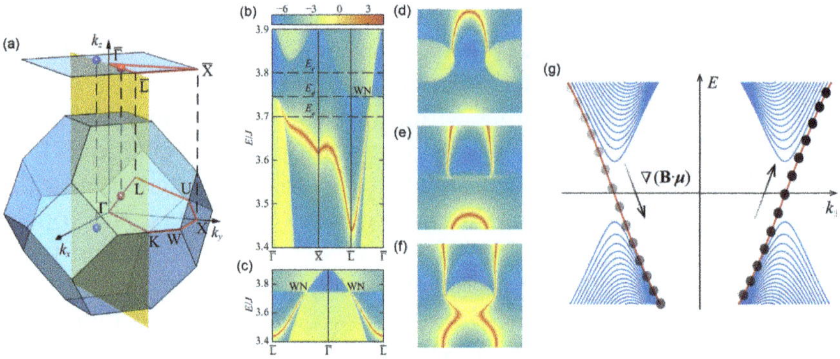

Figure 4.18. (a) Illustrations of the first bulk Brillouin zone and the first surface (001) Brillouin zone of the pyrochlore lattice. Here, the blue and red dots denote the Weyl nodes having opposite topological charges. The density diagrams of the magnon wave-function on the top surface (001) along (b) $\overline{\Gamma X L \Gamma}$ and (c) $\overline{L \Gamma L}$. (d)–(f) The density diagrams of the magnon wavefunction on the top surface (001) for fixed energies of E_c, E_d, and E_e, as marked in (b). (g) A sketch map of the magnon Landau levels. Here, the red and blue curves represent the zeroth and higher magnon Landau levels, respectively. The black dots denote the magnons in the zeroth magnon Landau level. The black arrows indicate the direction of magnon motion induced by magnetic field **B**. (Reproduced with permission from [64]. Copyright 2017 The American Physical Society.)

which the magnon arcs are clearly identified. Su *et al* [64] further showed that when the mutually perpendicular inhomogeneous electric and magnetic fields are applied, one can observe the magnonic chiral anomaly. Here, the electric field is used to induce the magnonic Landau level based on the Aharonov–Casher (AC) effect [75]: assuming a magnon propagating along route P in the electric field $\mathbf{E}$, the magnon will gain an AC phase $\phi_{\mathrm{AC}} = (1/\hbar c^2)\int_P (\mathbf{E} \times \mu) \cdot \mathbf{dr}$ with μ being the magnetic moment of the magnon, and the magnetic field is used to drive the magnon (see the sketch map plotted in figure 4.18(g)). When magnons driven by the magnetic field move from one WN to the other via the zeroth magnonic Landau level, the balance of the chirality is broken, which represents the typical feature for the magnonic chiral anomaly.

In addition, Su *et al* [65] demonstrated that a ferromagnet with a stacked honeycomb lattice can also exhibit Weyl magnons, with the schematic diagram of crystal structure shown in figure 4.19(a). The Hamiltonian can be expressed as

$$\begin{aligned} H = & -J \sum_{\langle i,j \rangle, l} \mathbf{S}_{i,l} \cdot \mathbf{S}_{j,l} - \sum_{i,\, \langle l,\, l' \rangle} J_i \mathbf{S}_{i,l} \cdot \mathbf{S}_{i,l'} - \sum_{i,l} K_i (S^z_{i,\, l})^2 \\ & + D \sum_{\langle\langle i,j \rangle\rangle, l} \nu_{ij} \hat{\mathbf{z}} \cdot (\mathbf{S}_{i,l} \times \mathbf{S}_{j,l}) - g\mu_{\mathrm{B}} B_0 \sum_{i,l} S^z_{i,\, l}, \end{aligned} \tag{4.19}$$

where i and j denote the spin sites within one layer, while l and l' represent the layer indices. $\langle i, j \rangle$ and $\langle\langle i, j \rangle\rangle$ denote the NN and NNN intralayer lattice sites, respectively, and $\langle l, l' \rangle$ represents the NN layers. $\mathbf{S}_{i,l}$ is the spin at site i of the lth layer. The first and second terms are the intralayer and interlayer exchange interactions, respectively, where the exchange coefficients $J > 0$ and $J_i > 0$. The third term is the anisotropy energy with the easy-axis along the z-axis. The fourth and last terms are the DM interaction and Zeeman interaction, respectively, with B_0 the strength of the magnetic field along the z-direction.

By solving the Hamiltonian (4.19), one can obtain the phase diagram of the system which supports five distinct phases: the trivial phase, topological insulator phase, and three different topological Weyl semimetal phases. Interestingly, two of the Weyl semimetal phases have only one pair of WNs located at a different **k** point, while one of the Weyl semimetal phases exhibits two pairs of WNs. To show the unique band structures of the magnonic Weyl semimetal phase, we choose two different parameters supporting one and two pairs of WNs. Figures 4.19(c) and (d) plot the dispersion relation on the k_x–k_z-plane with $k_y = 0$, from which one can identify the paired WNs. The blue and red balls denote the topological charges −1 and +1 of WNs, respectively. The corresponding Berry curvatures are presented in figures 4.19(e) and (f), where the black arrows represent the direction of Berry curvatures, and the background color denotes the divergence of Berry curvature with red and blue regions indicating positive and negative values, respectively. Apparently, the WNs emerging in figures 4.19(c) and (d) can be regarded as the monopoles of Berry curvature. Furthermore, figures 4.19(g) and (h) plot the band structure along $\overline{Z\Gamma Z}$ for semi-infinite systems. One can clearly see that the surface states (red color) connect the paired WNs and form magnon arcs at the surface of the system.

For Weyl magnons, although there exist a lot of theoretical studies, experimental observation is still rare. The most important reason is that the frequencies of the WNs are very high (about hundreds of gigahertz) because the WNs generally emerge far away from the center of the Brillouin zone. It is beyond the current microwave technology to excite magnons with such a high frequency. Recently, Zhang *et al* [76] reported the observation of Weyl magnons in multiferroic ferrimagnet Cu_2OSeO_3 by

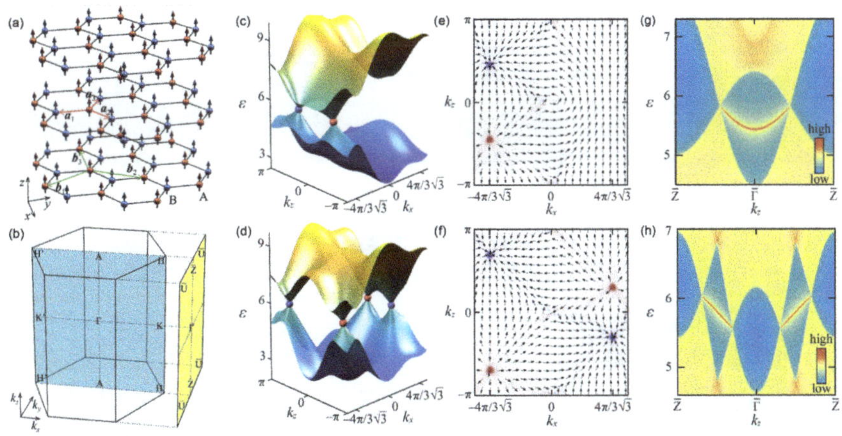

Figure 4.19. (a) A sketch map and (b) the corresponding first Brillouin zone of a stacked honeycomb ferromagnet. (c) and (d) The dispersion relations of Weyl magnons. Here, the red and blue dots denote the Weyl nodes with topological charges +1 and −1, respectively. (e) and (f) The spatial distribution of Berry curvatures for the lower bands considered in (c) and (d). The black arrows represent the direction of Berry curvature within the k_x–k_z-plane for $k_y = 0$. The background color describes the divergence of Berry curvature with the red and blue regions denoting the positive and negative values, respectively. (g) and (h) The density plots of the magnon wave-functions along $\overline{Z\Gamma Z}$ with the energy bands considered in (c) and (d). (Reproduced with permission from [65]. Copyright 2017 The American Physical Society.)

utilizing an inelastic neutron scattering technique. They showed that two pairs of magnon WNs with topological charge +2 and −2 emerge at the center and boundary of the Brillouin zone when the DM interaction is equal to zero. However, if the DM interaction is considered, the WNs will shift to other positions which depend on the strength and direction of the DM interaction vector. The measured and calculated magnon band structures are plotted in figure 4.20, which shows good agreement.

4.3.3 Magnon nodal-line semimetals

In the above two sections, we have discussed the magnon semimetals (Dirac and Weyl semimetals) with 0-dimensional (0D) band crossings, i.e. discrete nodes. However, in addition to discrete points, the band-crossing points can also form 1D loops in 3D momentum space, called topological nodal-line semimetals. In recent years, the magnon nodal-line semimetals have received a certain amount of attention [77–82]. Compared to Dirac and Weyl semimetals, the nodal-line semimetals exhibit several distinct features [83]: (i) the nodal-line semimetals have a 1D bulk Fermi surface, while the Fermi surfaces are 0D to the Dirac and Weyl semimetals; (ii) the Dirac and Weyl semimetals have linear band dispersion in all k-directions, 1D nodal lines exhibit linear band dispersion only along directions perpendicular to the nodal line; (iii) generally speaking, the monopole charges are used to characterize the Dirac and Weyl

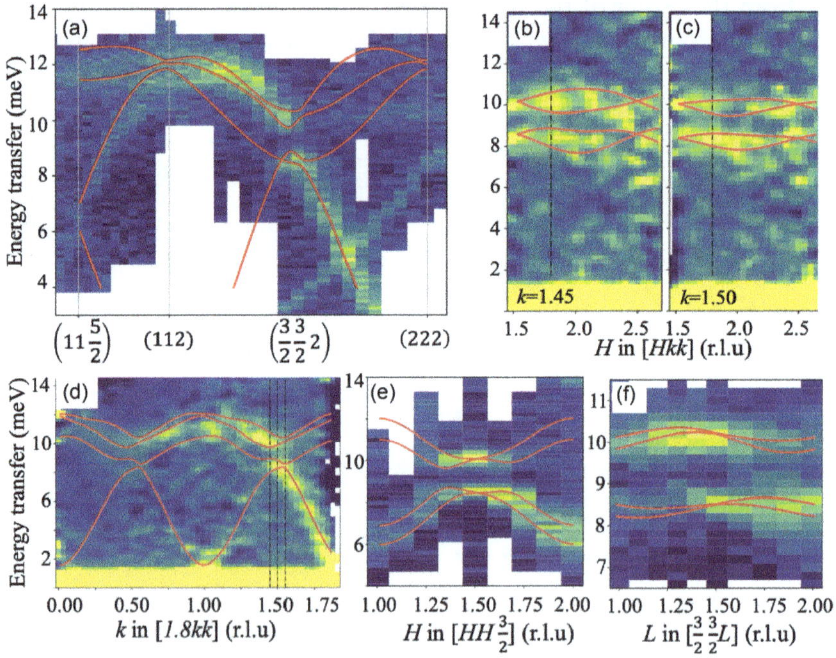

Figure 4.20. The momentum–energy cuts along different directions in the momentum space. Here, the conditions of temperature at 2 K and zero magnetic field are considered for experimental measurements. The data in (a), (e), and (f) were measured with the triple-axis spectrometer, while (b)–(d) are cuts from the time-of-flight data set. The red curves denote the magnon bands determined by the fitted data of the DM interaction. (Reproduced with permission from [76]. Copyright 2020 The American Physical Society.)

semimetals, however, for the nodal-line semimetals, the form of topological invariant which protects the nodal-line structure depends on the symmetry group; and (iv) crystallographic symmetries are necessary for stabilizing nodal lines and the bulk–edge correspondence may not be applicable. The surface states in nodal-line semimetals manifest as the 'drumhead' surface states and are not topologically protected.

Mook *et al* [77] reported the magnon nodal-line semimetals in anisotropic pyrochlore ferromagnets. The pyrochlore lattice consists of four interpenetrating face-centered cubic (fcc) lattices, resulting in a regular array of corner-sharing tetrahedra, as shown in figure 4.21(a). Along the [111] direction, kagome layers alternate with triangular layers. We consider two different exchange interactions: the nearest-neighbor interactions within the kagome layer (J_{N}, solid lines in figure 4.21(a)) and interactions in a triangular layer with sites in the adjacent kagome layers (J'_{N}, dashed lines in figure 4.21(a)). Assuming $0 < J'_{\mathrm{N}} < J_{\mathrm{N}}$, which leads to the anisotropic exchange interactions, the Hamiltonian of the system can be expressed as

$$H = -J_{\mathrm{N}} \sum_{\langle ij \rangle}^{\text{kagome}} \mathbf{s}_i \cdot \mathbf{s}_j - J'_{\mathrm{N}} \sum_{\langle ij \rangle}^{\text{triangular}} \mathbf{s}_i \cdot \mathbf{s}_j, \tag{4.20}$$

where $\mathbf{s}_i$ and $\mathbf{s}_j$ are the spins at sites i and j. By applying the truncated Holstein–Primakoff transformation [31], we can obtain the free-magnon Hamiltonian matrix

$$H_{\text{free}} = -2sJ_{\mathrm{N}} \begin{bmatrix} -2-\lambda & c(y,-z) & \lambda c(x,y) & c(z,-x) \\ c(y,-z) & -2-\lambda & \lambda c(x,z) & c(x,-y) \\ \lambda c(x,y) & \lambda c(x,z) & -3\lambda & \lambda c(y,z) \\ c(z,-x) & c(x,-y) & \lambda c(y,z) & -2-\lambda \end{bmatrix}, \tag{4.21}$$

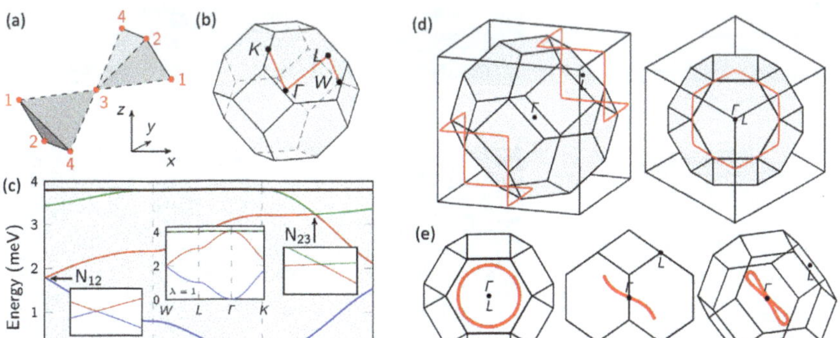

Figure 4.21. (a) Corner-sharing tetrahedra of the pyrochlore lattice with basis sites. Sites 1, 2, and 4 form a kagome lattice with nearest-neighbor interaction J_{N} (solid bonds); site 3 connects adjacent kagome layers by a weaker interaction J'_{N} (dashed bonds). (b) Brillouin zone of the fcc lattice with indicated high-symmetry points. (c) Magnon spectrum of the anisotropic pyrochlore ferromagnet along the red path depicted in (b). Band crossings are marked N_{12} and N_{23}, with insets showing magnifications. The central inset displays the spectrum for the isotropic case ($\lambda = 1$). Perspective views of the nodal lines (red lines) (d) N_{12} and (e) N_{23} in the fcc Brillouin zone. (Reproduced with permission from [77]. Copyright 2017 The American Physical Society.)

where $c(\alpha, \beta) = \cos[\mathrm{sgn}(\alpha)k_\alpha + \mathrm{sgn}(\beta)k_\beta]$ for $\alpha, \beta = x, y, z$ and $\lambda = J_N'/J_N$. The magnon dispersions can be determined by diagonalizing H_{free}. The magnon band structure along high-symmetry lines (see figure 4.21(b)) is plotted in figure 4.21(c) with parameters $s = 1/2$, $J_N = 1$ meV, and $\lambda = 0.8$. One can see that there exist the crossing N_{12} of band 1 (blue) and band 2 (red), and the crossing N_{23} of band 2 (red) and band 3 (green). Indeed, these band crossings are nodal-line, which are depicted in figures 4.21(d) and (e) with perspective views in 3D momentum space.

It is noted that the Hamiltonian (4.21) respects the time-reversal symmetry because there is no complex hopping. Meanwhile, the inversion symmetry is also present. Therefore, these magnon nodal lines are topologically protected by the coexistence of inversion and time-reversal symmetry and characterized by the Berry phase. The constant-energy cuts of the surface magnon spectrum at energies close to the nodal line N_{23} are plotted in figure 4.22(a) (the corresponding energies are indicated in figure 4.22(b)). One can see that, for the lowest energy, there is no surface state and only bulk states are visible (bottom constant-energy cut in figure 4.22(a)). At the energy for which only N_{23} is present, the projected bulk states form a closed ring in the surface Brillouin zone (top constant-energy cut in figure 4.22(a)). In addition, with decreasing energy, the drumhead surface states

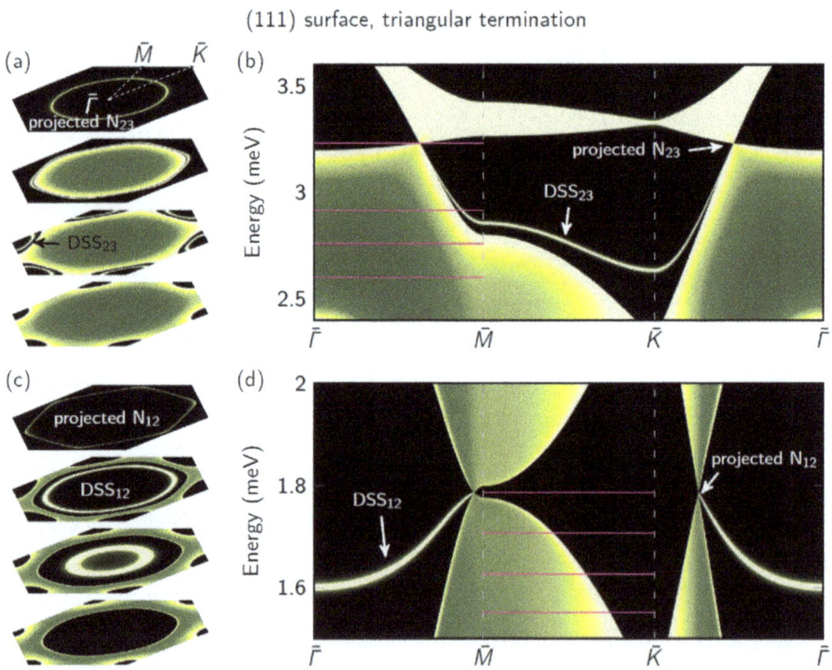

Figure 4.22. Magnons at the [111] surface of an anisotropic pyrochlore ferromagnet for triangular termination. Panels (a) and (c) show the constant-energy cuts through the entire surface Brillouin zone for energies indicated by magenta lines in (b) and (d), respectively. Panels (b) and (d) display the spectral densities along high-symmetry directions of the surface Brillouin zone. (Reproduced with permission from [77]. Copyright 2017 The American Physical Society.)

DSS_{23} associated with N_{23} produce a ring-like feature whose extension shrinks toward $\bar{K}$. We can perform a similar analysis for nodal line N_{12}, as shown in figures 4.22(c) and (d).

Furthermore, Hwang *et al* [81] reported topological magnons with nodal-line and triple-point degeneracies based on pyrochlore iridates with all-in-all-out (AIAO) antiferromagnet theoretically, as shown in figure 4.23(a). The antiferromagnetic (AFM) Heisenberg exchange and DM interaction between nearest-neighbor moments are considered in the Hamiltonian (see figure 4.23(b)), which can lead to the AIAO ordering. The corresponding first Brillouin zone is plotted in figure 4.23(c). By tuning the strength of the exchange interaction and DM interaction, the topological magnon bands can exhibit the triply degenerate cross-ings and nodal lines, as shown in figures 4.23(d) and (e), respectively. In addition, Liu *et al* [80] showed that the dipolar spin waves in uniaxial easy-axis antiferro-magnets are naturally in a topological nodal-line semimetal phase, and the topological surface modes display a chirality-momentum locking. The magnon nodal-line can exist in the quasi-2D quantum magnetic systems [79] with bilayer honeycomb ferromagnets as well.

4.4 Topological magnon polarons

In the above sections, we have showed that magnons can exhibit topological features based on various models. Meanwhile, it has been proved that phonons can support the Hall effect because of their topological nature [5, 84, 85]. In a magnetic system, when external excitation is applied, magnons and phonons are generally excited simultaneously. Remarkably, due to magnon–phonon coupling [86–88], magnons and phonons can be hybridized in the form of a magnetoelastic wave [89] or magnon polaron [90]. In previous studies, researchers discussed the topological phases separately for the magnon or phonon. Interestingly, very recently, the topological characteristics of a magnon polaron (or magnetoelastic wave) begin to attract people's interest. It was surprising that the magnon polaron can support topology even if the single magnon or phonon system is in its trivial phase. In this section, we introduce the topological properties of magnon polarons.

In 2016 Takahashi *et al* [91] theoretically investigated the Berry curvature of magnon polarons in which the magnon and phonon are hybridized through the dipole–dipole interaction. It was observed that an anticrossing emerges due to the magnon–phonon hybridization. In addition, it was found that the Berry curvature is highly anisotropic for longitudinal and transverse acoustic modes. On the one hand, the divergence of the Berry curvature emerges at which the hybridization disappears for the longitudinal acoustic mode. On the other hand, for the transverse acoustic mode, the divergence of the Berry curvature always appears even though there are no gapless points.

In addition to the long-range dipole–dipole interaction, Zhang *et al* [92] proved that short-range coupling can also induce the thermal Hall effect for magnon polarons. Utilizing symmetry analysis, they found that as long as the mirror symmetry along the direction of magnetization is broken, the short-range magnon–phonon coupling will

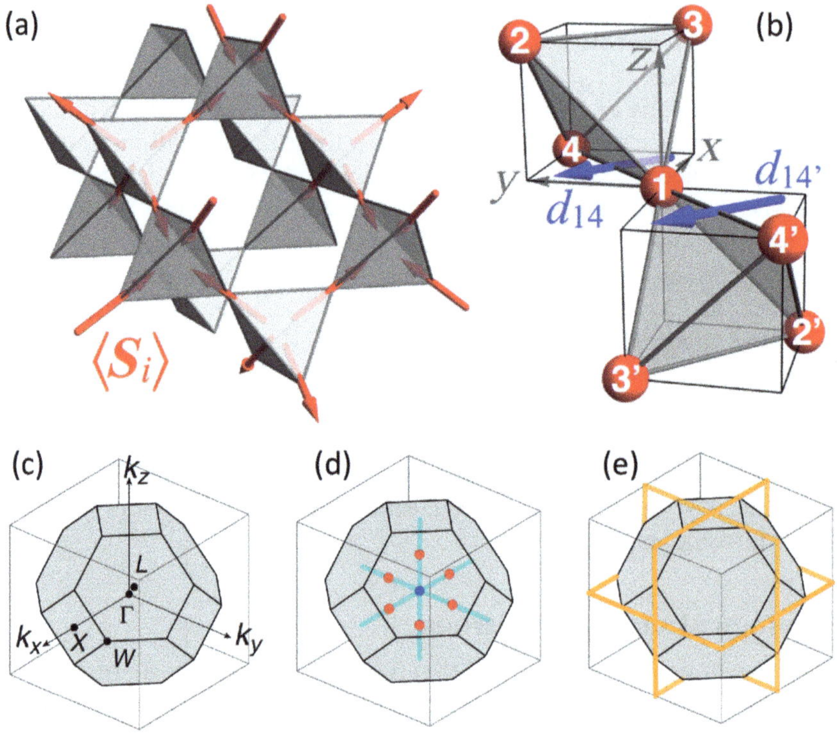

Figure 4.23. (a) The spin configuration of the AIAO state on the pyrochlore lattice. (b) The blue arrows represent the DM vectors and the numbers (1–4) denote the four sublattices of the pyrochlore crystal. (c) The first Brillouin zone of the pyrochlore lattice. Locations of the (d) triple points and (e) nodal lines in the Brillouin zone. (Reproduced with permission from [81]. Copyright 2020 The American Physical Society.)

appear. In addition, they showed that the hybridized magnon–phonon system can support the thermal Hall effect even if the single magnon or phonon system cannot. To verify the theoretical predictions, the authors chose a ferromagnet with a square lattice, as shown in figure 4.24(a). In this model, the perpendicular anisotropy and DM interaction are considered (see figure 4.24(b)). In general, the band structure of the magnon polaron can be divided into three parts: the magnon part, the phonon part, and the hybridized magnon–phonon part. On the one side, the magnon dispersion relation can be calculated via a Holstein–Primakoff transformation, and the dynamic matrix of the phonon can be determined by considering the NN and NNN couplings (see figure 4.24(c)). On the other side, the dynamics of the hybridized magnon–phonon part is described by the generalized Bogoliubov–de Gennes equation. A band structure exhibiting anticrossing points is the hallmark for a hybridized magnon–phonon system and these special points will result in Berry curvature hotspots which cause the thermal Hall conductivity. The dispersion relation of the magnon–phonon hybridized system is plotted in figure 4.24(d), from which one can clearly identify the anticrossing points. In addition, the distribution of Berry curvature hotspots can verify the emergence of anticrossing points, as shown in figures 4.24(e)–(g).

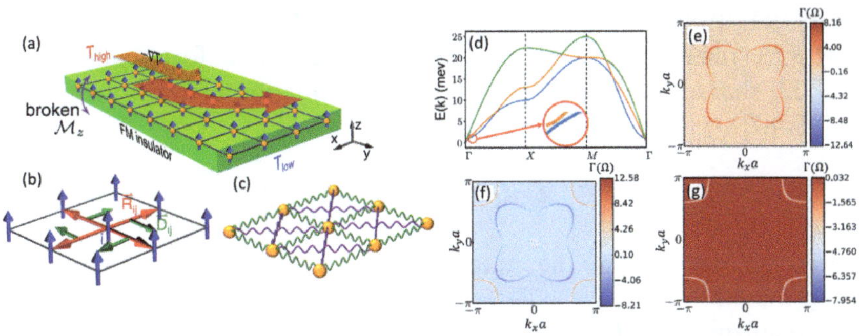

Figure 4.24. (a) A schematic diagram of the thermal Hall effect in a hybrid magnon–phonon system. (b) An illustration of a spin system, where the blue arrows denote the magnetization, the green arrows represent DM interaction vectors, and the red arrows connect the nearest-neighbor bonds. (c) An illustration of a phonon system, where the green and purple wavy lines represent the first- and second-nearest-neighbor interactions. (d) The dispersion relation of the hybrid magnon–phonon system. The spatial distribution of the Berry curvatures of (e) the lowest band, (f) the middle band, and (g) the highest band. (Reproduced with permission from [92]. Copyright 2019 The American Physical Society.)

In addition to the ferromagnet discussed above, very recently, Park *et al* [93] investigated the topological phase of a magnon–phonon hybrid based on antiferromagnets. Here, the coupling between magnons and phonons is realized through a magnetoelastic interaction which originates from the exchange striction. They showed that when an out-of-plane magnetic field is applied, although the isolated magnon and phonon band structures are trivial, the hybridized magnon–phonon band can support a nonzero Chern number. In addition, they observed a large Berry curvature which is induced by the anticrossing region.

It is worth mentioning that Go *et al* [94] presented a systematic investigation about the topological phases of a magnon–phonon hybridized system in ferromagnets, where the magnon and phonon are coupled via Kittel's magnetoelastic interaction. They further proved that the topology of the system can be induced by this magnetoelastic interaction, even though the dipolar interaction, DM interaction, or lattice symmetry are all vanishing. The ferromagnet with a square lattice (figure 4.25(a)) can be characterized by the following Hamiltonian

$$H = H_{\text{mag}} + H_{\text{ph}} + H_{\text{mp}}, \tag{4.22}$$

where the first term denotes the magnetic Hamiltonian which can be expressed as

$$H_{\text{mag}} = -J\sum_{\langle i,j \rangle}\mathbf{S}_i \cdot \mathbf{S}_j - \frac{K_z}{2}\sum_i S_{i,z}^2 - B\sum_i S_{i,z}, \tag{4.23}$$

with $J > 0$ the ferromagnetic exchange constant, $K_z > 0$ the perpendicular magnetocrystalline anisotropy coefficient, and B being the strength of the magnetic field. The second term in (4.22) represents the phonon Hamiltonian

$$H_{\text{ph}} = \sum_i \frac{\mathbf{p}_i^2}{2M} + \frac{1}{2}\sum_{i,j,\alpha,\beta} u_i^\alpha \Phi_{i,j}^{\alpha,\beta} u_j^\beta, \tag{4.24}$$

where $\mathbf{u}_i$ is the displacement vector of the ith ion, $\mathbf{p}_i$ is the conjugate momentum vector, M is the mass of ion, and $\Phi_{i,j}^{\alpha,\beta}$ is a matrix describing force constant. The last term in (4.22) characterizes the magnetoelastic interaction between the magnon and phonon [86, 87]

$$H_{\mathrm{mp}} = \kappa \sum_i \sum_{\mathbf{e}} (\mathbf{S}_i \cdot \mathbf{e})(u_i^z - u_{i+\mathbf{e}}^z), \tag{4.25}$$

where κ is the strength of the magnon–phonon coupling and $\mathbf{e}$ is the vector connecting the NN sites.

By diagonalizing the Hamiltonian (4.22), we can obtain the dispersion relation and topological invariant Chern number, with the results being shown in figures 4.25(b)–(d). When the magnon–phonon coupling is not considered, i.e. $\kappa = 0$, the system supports two bands which separately describe the magnon and phonon, and two band-crossing points naturally emerge (see figure 4.25(c)). However, if $\kappa \neq 0$, the band-crossing points will vanish and anticrossing points appear, as shown in figure 4.25(d). For a nonzero κ, by tuning the strength of effective field and exchange interaction, one can realize the phase transition between the topological phase with a Chern number equal to 1 (or 2) and a trivial phase with Chern number equal to 0, as plotted in figure 4.25(b). In addition, the dispersion relations of an infinite system for Chern number $|C| = 1$ are plotted in figures 4.25(e) and (f), where the parameters of CrI_3 are adopted [95, 96]. Again, we can clearly see that band-crossing points emerging for $\kappa = 0$ are replaced by anticrossing points for the case $\kappa = 5$ meV Å^{-1}. In addition, large Berry curvature is observed at the vicinity of the anticrossing point, as shown in figure 4.25(g). The topological properties of

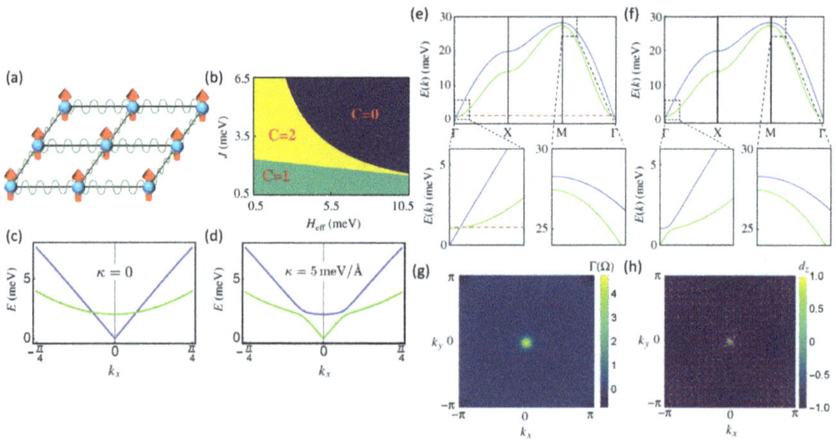

Figure 4.25. A sketch map (a) and the corresponding Chern number (b) of the hybrid magnon–phonon system. The band structures (c) without and (d) with magnetoelastic coupling. The dispersion relations of the system for (e) $\kappa = 0$ and (f) $\kappa = 5$ meV Å^{-1} with Chern number $|C| = 1$. (g) The spatial distribution of Berry curvatures for the upper band. (h) A schematic diagram of $\hat{\mathbf{d}}(\mathbf{k})$ for $\kappa = 5$ meV Å^{-1}, where the red arrows denote the in-plane components $(\hat{d}_x, \hat{d}_y)$. (Reproduced with permission from [94]. Copyright 2019 The American Physical Society.)

the magnon–phonon hybridized system can also described by the skyrmion number (Q):

$$Q_{\pm} = \frac{1}{2\pi}\int \mp\frac{1}{2}\hat{\mathbf{d}}(\mathbf{k})\cdot\left[\frac{\partial\hat{\mathbf{d}}(\mathbf{k})}{\partial k_x}\times\frac{\partial\hat{\mathbf{d}}(\mathbf{k})}{\partial k_y}\right]\mathrm{d}k_x\mathrm{d}k_y, \tag{4.26}$$

where $\hat{\mathbf{d}}(\mathbf{k}) = \mathbf{d}(\mathbf{k})/|\mathbf{d}(\mathbf{k})|$ with vector $\mathbf{d}(\mathbf{k})$ characterizing the band structures of magnons and phonons:

$$\mathbf{d}(\mathbf{k}) = \left(\tilde{\kappa}\sin k_y,\ \tilde{\kappa}\sin k_x,\ \frac{\hbar}{2}[\omega_m(\mathbf{k}) - \omega_p(\mathbf{k})]\right). \tag{4.27}$$

Here $\tilde{\kappa} = \kappa\sqrt{\hbar S/[M\omega_p(\mathbf{k})]}$, and $\omega_m(\mathbf{k})$ and $\omega_p(\mathbf{k})$ are the magnon and phonon dispersion relations, respectively. Figure 4.25(h) plots the distribution of $\hat{\mathbf{d}}(\mathbf{k})$ for $\kappa = 5$ meV $\mathring{\mathrm{A}}^{-1}$.

Moreover, Zhang *et al* [97] investigated the topological properties of magnon polarons in an antiferromagnet with a square lattice, as shown in figure 4.26(a). Here, the magnetoelastic coupling taking the form of (4.25) is considered. They showed that when $\kappa = 0$, the two magnon bands intersect with one phonon band, as shown in figures 4.26(b)–(d). However, if $\kappa \neq 0$, the anticrossing characteristic of the bands can be seen clearly, as shown in figures 4.26(e)–(g). Noticeably, the skyrmion number (4.26) can also be used to understand the band topology for antiferromagnetic magnon polarons. Since the topological magnon polaron can lead to the thermal Hall effect, one therefore can identify the topological properties of magnon polaron by detecting the thermal Hall conductance.

4.5 Higher-order topological magnons

In the above sections, the topological phases emerging in magnetic systems belong to first-order. In 2019 Li *et al* [98] reported a higher-order topological insulator (HOTI)

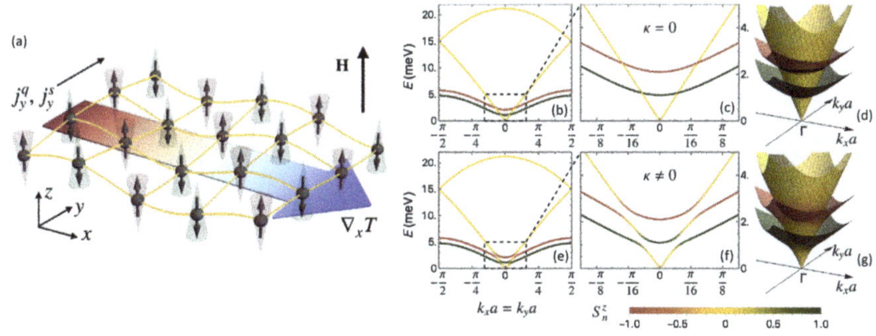

Figure 4.26. (a) The sketch map of hybrid magnon–phonon excitations based on a antiferromagnetic square lattice. The hybrid system shows the thermal Hall effect (the generation of a transverse heat current j_y^q) and the spin Nernst effect (the generation of a transverse spin current j_y^s). The band structures of the system in the absence (b)–(d) and in the presence (e)–(g) of magnetoelastic coupling. The green and red curves describe the magnon bands with $S^z = 1$ and $S^z = -1$, respectively, while the yellow curves depict the phonon bands. (Reproduced with permission from [97]. Copyright 2020 The American Physical Society.)

in a magnetic system for the first time (see chapter 7 for a detailed discussion). Subsequently, Sil *et al* [99] and Hirosawa *et al* [100] showed a second-order magnon TI in a two-dimensional system based on a breathing kagome lattice and anti-skyrmion crystals, respectively, where the robust corner states are clearly identified. Very recently, in three-dimensional van der Waals honeycomb ferromagnets, chiral hinge magnons emerging in the second-order TIs were discovered [101–103]. In this section we discuss the properties of magnonic HOTIs.

As shown in section 4.2.1, magnon-based FOTIs can be realized in two-dimensional kagome arrays. Sil *et al* [99] constructed a 'breathing' kagome lattice in a ferromagnet and found that the system can support second-order magnon TIs with appropriate parameters. The Hamiltonian of the system includes the ferromagnetic exchange interaction, DM interaction, and Zeeman energy. The crystal structure is plotted in figure 4.27(a), where the unit cell contains three sublattices, i.e. *A*, *B*, and *C*, the red and blue arrows denote the exchange interaction with strengths J_1 and J_2, respectively, the DM interaction vector with strength D_1 (D_2) along the z ($-z$) direction for upward (downward) triangle regions, and the magnetic field is applied along the z-axis to saturate the magnetization. The Hamiltonian is expressed as

$$\begin{aligned} H = &- J_\alpha \sum_{\langle n, n' \rangle} (\mathbf{S}_\mathbf{n}^\mathbf{a} \cdot \mathbf{S}_{\mathbf{n}'}^\mathbf{b} + \mathbf{S}_\mathbf{n}^\mathbf{b} \cdot \mathbf{S}_{\mathbf{n}'}^\mathbf{c} + \mathbf{S}_\mathbf{n}^\mathbf{c} \cdot \mathbf{S}_{\mathbf{n}'}^\mathbf{a}) - h \sum_n (S_n^{az} + S_n^{bz} + S_n^{cz}) \\ &- D_\alpha \sum_{\langle n, n' \rangle} (\mathbf{S}_\mathbf{n}^\mathbf{a} \times \mathbf{S}_{\mathbf{n}'}^\mathbf{b} + \mathbf{S}_\mathbf{n}^\mathbf{b} \times \mathbf{S}_{\mathbf{n}'}^\mathbf{c} + \mathbf{S}_\mathbf{n}^\mathbf{c} \times \mathbf{S}_{\mathbf{n}'}^\mathbf{a}) \cdot \hat{z}, \end{aligned} \tag{4.28}$$

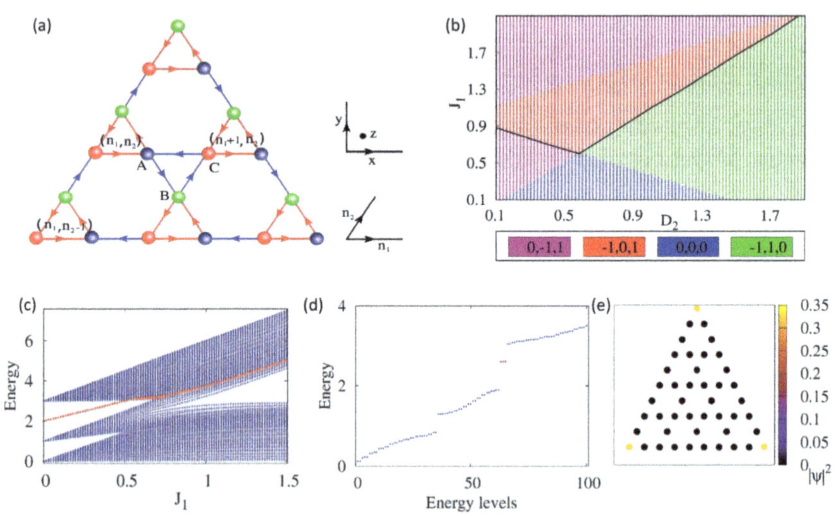

Figure 4.27. (a) Schematic diagram of a breathing kagome lattice based on a ferromagnet. (b) The phase diagram of the system for different combinations of parameters J_1 and D_2, with $J_2 = 1.0$ and $D_1 = 0.1$. Different colors represent different phases and the lower panel shows the corresponding Chern numbers C_1, C_2, C_3. (c) The energy eigenvalues of the breathing kagome lattice for different values of J_1 with $J_2 = 1.0$ and $D_1 = 0$. (d) The energy eigenvalues of the finite system with parameters $J_1 = 0.3$, $J_2 = 1.0$, and $D_1 = 0$. (e) The probability distribution of the corner state for $J_1 = 0.2$. (Reproduced with permission from [99]. Copyright 2020 IOP Publishing.)

where $\alpha = 1, 2, \langle n, n' \rangle$ represents the NN sites, $h = g\mu_B H_0$ with H_0 the magnetic field strength, and $\mathbf{S}_\mathbf{n}^\mathbf{a}$, $\mathbf{S}_\mathbf{n}^\mathbf{b}$, and $\mathbf{S}_\mathbf{n}^\mathbf{c}$ denote the spin operators for different sites.

It has been proved that when the DM interaction is absent, the ferromagnet with a kagome lattice ($J_1 = J_2$) is trivial [20]. Sil *et al* [99] showed that the breathing kagome lattice ($J_1 \neq J_2$) allows a second-order TI. The spectra of a finite system for different J_1 (J_2 fixed to 1) are plotted in figure 4.27(c), from which we can see that when $0 < J_1 < 0.5$, the system supports many degenerate modes. To further identify these degenerate modes, we choose $J_1 = 0.3$ as an example. Figure 4.27(d) shows the corresponding spectrum and one can clearly see three degenerate states as marked by yellow dots. These degenerate states are actually corner states with oscillations being highly confined at three corners, as shown in figure 4.27(e). Remarkably, when the nonzero DM interaction is considered, the breathing kagome lattice shows various phases which contain FOTI and HOTI as well as their coexistence. Figure 4.27(b) plots the phase diagram with $J_2 = 1$ and $D_1 = 0.1$. On the one side, the Chern number is calculated to describe the FOTI phase, four different phases are identified as marked with different colors (see figure 4.27(b)). On the other side, the bulk polarization is determined to characterize the HOTI phase. It has been found that only when the parameters are chosen in the regions beneath the black line, can the system support the higher-order (second-order) TI phase. Therefore, we have the conclusions that the red and magenta regions above the black line denote the FOTI phase, the magenta and green regions below the black line allow the coexistence of FOTI and second-order TI phases, and the blue region only supports the second-order TI phase. Noticeably, the second-order TI phases emerging in the breathing kagome lattice are protected by both the crystalline symmetry (for example, C_3 and mirror symmetries) and generalized chiral symmetry [104].

In addition, Hirosawa *et al* [100] have shown that the antiskyrmion crystals can realize a magnonic quadrupole topological insulator. The following two-dimensional spin lattice Hamiltonian is considered:

$$H = \frac{1}{2}\sum_{\langle \mathbf{r}, \mathbf{r}' \rangle} (-J_{\mathbf{r},\mathbf{r}'}\mathbf{S}_\mathbf{r} \cdot \mathbf{S}_{\mathbf{r}'} + D_{\mathbf{r},\mathbf{r}'} \cdot \mathbf{S}_\mathbf{r} \times \mathbf{S}_{\mathbf{r}'}) - g\mu_B B_z \sum_\mathbf{r} \mathbf{S}_\mathbf{r} \cdot \hat{z}, \tag{4.29}$$

where $\mathbf{S}_\mathbf{r}$ is the spin operator at site $\mathbf{r}$ in a square lattice with lattice constant a, $J_{\mathbf{r},\mathbf{r}'} = J(\delta_{\mathbf{r}-\mathbf{r}',\pm a\hat{x}} + \delta_{\mathbf{r}-\mathbf{r}',\pm a\hat{y}})$ is the ferromagnetic exchange coupling with $J > 0$, $D_{\mathbf{r},\mathbf{r}'} = D(\mp\hat{x}\delta_{\mathbf{r}-\mathbf{r}',\pm a\hat{x}} \pm \hat{y}\delta_{\mathbf{r}-\mathbf{r}',\pm a\hat{y}})$ is the DM interaction, B_z is the external magnetic field, and g and μ_B denote the g-factor and Bohr magneton, respectively. The phase diagram is determined by the value of $g\mu_B B_z/(JS)$ (see figure 4.28(c)). Interestingly, Hirosawa *et al* [100] found that the robust magnonic corner states emerge only when the external magnetic field is below a critical value, triggering the self-assembly of fractional antiskyrmions stabilized along the sample edges that restore the protecting symmetries. Specifically, when $g\mu_B B_z/(JS) \leqslant 0.40$, the fractional antiskyrmions can be stabilized along the sample edges. In this case, the system is in the second-order TI phase and magnonic topological corner states will appear (see figure 4.28(a)). However, if $g\mu_B B_z/(JS) > 0.40$, there are no fractional antiskyrmions (see figure 4.28(b)) and the system is in the trivial phase with no corner states.

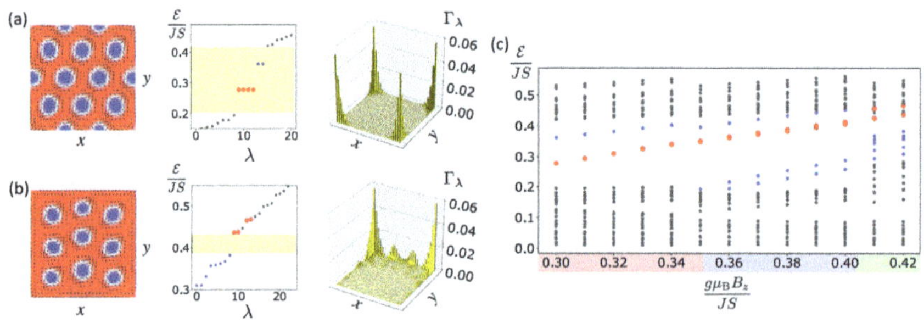

Figure 4.28. Formation of magnonic corner states in confined antiskyrmion crystals. (a) and (b) Characterization of confined antiskyrmion crystals at $g\mu_B B_z/(JS) = 0.3$ and $g\mu_B B_z/(JS) = 0.42$, respectively. Left: Ground-state of the system. Middle: Magnon spectrum showing corner states (red) and trivial edge states (blue). Right: Probability density of the corner states. (c) Magnon spectrum against the applied magnetic field with corner and edge-localized states. (Reproduced with permission from [100]. Copyright 2020 The American Physical Society.)

Although the modes (denoted by red dots) with significant probability density near the corners can still be identified, they spread over the edges and into the bulk. It is noted that the second-order topological phase here is protected by the combined $C_{2x}T$ and $C_{2y}T$ symmetries.

For the second-order TI phases of magnons, the corner states will emerge in 2D systems. If we consider a three-dimensional system, the topological edge states are chiral hinge magnons. Recently, Mook *et al* [101] theoretically studied the chiral hinge magnons based on a stack of honeycomb magnets, as shown in figure 4.29(a), where the red and blue spheres denote spins. The interactions between these spins are comprised in the following Hamiltonian

$$H = H_{\parallel} + H_{\perp} + H_{\perp}^{\delta} + H_Z. \tag{4.30}$$

Here, $H_{\parallel} = \sum_l h_{\parallel}^{(l)}$ (l is the layer index) is the intralayer interaction with

$$\sum_l h_{\parallel}^{(l)} = -\frac{J}{2}\sum_{\langle ij\rangle}\mathbf{S}_i^{(l)} \cdot \mathbf{S}_j^{(l)} + \frac{(-1)^l D}{2}\sum_{\langle\langle ij\rangle\rangle} v_{ij}\hat{z} \cdot \mathbf{S}_i^{(l)} \times \mathbf{S}_j^{(l)}, \tag{4.31}$$

where the first term is the nearest-neighbor exchange interaction and the second term is the next-nearest-neighbor DMI. $H_{\perp}$ is AA-stacked interlayer interaction

$$H_{\perp} = -J_{\perp}\sum_l \sum_i \mathbf{S}_i^{(l)} \cdot \mathbf{S}_i^{(l+1)}. \tag{4.32}$$

In addition, to obtain the second-order topological phase, an extra term $H_{\perp}^{\delta}$ is considered to break the mirror symmetry

$$H_{\perp}^{\delta} = \delta J_{\perp}\sum_l (-1)^l \left(\sum_{i\in\mathrm{A}}\mathbf{S}_i^{(l)} \cdot \mathbf{S}_i^{(l+1)} - \sum_{i\in\mathrm{B}}\mathbf{S}_i^{(l)} \cdot \mathbf{S}_i^{(l+1)}\right), \tag{4.33}$$

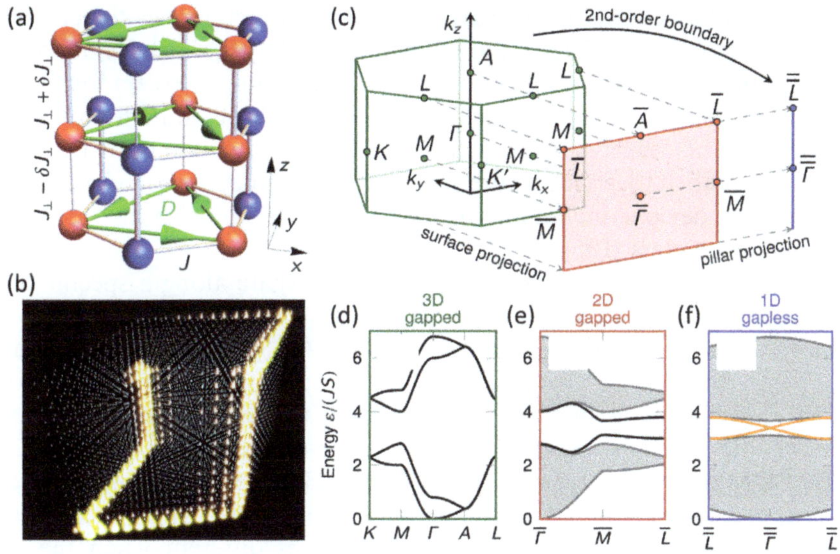

Figure 4.29. (a) A stack of honeycomb layers with indicated magnetic interactions. The blue and red dots denote the A and B sublattices, respectively. (b) Snapshot of a chiral hinge magnon in a second-order topological magnon insulator. The size of the cones represents the strength of the spin excitation. (c) The 3D Brillouin zones and high-symmetry points. The projected 2D and 1D Brillouin zones are also shown. (d) Gapped 3D spectrum showing four magnon branches. (e) Gapped 2D spectrum, with gray areas indicating the bulk continuum projected onto the xz surface and black lines indicating surface states. (f) Gapless 1D spectrum with chiral hinge magnons crossing the band gap (orange lines). (Reproduced with permission from [101]. Copyright 2021 The American Physical Society.)

which is opposite for the A and B sublattices of the honeycomb magnet (see figure 4.29(a)). At last, H_Z is the Zeeman Hamiltonian with

$$H_Z = \sum_l \sum_i B^z S_i^{(l),z}. \tag{4.34}$$

The Brillouin zone convention of the system is shown in figure 4.29(c). When the parameters are chosen for proper values, the system can support the second-order TI phase. In this situation, the three-dimensional bulk magnon spectrum shows the band gap between the lower and upper pair of bands (see figure 4.29(d)). Meanwhile the projected two-dimensional band structures are also gapped, in contrast to the gapless dispersion relation for the one-dimensional case (see figures 4.29(e) and (f)). The representative hinge magnon mode is plotted in figure 4.29(b) with energy being localized at the edge of the system. In addition, Park *et al* [102] and Li *et al* [103] investigated the second-order TI phase of magnons in van der Waals ferromagnets with noncollinear and collinear antiferromagnetic interlayer coupling, respectively. The chiral topological hinge magnon states are also identified.

4.6 Topological magnon devices

In magnonics or magnon spintronics, the manipulation of magnon states in a desired manner is a central topic. On the one hand, the ordinary magnons are sensitive to various perturbations, such as disorder and defects, which makes the conventional magnon spintronic devices fragile and inflexible. On the other hand, it is challenging to realize the unidirectional propagation of ordinary magnons except for the MSSW. Topological magnons become essential to design robust and tunable magnonic devices. Wang *et al* [59] have investigated the topological magnons in a ferromagnet with a honeycomb lattice (see figure 4.17). Further, they designed various topological magnon devices including magnon diode, magnon beam splitters, and magnon interferometers [105].

As shown in [59], the topological magnons propagate along a specific direction which depends on the direction of the magnetic moment, i.e. counterclockwise (clockwise) for a magnetic moment along the $+z$ ($-z$) direction, as plotted in figure 4.30(a). Because of the unidirectional propagation feature of topological magnons, the system boundary can be used to design the magnon diode. Considering the magnetic moment along the $+z$-direction and $D = 0$ (D is the strength of DM interaction), as shown in figure 4.30(b), when the magnon exciting source is applied at position ①, no magnons can be probed at position ②, and the system is in an *off* state. However, the situation is different when the magnon exciting source is applied at position ②, the topological magnons can propagate to position ①, and an *on* state therefore appears. Furthermore, Wang *et al* [105] showed that by tuning the DM interaction, the system can change from one topological phase to another in which the propagation direction of the topological magnon is reversed, so the *on* and *off* states can also be swapped. Figure 4.30(c) shows the switching of *on* and *off* states for $D = 0.4$.

Because the topological magnons propagate along opposite directions for the magnetic moment along the $+z$- and $-z$-directions, the domain wall can be used to design magnon beam splitters, as shown in the right-hand panel of figure 4.30(a), where the pink and cyan regions denote the magnetic moment along the $+z$- and $-z$-directions, respectively. As a result, a beam of magnons is finally divided into two beams, i.e. 1:2 beam splitters. Furthermore, Wang *et al* [105] designed a 1:4 magnon

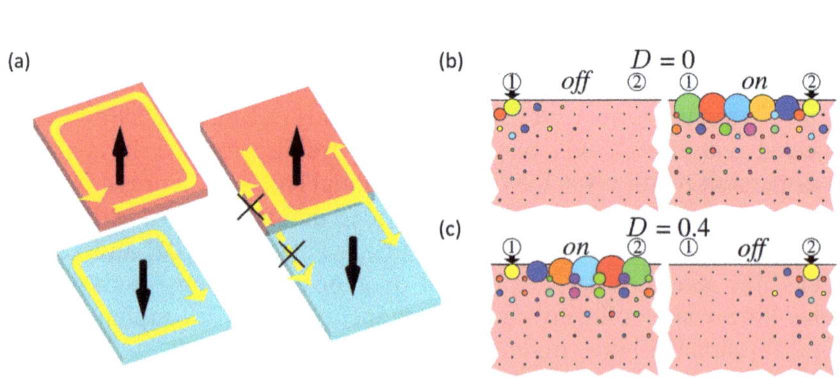

Figure 4.30. (a) Schematic diagram of the chiral magnon propagation. The cyan and red regions represent the magnetization along the $-z$- and $+z$-axes, respectively. The yellow arrows represent the propagation direction of the chiral magnon. Schematic diagrams of a magnon diode for (b) $D = 0$ and (c) $D = 0.4$. (Reproduced with permission from [105]. Copyright 2018 The American Physical Society.)

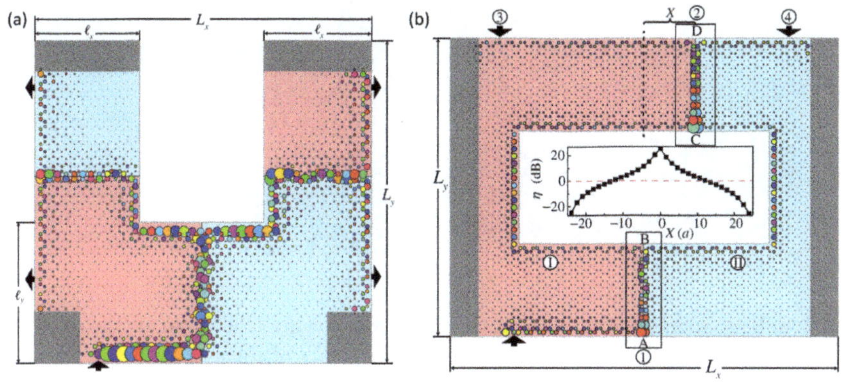

Figure 4.31. Sketch maps of (a) magnon beam splitters and (b) magnon interferometers. (Reproduced with permission from [105]. Copyright 2018 The American Physical Society.)

beam splitter, as shown in figure 4.31(a), where three domain walls are used to separate different domains. The magnon exciting source is applied at the bottom boundary as marked by the black inward arrow. One can clearly see that a magnon beam splits ultimately into four beams.

In addition, it is well known that the magnon interferometer is also a key element in spintronic devices. With the help of topological magnons, one can design a stable and reconfigurable magnon interferometer. Figure 4.31(b) plots a sketch map of a Mach–Zehnder–type magnon interferometer consisting of two different domains which are separated by a domain wall. The working principle of magnon interferometer can be explained as follows: first, the topological magnons are excited at the bottom boundary, as marked by the black arrow. Then the magnon propagates along the domain wall *AB*, and splits into beam Ⓘ and Ⓘⅰ. After propagating over a certain distance, the two magnon beams recombine and pass through the domain wall *CD*. After that, the magnons can reach the areas ③ and ④ with the strength depending on the interference of the two magnon beams in the domain wall *CD*. Noticeably, by adjusting the position or length of the domain wall *CD*, one can tune the relative phase of two interfered magnon beams.

Remarkably, Shindou *et al* [106] proved that the magnonic crystal (MC) can also support topological magnons. Here, the MC consists of YIG and Fe, where the YIG film has periodic holes which are filled by Fe, as plotted in figure 4.32. By utilizing the topological magnon edge states emerging in this MC, one can design the magnon splitter and interferometer. Figure 4.32(a) shows a schematic diagram of the magnon splitter, where regions II and III denote two different topological phases with Chern numbers $C = 2$ and 1, respectively, and $r = a_y/a_x$ describes the shape of the unit cell, with a_x and a_y representing the length of the unit cell in the x- and y-directions, respectively. The two topological magnon edge modes are excited in region II and propagate along the boundary. When they pass through the interface, one mode propagates along the boundary of region II while the other one propagates along the boundary of region III. We therefore realize a magnon splitter. In addition, the Fabry–Perot interferometer can be designed by using a couple of topological

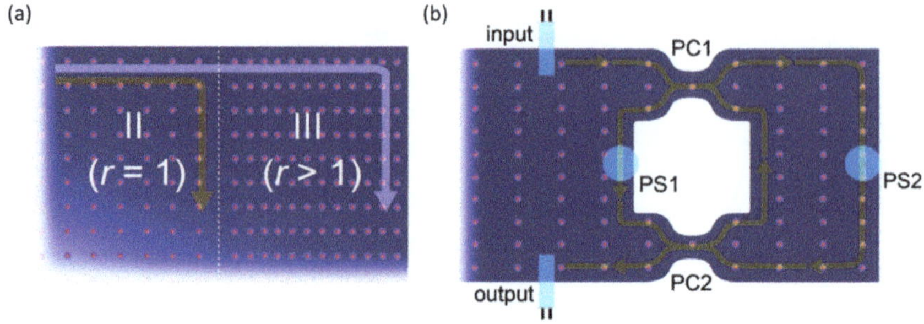

Figure 4.32. (a) Illustration of the magnon splitter. (b) A sketch map of the magnonic analog of the Fabry–Perot interferometer. (Reproduced with permission from [106]. Copyright 2013 The American Physical Society.)

magnon edge states in a single MC, as shown in figure 4.32(b). A topological magnon beam is excited in the antenna 'input', then splits into two beams at the point contact (PC1). After that, the two magnon beams propagate along separate paths and recombine at the other point contact (PC2). The superposed magnon beam shows destructive or constructive interference, which can be probed via an electric measurement from the antenna 'output'. Noticeably, by introducing the external magnetic field (PS1 and PS2), one can modify the propagation velocities of magnon beams.

References

[1] Hall E H 2010 *Am. J. Math.* **2** 287–92
[2] Nagaosa N *et al* 2010 *Rev. Mod. Phys.* **82** 1539
[3] Katsura H, Nagaosa N and Lee P A 2010 *Phys. Rev. Lett.* **104** 066403
[4] Onose Y *et al* 2010 *Science* **329** 297–9
[5] Strohm C, Rikken G L J A and Wyder P 2005 *Phys. Rev. Lett.* **95** 155901
[6] Sheng L, Sheng D N and Ting C S 2006 *Phys. Rev. Lett.* **96** 155901
[7] Kagan Y and Maksimov L A 2008 *Phys. Rev. Lett.* **100** 145902
[8] Matsumoto R and Murakami S 2011 *Phys. Rev. Lett.* **106** 197202
[9] Matsumoto R and Murakami S 2011 *Phys. Rev.* B **84** 184406
[10] Zhang L *et al* 2013 *Phys. Rev.* B **87** 144101
[11] Haug H and Jauho A P 1996 *Quantum Kinetics in Transport and Optics of Semiconductors* (Berlin: Springer) https://doi.org/10.1007/978-3-540-73564-9
[12] Hirschberger M *et al* 2015 *Phys. Rev. Lett.* **115** 106603
[13] Marcipar L *et al* 2009 *Phys. Rev.* B **80** 132402
[14] Hirschberger M *et al* 2015 *Science* **348** 106–9
[15] Tanabe K *et al* 2016 *Phys. Status Solidi* B **253** 783–7
[16] Matsumoto R, Shindou R and Murakami S 2014 *Phys. Rev.* B **89** 054420
[17] Hoogdalem K A V, Tserkovnyak Y and Loss D 2013 *Phys. Rev.* B **87** 024402
[18] Mochizuki M *et al* 2014 *Nat. Mater.* **13** 241–6
[19] Mook A *et al* 2017 *Phys. Rev.* B **95** 020401(R)
[20] Mook A, Henk J and Mertig I 2014 *Phys. Rev.* B **90** 024412
[21] Heisenberg W 1928 *Z. Phys.* **49** 619–36

[22] Mook A, Henk J and Mertig I 2014 *Phys. Rev.* B **89** 134409
[23] Shindou R *et al* 2013 *Phys. Rev.* B **87** 174402
[24] Tamm I 1932 *Z. Phys.* **76** 849–50
[25] Shockley W 1939 *Phys. Rev.* **56** 317
[26] Nakata K *et al* 2017 *Phys. Rev.* B **96** 224414
[27] Kawano M and Hotta C 2019 *Phys. Rev.* B **99** 054422
[28] Nakata K and Kim S K 2021 *J. Phys. Soc. Jpn.* **90** 081004
[29] Li Y M, Xiao J and Chang K 2018 *Nano Lett.* **18** 3032–7
[30] Roldán-Molina A, Nunez A S and Fernández-Rossier J 2016 *New J. Phys.* **18** 045015
[31] Holstein T and Primakoff H 1940 *Phys. Rev.* **58** 1098
[32] Díaz S A *et al* 2020 *Phys. Rev. Res.* **2** 013231
[33] Hamamoto K, Ezawa M and Nagaosa N 2015 *Phys. Rev.* B **92** 115417
[34] Wang X G *et al* 2020 *J. Appl. Phys.* **128** 063901
[35] Xie K, Zhang L and Ma F 2021 *J. Appl. Phys.* **130** 153901
[36] Díaz S A, Klinovaja J and Loss D 2019 *Phys. Rev. Lett.* **122** 187203
[37] Kim S K *et al* 2019 *Phys. Rev. Lett.* **122** 057204
[38] Weber T *et al* 2022 *Science* **375** 1025–30
[39] Neto A H C *et al* 2009 *Rev. Mod. Phys.* **81** 109
[40] Wehling T O, Black-Schaffer A M and Balatsky A V 2014 *Adv. Phys.* **63** 1–76
[41] Ozawa T *et al* 2019 *Rev. Mod. Phys.* **91** 015006
[42] Lu L, Joannopoulos J D and Soljačić M 2014 *Nat. Photon.* **8** 821–9
[43] Huang X *et al* 2011 *Nat. Mater.* **10** 582–6
[44] Hu C *et al* 2018 *Phys. Rev. Lett.* **121** 024301
[45] Weick G *et al* 2013 *Phys. Rev. Lett.* **110** 106801
[46] Pietro P D *et al* 2013 *Nat. Nanotechnol.* **8** 556–60
[47] Politano A *et al* 2018 *Phys. Rev. Lett.* **121** 086804
[48] Chen Z J *et al* 2021 *Phys. Rev. Lett.* **126** 185301
[49] Liu F, Huang X and Chan C T 2012 *Appl. Phys. Lett.* **100** 071911
[50] Jin Y, Wang R and Xu H 2018 *Nano Lett.* **18** 7755–60
[51] Banerjee S *et al* 2016 *Phys. Rev.* B **93** 134502
[52] Chen L *et al* 2018 *Phys. Rev.* X **8** 041028
[53] Owerre S A 2016 *J. Phys.: Condens. Matter* **28** 386001
[54] Fransson J, Black-Schaffer A M and Balatsky A V 2016 *Phys. Rev.* B **94** 075401
[55] Boyko D, Balatsky A V and Haraldsen J T 2018 *Phys. Rev.* B **97** 014433
[56] Pershoguba S S *et al* 2018 *Phys. Rev.* X **8** 011010
[57] Costa A T *et al* 2020 *2D Mater.* **7** 045031
[58] Kondo H, Akagi Y and Katsura H 2019 *Phys. Rev.* B **99** 041110(R)
[59] Wang X S, Su Y and Wang X R 2017 *Phys. Rev.* B **95** 014435
[60] Shekhtman L, Entin-Wohlman O and Aharony A 1992 *Phys. Rev. Lett.* **69** 836
[61] Jackeli G and Khaliullin G 2009 *Phys. Rev. Lett.* **102** 017205
[62] Yuan B *et al* 2020 *Phys. Rev.* X **10** 011062
[63] Mook A, Henk J and Mertig I 2016 *Phys. Rev. Lett.* **117** 157204
[64] Su Y, Wang X S and Wang X R 2017 *Phys. Rev.* B **95** 224403
[65] Su Y and Wang X R 2017 *Phys. Rev.* B **96** 104437
[66] Zyuzin V A and Kovalev A A 2018 *Phys. Rev.* B **97** 174407
[67] Owerre S A 2018 *Sci. Rep.* **8** 10098

[68] Li F Y *et al* 2016 *Nat. Commun.* **7** 12691
[69] Wan X *et al* 2011 *Phys. Rev.* B **83** 205101
[70] Xu S Y *et al* 2015 *Nat. Phys.* **11** 748–54
[71] Hosur P and Qi X 2013 *C. R. Phys.* **14** 857–70
[72] Nielsen H B and Ninomiya M 1983 *Phys. Lett.* B **130** 389–96
[73] Lv B Q *et al* 2015 *Phys. Rev.* X **5** 031013
[74] Xu S Y *et al* 2015 *Science* **349** 613–7
[75] Aharonov Y and Casher A 1984 *Phys. Rev. Lett.* **53** 319
[76] Zhang L C *et al* 2020 *Phys. Rev. Res.* **2** 013063
[77] Mook A, Henk J and Mertig I 2017 *Phys. Rev.* B **95** 014418
[78] Li K *et al* 2017 *Phys. Rev. Lett.* **119** 247202
[79] Owerre S A 2017 *Sci. Rep.* **7** 6931
[80] Liu J, Wang L and Shen K 2020 *Phys. Rev. Res.* **2** 023282
[81] Hwang K, Trivedi N and Randeria M 2020 *Phys. Rev. Lett.* **125** 047203
[82] Scheie A *et al* 2022 *Phys. Rev. Lett.* **128** 097201
[83] Lv B Q, Qian T and Ding H 2021 *Rev. Mod. Phys.* **93** 025002
[84] Zhang L *et al* 2010 *Phys. Rev. Lett.* **105** 225901
[85] Qin T, Zhou J and Shi J 2012 *Phys. Rev.* B **86** 104305
[86] Kittel C 1949 *Rev. Mod. Phys.* **21** 541
[87] Kittel C 1958 *Phys. Rev.* **110** 836
[88] Callen E R and Callen H B 1963 *Phys. Rev.* **129** 578
[89] Ogawa N *et al* 2015 *Proc. Natl. Acad. Sci. USA* **112** 8977–81
[90] Kikkawa T *et al* 2016 *Phys. Rev. Lett.* **117** 207203
[91] Takahashi R and Nagaosa N 2016 *Phys. Rev. Lett.* **117** 217205
[92] Zhang X *et al* 2019 *Phys. Rev. Lett.* **123** 167202
[93] Park S and Yang B J 2019 *Phys. Rev.* B **99** 174435
[94] Go G, Kim S K and Lee K J 2019 *Phys. Rev. Lett.* **123** 237207
[95] Huang B *et al* 2017 *Nature* **546** 270–3
[96] Lado J L and Fernández-Rossier J 2017 *2D Mater.* **4** 035002
[97] Zhang S *et al* 2020 *Phys. Rev. Lett.* **124** 147204
[98] Li Z X *et al* 2019 *npj Comput. Mater.* **5** 107
[99] Sil A and Ghosh A K 2020 *J. Phys.: Condens. Matter* **32** 205601
[100] Hirosawa T *et al* 2020 *Phys. Rev. Lett.* **125** 207204
[101] Mook A *et al* 2021 *Phys. Rev.* B **104** 024406
[102] Park M J, Lee S B and Kim Y B 2021 *Phys. Rev.* B **104** L060401
[103] Li Y M *et al* 2022 *Phys. Rev.* B **106** 054403
[104] Ni X *et al* 2019 *Nat. Mater.* **18** 113–20
[105] Wang X S, Zhang H W and Wang X R 2018 *Phys. Rev. Appl.* **9** 024029
[106] Shindou R *et al* 2013 *Phys. Rev.* B **87** 174427

IOP Publishing

Topology in Collective Magnetization Dynamics

Zhixiong Li, Yunshan Cao and Peng Yan

Chapter 5

Twisted magnons

This chapter is devoted to the investigation of magnons carrying orbital angular momentum. Such a magnon state is dubbed as a twisted magnon (TM). We first introduce the concept of TMs in three-dimensional magnetic nanocylinders, by deriving the dispersion relation and simulating the propagation length. Then we focus on the dynamics of TMs in two-dimensional magnetic nanodisks. The method to generate TMs in single nanodisk and the propagation characteristic of TMs in one-dimensional nanodisk arrays are also discussed. We finally introduce the topological phases of TMs in a constructed one-dimensional Su–Schrieffer–Heeger model.

5.1 Intrinsic dynamics of twisted magnons

The quantized orbital angular momentum (OAM) states of (quasi-) particles were rarely explored until the pioneering work by Allen *et al* [1] in 1992—the theoretical prediction of the OAM of light. The OAM is associated with the rotation of a (quasi-)particle about a fixed axis and is encoded in the spiral phase profile of the particle's wave function characterized by an azimuthal ϕ phase dependence $e^{il\phi}$ with a nonzero topological charge l (an integer) and a vanishing field at the core. The OAM component along the direction of the rotational axis has a quantized value $l\hbar$ with $\hbar$ the reduced Planck constant, in contrast to the spin angular momentum (SAM) originating from the wave polarization. Such twisted OAM states have a phase dislocation on the axis that is sometimes referred to as a vortex. Ever since its prediction in an optical system [2–5], the OAM states have been studied in a broad field of electronics [6–10], acoustics [11–15], neutron sciences [16–18], and very recently in magnonics [19–25]. The OAM state of the elementary excitations (spin waves or magnons) in a magnetic system is called twisted magnons (TMs). In this section, we discuss the dynamic properties of TMs in detail.

doi:10.1088/978-0-7503-5082-2ch5

5.1.1 Twisted magnons in cylindrical nanowires

We consider an infinitely long cylindrical ferromagnetic nanowire with radius R. A static external field $\mathbf{H}$ along the z-axis (the symmetry axis of the cylinder) is applied to saturate the magnetization, as shown in figure 5.1. The dynamics of magnetization in the cylindrical nanowire are described by the Landau–Lifshitz–Gilbert (LLG) equation [26]

$$\frac{\partial \mathbf{m}}{\partial t} = -\gamma\mu_0 \mathbf{m} \times \mathbf{H}_{\text{eff}} + \alpha \mathbf{m} \times \frac{\partial \mathbf{m}}{\partial t}, \tag{5.1}$$

where α is the Gilbert damping constant, γ is the gyromagnetic ratio, μ_0 is the vacuum permeability, $\mathbf{m}$ is unit vector along local magnetization, and

$$\mathbf{H}_{\text{eff}} = H_0\hat{z} + \mathbf{h}(\mathbf{r}, t) + \frac{2A}{\mu_0 M_s}\nabla^2 \mathbf{m}(\mathbf{r}, t), \tag{5.2}$$

represents the total effective field. The first, second, and last terms on the right side of (5.2) denote the external magnetic field, the dynamical dipolar field, and the exchange field, respectively. Here A is the exchange constant and M_s is saturation

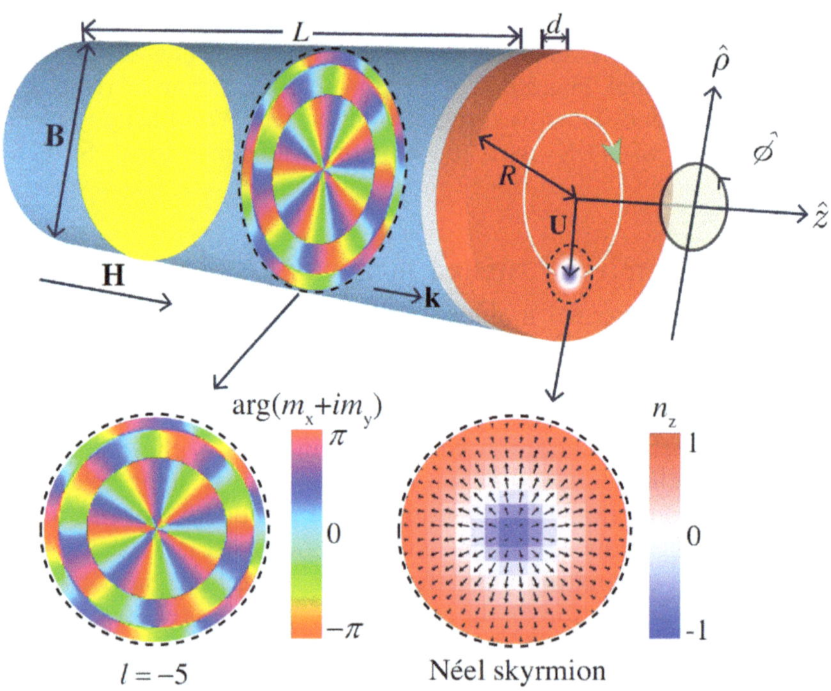

Figure 5.1. The schematic diagram of a heterostructured nanocylinder exchange-coupled to a chiral magnetic nanodisk hosting a Néel-type skyrmion. An external magnetic field $\mathbf{H}$ is applied along the z-axis direction. A spin-wave beam with the wave vector $\mathbf{k}$ and OAM quantum number $l = -5$ is excited by a local microwave field $\mathbf{B}$, leading to a steady skyrmion gyration around the disk center. (Reproduced with permission from [19]. Copyright 2020 The American Physical Society.)

magnetization. In the long-wave approximation, the dynamical dipolar field $\mathbf{h}(\mathbf{r}, t)$ satisfies the magnetostatic Maxwell's equations

$$\nabla \times \mathbf{h}(\mathbf{r}, t) = 0, \tag{5.3}$$

and

$$\nabla \cdot [\mathbf{h}(\mathbf{r}, t) + M_s\mathbf{m}(\mathbf{r}, t)] = 0. \tag{5.4}$$

Therefore, we have $\mathbf{h}(\mathbf{r}, t) = -\nabla\Phi(\mathbf{r}, t)$ with $\Phi(\mathbf{r}, t)$ denoting the magnetostatic potential.

Considering the magnetization precession around equilibrium position with small angle, i.e. $\mathbf{m} = (m_x, m_y, 1)$ with $|m_x|^2 + |m_y|^2 \ll 1$, then assuming $m_{x(y)}(\mathbf{r}, t) = m_{x(y)}(\mathbf{r})\mathrm{e}^{-\mathrm{i}\omega t}$ and $\Phi(\mathbf{r}, t) = \Phi(\mathbf{r})\mathrm{e}^{-\mathrm{i}\omega t}$, we can linearize the LLG equation as follows

$$\mathrm{i}\varpi m_x = (H_0 - D\nabla^2)m_y + \frac{\partial\Phi}{\partial y}, \tag{5.5}$$

$$-\mathrm{i}\varpi m_y = (H_0 - D\nabla^2)m_x + \frac{\partial\Phi}{\partial x}, \tag{5.6}$$

$$\nabla^2\Phi = M_s\left(\frac{\partial m_x}{\partial x} + \frac{\partial m_y}{\partial y}\right), \tag{5.7}$$

where $\varpi = \omega/(\gamma\mu_0)$ and $D = 2A/(\mu_0 M_s)$. To derive the differential equation about Φ, we proceed as follows. First, we differentiate equations (5.5) and (5.6) with respect to y and x, respectively. Then we obtain

$$\mathrm{i}M_s\varpi\left(\frac{\partial m_x}{\partial y} - \frac{\partial m_y}{\partial x}\right) = (H_0 - D\nabla^2)\nabla^2\Phi + M_s\left(\frac{\partial^2\Phi}{\partial y^2} + \frac{\partial^2\Phi}{\partial x^2}\right). \tag{5.8}$$

Similarly, if we differentiate the equations (5.5) and (5.6) with respect to x and y, respectively, we can obtain another equation

$$-\mathrm{i}\varpi\nabla^2\Phi = M_s(H_0 - D\nabla^2)\left(\frac{\partial m_x}{\partial y} - \frac{\partial m_y}{\partial x}\right). \tag{5.9}$$

By combining equations (5.7)–(5.9), we obtain the following homogeneous equation that the magnetic potential Φ follows,

$$[(H_0 - D\nabla^2)(H_0 + M_s - D\nabla^2) - \varpi^2]\nabla^2\Phi + M_s(D\nabla^2 - H_0)\frac{\partial^2\Phi}{\partial z^2} = 0. \tag{5.10}$$

By using cylindrical polar coordinates, the magnetic potential in a cylindrical nanowire has the form

$$\Phi(\rho, \phi, z) = J_n(\kappa\rho)\exp[\mathrm{i}(n\phi + kz)], \tag{5.11}$$

where $J_n(\kappa\rho)$ is a Bessel function of the first-kind with order n, and k is the wave vector in the direction parallel to the z-axis. By using the relation

$$\nabla^2\Phi = \frac{1}{\rho}\frac{\partial\Phi}{\partial\rho} + \frac{\partial^2\Phi}{\partial\rho^2} + \frac{1}{\rho^2}\frac{\partial^2\Phi}{\partial\phi^2} + \frac{\partial^2\Phi}{\partial z^2}, \tag{5.12}$$

we obtain $\nabla^2\Phi = -(\kappa^2 + k^2)\Phi$, then equation (5.10) becomes

$$\begin{aligned} &D^2(\kappa^2 + k^2)^3 + D(2H_0 + M_s)(\kappa^2 + k^2)^2 \\ &\quad + [H_0(H_0 + M_s) - \bar{\omega}^2 - DM_sk^2](\kappa^2 + k^2) - H_0M_sk^2 = 0, \end{aligned} \tag{5.13}$$

which is the dispersion relation of TMs in cylindrical nanowire. Since the above equation is cubic in κ^2, for any combination of n and k, there exist three linearly independent solutions. The magnetic potential within the cylinder thus can be written as [27]

$$\Phi^{\text{wire}}(\rho, \phi, z) = \sum_{j=1}^{3} C_jJ_n(\kappa_j\rho)\exp[\text{i}(n\phi + kz)]. \tag{5.14}$$

Outside the cylindrical wire, the magnetic potential satisfies the Laplace equation,

$$\Phi^{\text{out}}(\rho, \phi, z) = C_4K_n(k\rho)\exp[\text{i}(n\phi + kz)], \tag{5.15}$$

where $K_n(k\rho)$ is the modified Bessel function of the second-kind with order n.

To determine the four coefficients C_1, C_2, C_3, and C_4, we consider the boundary conditions at the surface of the wire that $\mathbf{m}(\mathbf{r})$ and $\mathbf{h}(\mathbf{r})$ have to fulfill. For the boundary condition of the continuity of Φ, we have

$$\sum_{j=1}^{3} C_jJ_n(\kappa_jR) = C_4K_n(kR), \tag{5.16}$$

and for the continuity of the normal component of the magnetic field induction we have

$$\left. -\frac{\partial\Phi^{\text{wire}}}{\partial\rho} + M_sm_\rho \right|_{\rho=R} = -\frac{\partial\Phi^{\text{out}}}{\partial\rho}. \tag{5.17}$$

In addition, $\mathbf{m}(\mathbf{r})$ should satisfy the exchange boundary conditions [28]

$$\left.\left(\frac{\partial m_\phi}{\partial\rho}\right)\right|_{\rho=R} = 0, \tag{5.18}$$

and

$$\left.\left(\frac{\partial m_\rho}{\partial\rho}\right)\right|_{\rho=R} - \zeta m_\rho\Big|_{\rho=R} = 0, \tag{5.19}$$

where ζ is the pinning parameter.

To evaluate these boundary conditions, we need the explicit expressions for the dynamical radial and azimuthal components of the magnetization. Assuming $m_+ = m_x + \mathrm{i}m_y$ and $h_+ = h_x + \mathrm{i}h_y$, equations (5.5) and (5.6) can be simplified as

$$(H_0 - D\nabla^2 + \varpi)m_+ = h_+. \tag{5.20}$$

We assume that

$$m_+ = \sum_{j=1}^{3} Q_j J_{n+1}(\kappa_j\rho)\exp[\mathrm{i}(n+1)\phi + \mathrm{i}kz], \tag{5.21}$$

where Q_j are constants to be determined. Then we have

$$(H_0 - D\nabla^2 + \varpi)m_+ = \sum_{j=1}^{3} Q_j[\varpi + H_0 + D(\kappa_j^2 + k^2)]J_{n+1}(\kappa_j\rho) \exp[\mathrm{i}(n+1)\phi + \mathrm{i}kz]. \tag{5.22}$$

On the other hand, we have

$$h_+ = -\mathrm{e}^{\mathrm{i}\phi}\left(\frac{\partial}{\partial\rho} + \frac{\mathrm{i}}{\rho}\frac{\partial}{\partial\phi}\right)\Phi^{\mathrm{wire}} = \sum_{j=1}^{3} C_j\kappa_j J_{n+1}(\kappa_j\rho)\exp[\mathrm{i}(n+1)\phi + \mathrm{i}kz]. \tag{5.23}$$

Then, we obtain the relation

$$Q_j = \frac{C_j\kappa_j}{H_0 + \varpi + D(\kappa_j^2 + k^2)}. \tag{5.24}$$

Thus, we have

$$m_+ = \sum_{j=1}^{3} \frac{C_j\kappa_j}{H_0 + \varpi + D(\kappa_j^2 + k^2)} J_{n+1}(\kappa_j\rho)\exp[\mathrm{i}(n+1)\phi + \mathrm{i}kz]. \tag{5.25}$$

Similarly, we can define $m_- = m_x - \mathrm{i}m_y$ and $h_- = h_x - \mathrm{i}h_y$, then we obtain

$$m_- = -\sum_{j=1}^{3} \frac{C_j\kappa_j}{H_0 - \varpi + D(\kappa_j^2 + k^2)} J_{n-1}(\kappa_j\rho)\exp[\mathrm{i}(n-1)\phi + \mathrm{i}kz]. \tag{5.26}$$

Finally, we obtain the dynamical radial and azimuthal components of the magnetization

$$\begin{aligned} m_\rho(\rho, \phi, z) &= \frac{1}{2}(\mathrm{e}^{-\mathrm{i}\phi}m_+ + \mathrm{e}^{\mathrm{i}\phi}m_-) \\ &= \frac{1}{2}\sum_{j=1}^{3} C_j\kappa_j\left[\frac{J_{n+1}(\kappa_j\rho)}{H_0 + \varpi + D(\kappa_j^2 + k^2)} - \frac{J_{n-1}(\kappa_j\rho)}{H_0 - \varpi + D(\kappa_j^2 + k^2)}\right] \\ &\quad \exp(\mathrm{i}n\phi + \mathrm{i}kz), \end{aligned} \tag{5.27}$$

and

$$
\begin{aligned}
m_\phi(\rho, \phi, z) &= -\frac{i}{2}(\mathrm{e}^{-\mathrm{i}\phi}m_+ - \mathrm{e}^{\mathrm{i}\phi}m_-) \\
&= -\frac{i}{2}\sum_{j=1}^{3} C_j\kappa_j\left[\frac{J_{n+1}(\kappa_j\rho)}{H_0 + \varpi + D(\kappa_j^2 + k^2)}\right. \\
&\quad \left. + \frac{J_{n-1}(\kappa_j\rho)}{H_0 - \varpi + D(\kappa_j^2 + k^2)}\right] \\
&\quad \exp(in\phi + \mathrm{i}kz).
\end{aligned}
\tag{5.28}
$$

Therefore, we obtain four equations by considering four different boundary conditions (here we assume the free boundary conditions, i.e. $\zeta = 0$)

$$
\begin{aligned}
&J_n(\kappa_1 R)C_1 + J_n(\kappa_2 R)C_2 + J_n(\kappa_3 R)C_3 - K_n(kR)C_4 = 0, \\
&F_n(\kappa_1 R)C_1 + F_n(\kappa_2 R)C_2 + F_n(\kappa_3 R)C_3 + \frac{1}{2}k[K_{n-1}(kR) - K_{n+1}(kR)]C_4 = 0, \\
&P'_{n,-}(\kappa_1 R)C_1 + P'_{n,-}(\kappa_2 R)C_2 + P'_{n,-}(\kappa_3 R)C_3 = 0, \\
&-\mathrm{i}P'_{n,+}(\kappa_1 R)C_1 - \mathrm{i}P'_{n,+}(\kappa_2 R)C_2 - \mathrm{i}P'_{n,+}(\kappa_3 R)C_3 = 0,
\end{aligned}
\tag{5.29}
$$

where

$$
P_{n,\pm}(\kappa_j R) = \frac{\kappa_j}{2}\left[\frac{J_{n+1}(\kappa_j R)}{H_0 + \varpi + D(\kappa_j^2 + k^2)} \pm \frac{J_{n-1}(\kappa_j R)}{H_0 - \varpi + D(\kappa_j^2 + k^2)}\right], \tag{5.30}
$$

$$
F_n(\kappa_j R) = -\frac{1}{2}\kappa_j[J_{n-1}(\kappa_j R) - J_{n+1}(\kappa_j R)] + M_s P_{n,-}(\kappa_j R), \tag{5.31}
$$

$$
\begin{aligned}
P'_{n,\pm}(\kappa_j R) &= \left.\frac{\partial P_{n,\pm}(\kappa_j\rho)}{\partial\rho}\right|_{\rho=R} \\
&= \frac{\kappa_j^2}{4}\left[\frac{J_n(\kappa_j R) - J_{n+2}(\kappa_j R)}{H_0 + \varpi + D(\kappa_j^2 + k^2)} \pm \frac{J_{n-2}(\kappa_j R) - J_n(\kappa_j R)}{H_0 - \varpi + D(\kappa_j^2 + k^2)}\right].
\end{aligned}
\tag{5.32}
$$

Considering that equation (5.29) has nontrivial solutions, we obtain

$$
|M(\varpi, \kappa_i)| = 0, \tag{5.33}
$$

where $M(\varpi, \kappa_i)$ is a 4×4 coefficient matrix. In what follows, we use parameter l to represent the azimuthal quantum (OAM) number which counts the number of nodes along the azimuthal direction of half the nanodisk [24]. In addition, from Noether's theorem, one can derive the OAM of TMs

$$J_z = \frac{\hbar}{S} \iint (\nabla\psi \times \mathbf{r})_z \mathrm{d}x\mathrm{d}y, \tag{5.34}$$

where $S = \pi r^2$ is the cross-section area of nanocylinder and $\psi(\mathbf{r}) = \arg(m_x + \mathrm{i}m_y)$ [29]. Substituting the dynamical magnetization profile into equation (5.34), we obtain $J_z = l\hbar$ $(l = n - 1)$ denoting the quantized OAM per TM.

We can calculate the eigenfrequencies of TMs in a nanocylinder with the following processes. First, we fix the value of l and consider a frequency range (for example, 0 to 100 GHz). For each trial frequency in the given range, one can determine the transverse wave number κ using equation (5.13). Afterwards, we judge whether the determinant $\det[M(\bar{\omega}, \kappa_i)]$ is equal to 0, if so, this trial frequency is indeed the eigenfrequency of TMs. Noticeably, the determinant cannot be strictly equal to 0 due to the accuracy of calculations, determinant minima therefore can be regarded as the solutions.

A yttrium iron garnet (YIG) nanocylinder with length $L = 2$ μm and radius $R = 60$ nm is considered. The external magnetic field is set to $\mu_0 H_0 = 0.4$ T and we choose $l = -5$ as an example. Figure 5.2(a) plots the magnitude of determinant M varying with trial frequencies with wave vector $k = 1 \times 10^5$ cm^{-1}. One can clearly see that in the range of 0~100 GHz, there exist five frequencies at 11.3, 25, 42.5, 63.5,

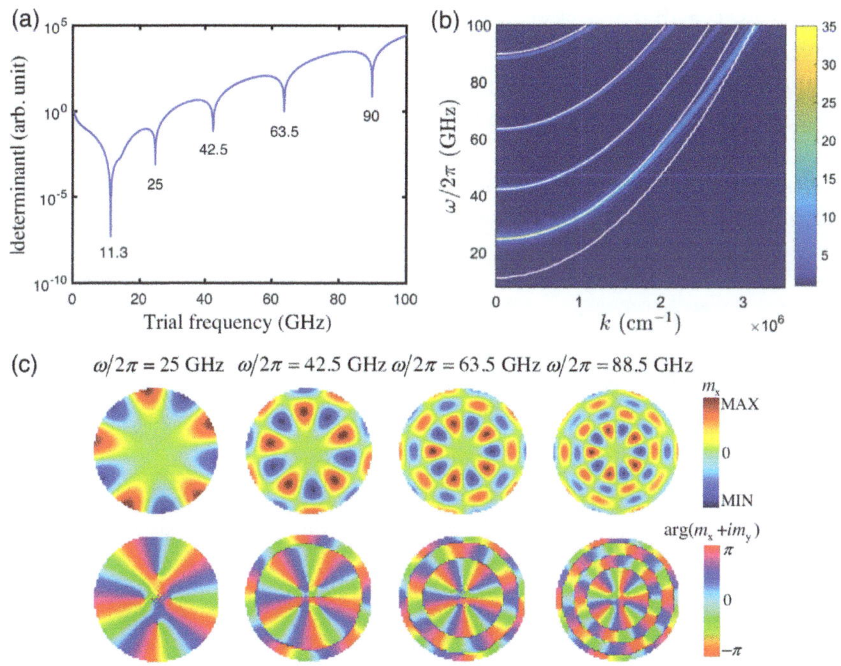

Figure 5.2. (a) Boundary-value determinant versus trial frequencies for a longitudinal wave vector $k = 1 \times 10^5$ cm^{-1} and $l = -5$. (b) Spin-wave spectrum in the nanocylinder. The solid curves are theoretical results, comparing with the micromagnetic simulations. (c) Cross-sectional distribution of spin-wave beams with 0, 1, 2, and 3 radial nodes (from left to right). (Reproduced with permission from [19]. Copyright 2020 The American Physical Society.)

and 90 GHz which satisfy the equation (5.33). Similarly, we can obtain the eigenfrequency of TMs for other wave vectors. The calculated full dispersion relations are plotted in figure 5.2(b) with white curves. In addition, the determined band structures of TMs from micromagnetic simulations are also shown in figure 5.2(b). One can see that simulation results compare well with theoretical curves. The oscillations distribution of different TM modes with $l = -5$ in the cross-section are plotted in figure 5.2(c), from which one can identify the various TM modes with radial nodes 0, 1, 2, and 3.

By utilizing the magnonic potential well between two adjacent nodes of TM, one can trap the topological spin texture. However, only when the width between two adjacent nodes is comparable to or larger than the size of the spin textures, is the trapping effect efficient. We therefore consider the TMs with two nodes. To show how the TMs manipulate the spin textures, we consider a heterostructure shown in figure 5.1, where the nanodisk is a chiral ferromagnet (Co|Pt) hosting a Néel-type skyrmion. The TMs are excited with a Laguerre–Gaussian type microwave field in the YIG nanocylinder and then propagate to the nanodisk, inducing the motion of the skyrmion.

The steady gyrotropic motion of skyrmion with the help of TMs is plotted in figure 5.3(b). The skyrmion initial displacement is set as $\mathbf{U}(t = 0) = (-22, 0)$ nm, where we assume the coordinate origin is located at the center of the nanodisk. For $l = -5$, we observe that the skyrmion exhibits a clockwise circular motion with an average velocity of 15.2 m s^{-1} and orbit radius of 31.9 nm. To demonstrate that the circular motion of the skyrmion is induced by the OAM transfer from TMs to the skyrmion, we switch the sign of the OAM of the TMs to $l = +5$ while keeping the rest of the parameters unchanged. Then we find that the skyrmion shows an expected counterclockwise circular motion. However, the average velocity of the skyrmion is reduced to 7.82 m s^{-1} and the orbit radius is decreased to 19.7 nm (see figures 5.3(b)).

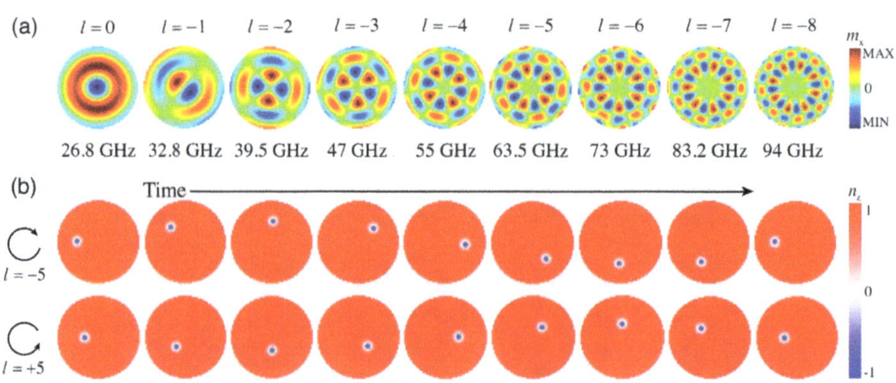

Figure 5.3. (a) Cross-sectional view of spin-wave beams carrying different OAMs for a fixed radial node. (b) Time evolution of Néel skyrmions driven by twisted magnons with OAM $l = \mp 5$. The time intervals between successive snapshots are 1.65 and 1.98 ns, respectively. A negative (positive) OAM induces a clockwise (counterclockwise) skyrmion gyration. (Reproduced with permission from [19]. Copyright 2020 The American Physical Society.)

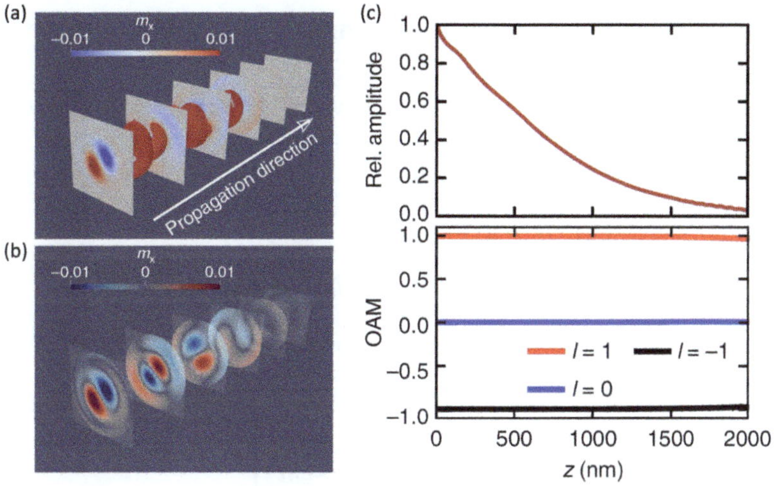

Figure 5.4. A spiraling spin-wave propagating along a cylindrical micro-scale waveguide of the insulting magnet YIG. (a) Snapshot of the magnon beams for the x-component of the triggered magnetization. (b) Vortex configuration of the excitation modes along the waveguide. (c) The amplitude and the OAM of the spin waves. (Reproduced with permission from [20]. Copyright 2019 the Authors.)

These results can be easily explained as the symmetry breaking of $l = -5$ and $l = 5$: because of the topological structure of the skyrmion, for a given topological charge ($Q = -1$ in our model), the intrinsic gyrotropic motion is along clockwise direction.

As we know, the Gilbert damping of magnetization procession inevitably leads to the time-decaying magnon density, which becomes a major obstacle for long distance transmission of information in magnonics. Very recently, Jia *et al* [20] theoretically showed that the topological charge (OAM quantum number) of twisted magnons is robust against magnetic damping. By utilizing this property of the twisted magnons, one can realize a robust transmission of information without or with little attenuation. Figures 5.4(a) and (b) show the snapshot of the twisted magnon beams in YIG tube. Although the finite damping leads to a decaying magnon density along the YIG cylinder, the OAM quantum number l is robust to the propagation distance, as shown in figure 5.4(c).

5.1.2 Twisted magnons in nanodisks

In previous section, we have discussed the TMs in ferromagnetic cylindrical nanowire. While the TMs can also exist in a nanodisk, in this section we introduce the properties of TMs in a magnetic nanodisk.

A magnetic nanodisk with radius r and thickness d is considered, and the magnetizations are perpendicularly magnetized through an out-of-plane uniform magnetic field H_0. Similar to a cylindrical nanowire, we use the LLG equation (equation (5.1)) to describe the magnetization dynamics. Because of the existence of surface magnetic charges, the effective field should be modified with

$$\mathbf{H}_{\text{eff}} = H_0\hat{z} + \mathbf{h}(\mathbf{r}, t) + \frac{2A}{M_s}\nabla^2\mathbf{m}(\mathbf{r}, t) - \mu_0 N_z M_z \hat{z}. \tag{5.35}$$

The last term on the right side of (5.35) denotes the static demagnetization field, where N_z is the demagnetizing factor along the z-axis. Here we consider the limit of thin-film, i.e. $d \ll r$, so we have $N_z = 1$.

By using the similar method adopted in a previous section, one can obtain the dispersion relation of TMs in a nanodisk

$$D^2\kappa^4 + (2H_0 - \mu_0 M_s)D\kappa^2 + [H_0(H_0 - \mu_0 M_s) - \varpi^2] = 0. \tag{5.36}$$

Quite straightforwardly, we can write the explicit expressions of the magnetization components

$$m_\rho = \frac{1}{2}\sum_{j=1}^{2} B_j\kappa_j\left[\frac{J_{n+1}(\kappa_j\rho)}{\Omega + \varpi} - \frac{J_{n-1}(\kappa_j\rho)}{\Omega - \varpi}\right]\exp(\mathrm{i}n\phi), \tag{5.37}$$

and

$$m_\phi = -\frac{\mathrm{i}}{2}\sum_{j=1}^{2} B_j\kappa_j\left[\frac{J_{n+1}(\kappa_j\rho)}{\Omega + \varpi} + \frac{J_{n-1}(\kappa_j\rho)}{\Omega - \varpi}\right]\exp(\mathrm{i}n\phi), \tag{5.38}$$

with $\Omega = H_0 - \mu_0 M_s + D\kappa_j^2$. Here, coefficients B_1 and B_2 can be determined by the following equations when one considers the free boundary conditions

$$\begin{aligned} T_{n,-}(\kappa_1 r)B_1 + T_{n,-}(\kappa_2 r)B_2 = 0,\\ -\mathrm{i}T_{n,+}(\kappa_1 r)B_1 - \mathrm{i}T_{n,+}(\kappa_2 r)B_2 = 0, \end{aligned} \tag{5.39}$$

where

$$T_{n,\pm}(\kappa_j r) = \frac{\kappa_j^2}{4}\left[\frac{J_n(\kappa_j r) - J_{n+2}(\kappa_j r)}{\Omega + \varpi} \pm \frac{J_{n-2}(\kappa_j r) - J_n(\kappa_j r)}{\Omega - \varpi}\right]. \tag{5.40}$$

To guarantee equation (5.39) has nontrivial solutions, we need $\det[M(\omega, \kappa_i)] = 0$ with $M(\omega, \kappa_i)$ denoting the coefficient matrix.

In the theoretical calculations, we use the following parameters based on a YIG nanodisk: radius of the disk $r = 50$ nm, thickness $d = 2$ nm, saturation magnetization $M_s = 1.92 \times 10^5$ A m^{-1}, exchange constant $A = 3.1 \times 10^{-12}$ J m^{-1}, and external field $H_0 = 400$ mT along the z-axis direction. The magnitude of the determinant $\det(M)$ (l is fixed to 3) as a function of trial frequencies is plotted in figure 5.5(a), from which we find five minimum points. In addition, by analyzing the oscillation distribution of different modes, we identify four eigenmodes (the first minimum point corresponds to a localized mode because of the boundary effect), as plotted in figure 5.5(b). Here we use s to represent the radial quantum number which counts the number of nodes along the radial direction.

We can perform the micromagnetic simulations to verify the theoretical predictions. Here the package MUMAX3 [30] is used to simulate the dynamics of

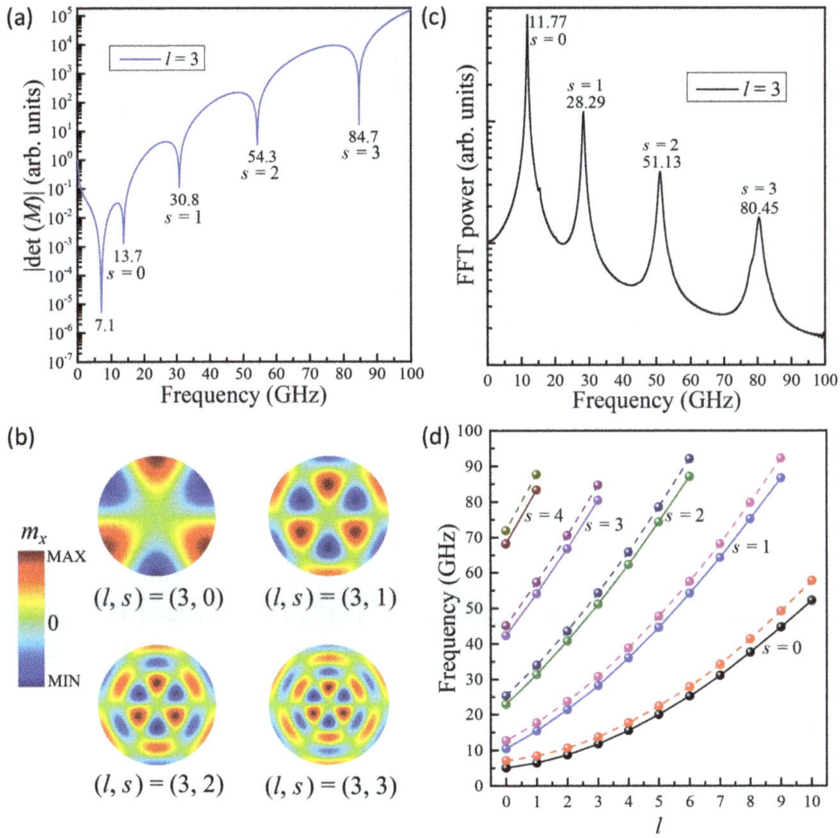

Figure 5.5. (a) The magnitude of the coefficient determinant for different trial frequencies with $l = 3$. (b) The FFT strength distribution of TM modes in a nanodisk for different radial quantum numbers $s = 0, 1, 2, 3$ with $l = 3$. (c) The Fourier spectrum of the oscillation of magnetization component m_x. (d) The dependence of the eigenfrequencies of TMs on l with different s. The dashed and solid curves represent the theoretical results and micromagnetic simulations, respectively. (Reproduced with permission from [32]. Copyright 2022 The American Physical Society.)

magnetization. To obtain the dispersion relation of TMs, we apply a sinc-function magnetic field to the whole disk

$$\mathbf{H}(t) = H_1 \frac{\sin[2\pi f_0 (t - t_0)]}{2\pi f_0 (t - t_0)} [\cos(l\phi), \sin(l\phi), 0], \tag{5.41}$$

where $H_1 = 10$ mT, $f_0 = 100$ GHz, $t_0 = 1$ ns, and $l = 3$. Figure 5.5(c) plots the Fourier spectrum for the oscillation of m_x, from which we can easily identify four TM eigenmodes. Similarly, we can obtain all eigenfrequencies of TMs below 100 GHz by varying parameter l, the results are shown in figure 5.5(d). The theoretical calculations (dashed lines) compare well with micromagnetic simulations (solid lines).

In the above calculations, the exciting field used to generate TMs has very complicated profile, which may hinder the application of TMs. Recently, some other methods have been proposed to generate TMs, such as using the Aharanov–Casher effect [20], a magnonic spiral phase plate in the heterostructure [22], a skyrmion-textured domain wall [25], etc. However, these methods are relatively too complex and cannot be extended easily to the cases with high OAM. Here we introduce a simple method to generate TMs with arbitrary l and s. The schematic diagram is shown in figure 5.6(a), where the nanodisk contacts a strip with width $W = 60$ nm and length $L = 1000$ nm. To generate TMs in the nanodisk, we apply a simple sinusoidal magnetic field $\mathbf{h}(t) = h_0 \sin(2\pi f t)\hat{y}$ localized at one end of the strip (marked by a red arrow) to excite the planar magnons. As an example, we set $f = 28.29$ GHz, which equals the frequency of the TM with $l = 3$ and $s = 1$ (see figure 5.5(d)). When the planar magnons carrying spin angular momentum propagate to the disk–strip contact point, the eigenmode of the TM with $(l, s) = (3, 1)$ in the nanodisk is resonantly excited. As a result, the planar magnons without any twisting in the nanostrip are converted to TMs carrying OAM in the nanodisk. We call this process the spin-to-orbital angular momentum conversion. The generated TMs with counterclockwise rotation can exist stably, the whole process is plotted in figure 5.6. Interestingly, if the exciting field is applied to the other end of the strip, the rotation direction of the generated TMs will reverse

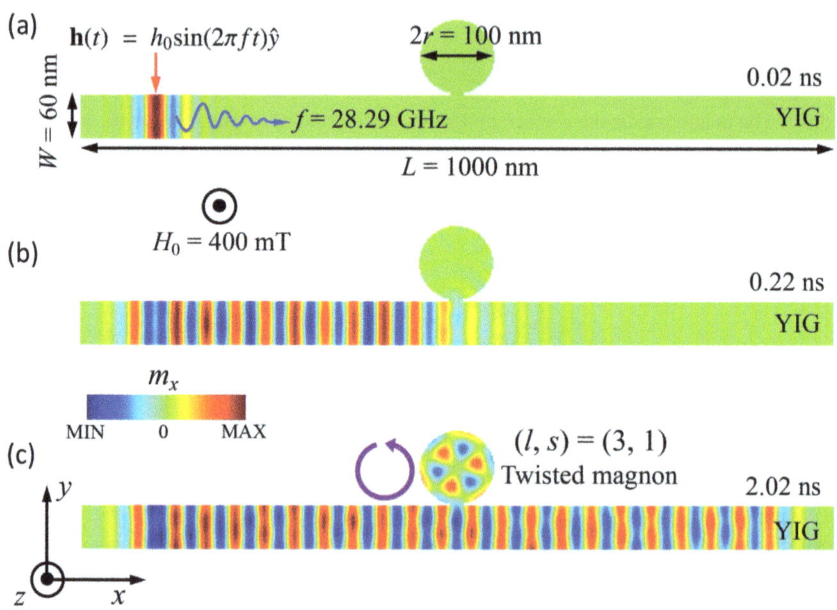

Figure 5.6. A sketch map of the generation of TM in a nanodisk through the spin-to-orbital angular momentum conversion at time (a) $t = 0.02$ ns, (b) 0.22 ns, and (c) 2.02 ns. The red arrow indicates the position of the excitation field, the purple arrow denotes the rotation direction of the emerging TM in the nanodisk. (Reproduced with permission from [32]. Copyright 2022 The American Physical Society.)

(clockwise). Similarly, by choosing different frequencies for the exciting fields, we can obtain TMs with arbitrary l and s.

To determine the conversion rate from planar magnons to TMs, we adopt the quantity $\sqrt{m_x^2 + m_y^2}$ [31] to characterize the amplitude of a magnon. The amplitude of the generated TMs (averaged over the whole disc) as a function of time for different s is plotted in figure 5.7(a), where l is fixed to 2. One can see that after applying the excitation field, the TM amplitude quickly reaches a stable value. In addition, the larger s we choose, the faster the stable value is reached. In addition, the amplitude of the TMs gradually decreases with the increase of s. Similarly, we find that with the increase of l, the amplitude of the TMs also decreases, with the results shown in figure 5.7(b), where we fix $s = 1$. The stable amplitude of the TMs for all combinations of s and l is plotted in figure 5.7(c), from which we can clearly see the trend of the TM amplitude varying with s or l. Then we calculate the

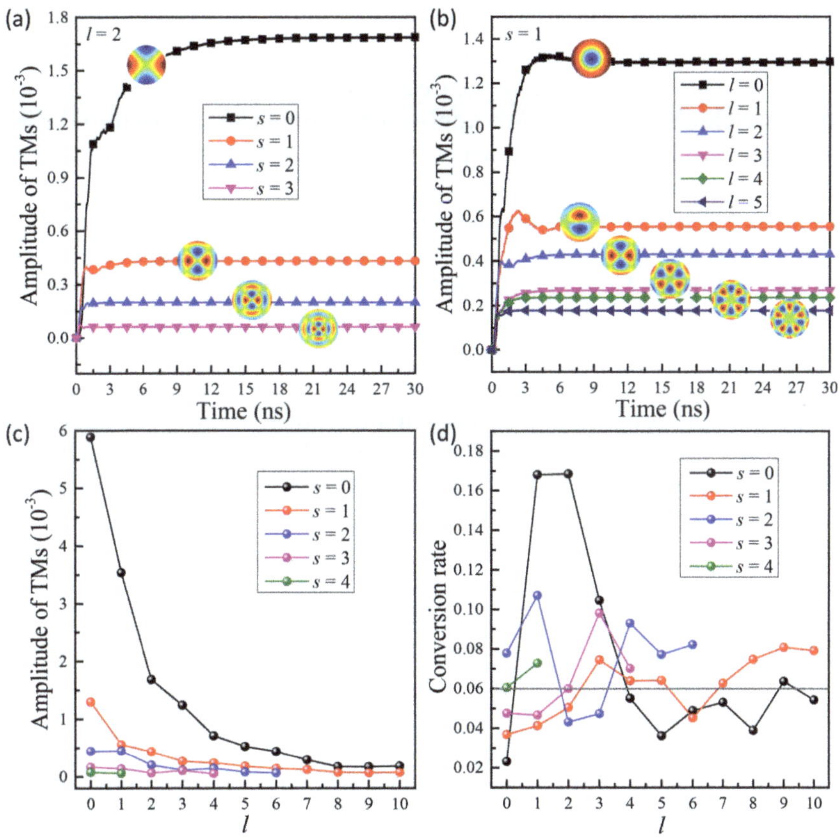

Figure 5.7. The TM amplitude over time for (a) different s with a fixed l ($l = 2$) and (b) different l with a fixed s ($s = 1$). The insets show the various eigenmodes of TM. (c) The amplitudes of generating TM over time with different l and s. (d) The dependence of the conversion rate from the planar magnons to TMs on l for different s. The gray line denotes the universal value 6%. (Reproduced with permission from [32]. Copyright 2022 The American Physical Society.)

conversion rate from planar magnons to TMs with the following processes. First, we choose two small areas which are located on the left and right sides of the contact point in the strip. Then we determine the amplitude of planar magnons in these two areas, with P_1 and P_2 representing the corresponding values in the left and right areas, respectively. Finally, the quantity of $1 - P_2/P_1$ can be used to denote the conversion rate, with the results being shown in figure 5.7(d). Interestingly, one can identify that the conversion rate has a universal number ~6%, although this value has some fluctuations when l and s change.

With the explosive growth of information, how to improve the bit rate of information has become the primary consideration for engineers and scientists. A promising solution is to use multiple channels to handle such huge amounts of information. For instance, by utilizing the different profiles of twisted photons, one can realize OAM-based spatial-division multiplexing (SDM) [33–37]. In a magnetic system, owing to the existence of the OAM degree of freedom, TMs have great potentials in future high-capacity communication devices. Here we propose the concept of magnonic OAM-based frequency-division multiplexing (FDM) in a one-dimensional nanodisk array. Our model is plotted in figure 5.8(a). Here h denotes the distance between two nearest-neighbor nanodisks. In the following simulations we choose $h = 6$ nm. We apply an exciting field localized to the left-most nanodisk with a specific TM frequency by utilizing the spin-to-orbit conversion mechanism mentioned above, then the generated TM will propagate in the nanodisk array because of the dipolar interaction between nanodisks. Within the linear regime, the TMs carrying different l or s do not interact with each other. Therefore, if we excite various TMs with different l or s simultaneously, the FDM based on TMs can be realized since these TMs spread in the lattice independently. Naturally, it is very important to clarify the propagation characteristics of TMs in this one-dimensional nanodisk array.

To investigate the propagation of TMs in a nanodisk array, as an example, we fix $l = 4$, and plot the dependence of the amplitude of TMs with different s on propagation distance in figure 5.8(b). Here the colored circles represent the results of simulation, while the solid curves denote the analytical fitting with formula $I = a + be^{-z/\lambda}$, where a and b are fitting coefficients, λ and z denote the attenuation length and propagation distance, respectively. One can clearly see that the decay of TMs satisfies the exponential function well. In addition, with the increase of s, the amplitude of TMs attenuate more quickly. Similarly, we also study the propagation of TMs for different l, where we fix $s = 3$. The corresponding results of simulation and analytical fitting are shown in figure 5.8(c). One can see that the TMs decay faster with the increase of l. The dependence of attenuation length λ on l and s is presented in figure 5.8(d), which clearly shows the feature that λ gradually decreases when l or s increases. To show the detailed propagation characteristic of TMs, we plot the intensity distribution of TMs in a nanodisk lattice with four representative modes, as shown in figure 5.8(e). Here we rotate the visual angle by 90 degrees for better observation. We can see that the OAM carried by TMs is very robust against the damping, although the amplitude of TMs has an obvious attenuation in the process of propagation.

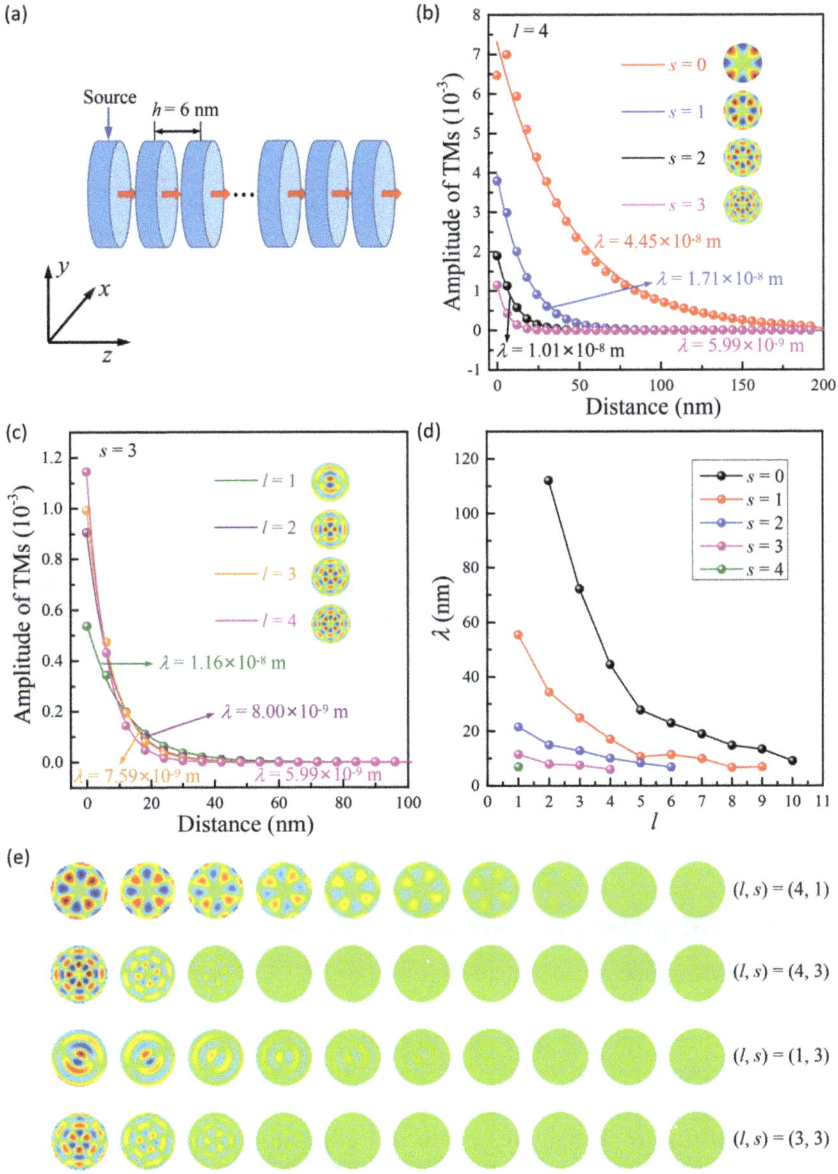

Figure 5.8. (a) A schematic diagram of a one-dimensional nanodisk lattice. The blue arrow denotes the position of the excitation field. The TM amplitudes over the propagation distance for (b) different s with $l = 4$ and (c) different l with $s = 3$. The small dots and solid lines denote the simulation results and analytical fittings, respectively. (d) Dependence of the attenuation length λ on l for different s. (e) The intensity distribution of TMs in a nanodisk lattice with four representative modes. (Reproduced with permission from [32]. Copyright 2022 The American Physical Society.)

In addition to the application of FDM, very recently Wang *et al* [38] showed that TMs can be used to realize a frequency comb, which is called a twisted magnon frequency comb (TMFC). An illustration is shown in figure 5.9. Because of the

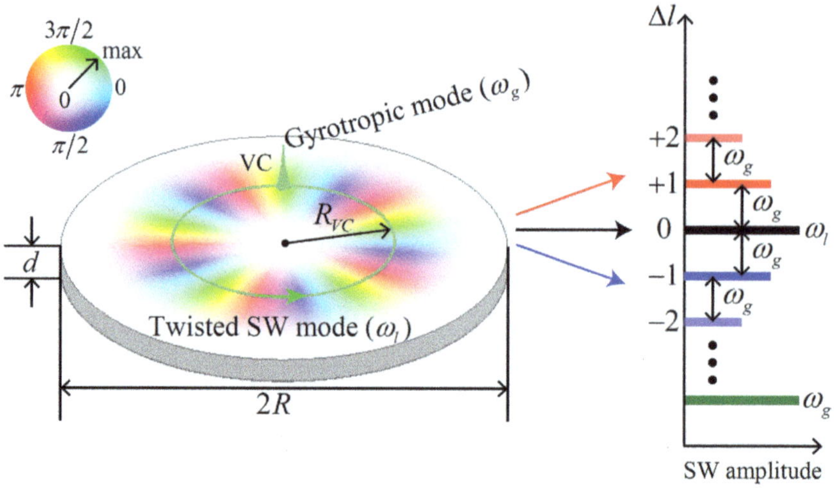

Figure 5.9. Schematic illustration of nonlinear interactions between the vortex core and TMs and the resulting TMFC. The green circle with an arrow indicates the vortex core gyration direction and trajectory. The inset is the color palette of TMs with the color indicating the phase and intensity representing the amplitude. (Reproduced with permission from [38]. Copyright 2022 The American Physical Society.)

quantization effect of the nonlinear magnon–vortex interaction in ferromagnetic nanodisks, the three-magnon confluence and splitting scattering of twisted magnons off the gyrating vortex core generates a set of discrete and equally spaced spectral lines, i.e. frequency comb. The mode spacing of the TMFC is the vortex gyration frequency and the OAM between neighboring spectral lines differ by one quantum number.

5.2 Topological phases of twisted magnons

As shown in previous sections (see e.g. figure 5.6), the TMs generated by the spin-to-orbital angular momentum conversion are usually fragile to small internal and/or external disturbances, such as defects and disorder. For practical applications, it is meaningful to seek a robust source of TMs. In this section, we discuss the Su–Schrieffer–Heeger (SSH) [39] model constructed in one-dimensional nanodisk arrays. The emerging topologically protected edge states can act as a robust TM source. Figure 5.10(b) shows a sketch map of the SSH model which contains 22 uniformly magnetized nanodisks with the alternating bond lengths being d_1 and d_2. According to the theory of the SSH model [39], when $d_1/d_2 > 1$, the system can support topologically protected edge states. As an example, we set $d_1 = 20$ nm and $d_2 = 10$ nm. A sinc-function magnetic field with the same form as equation (5.41) is applied to all nanodisks for 100 ns with $l = 2$. We plot the spectra of magnetization component oscillations (m_x) for different nanodisks in figure 5.10(c). The black and blue curves represent the position of the edge (disk 1) and bulk (disk 11) bands, respectively. We see that near the eigenfrequencies of TMs with $l = 2$, the edge spectrum has obvious strong peaks, which does not happen for the bulk band. We

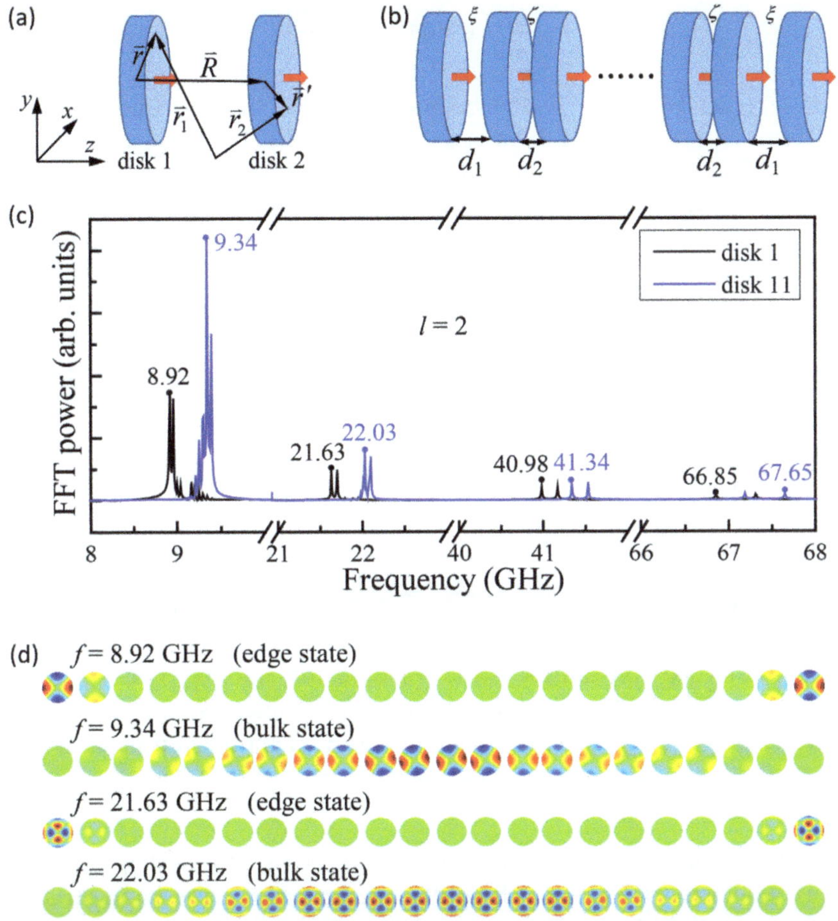

Figure 5.10. (a) Illustration of the two-nanodisk model considering magnetostatic interaction. **R** is the position vector connecting two nanodisk centers, **r** and **r**′ are the local nanodisk coordinates. The red arrows denote the direction of magnetization. (b) The SSH model of nanodisk arrays with alternating bond lengths $d_1 = 20$ nm and $d_2 = 10$ nm. (c) The Fourier spectra of the oscillation of magnetization component (m_x) at different nanodisks with $l = 2$. (d) The FFT intensity distribution for different states, as marked in (c).

therefore infer these modes to be edge states. Similarly, we obtain the frequency range of bulk modes. To visualize the difference between the bulk and edge states, we plot the FFT intensity distribution with four representative frequencies (as marked in figure 5.10(c)), as plotted in figure 5.10(d). We can see that the edge modes are confined at both ends of the nanodisk arrays while the bulk states can spread over the whole system. It is worth noting that if we change the value of l, we will obtain massive topological edge states with different l. At the frequency range of 0 to 100 GHz, we can identify almost 40 topological edge states. Below, we provide an analytical understanding of the topological features of TMs.

Based on the assumption of magnetic charges, the magnetostatic interaction energy between two magnetic nanodisks can be written as [40]

$$U_{\text{int}} = \frac{\mu_0}{4\pi} \int \mathrm{d}V \mathrm{d}V' \frac{\rho(\mathbf{r}_1)\rho(\mathbf{r}_2)}{|\mathbf{r}_1 - \mathbf{r}_2|}. \tag{5.42}$$

Here $\mathbf{r}_1 \in V$ ($\mathbf{r}_2 \in V'$) with V (V') belonging to the first (second) nanodisk (see figure 5.10(a)) and $\rho(\mathbf{r}) = -M_s \mathrm{div}\mathbf{m}$ is the density of the fictitious magnetic charges. The analytical calculation of equation (5.42) is quite cumbersome. To qualitatively understand the results of the micromagnetic simulation above, we propose a simple theoretical model. Without loss of generality, we use two coupling parameters ξ and ζ to qualitatively describe the magnetostatic interaction between two neighboring nanodisks (see figure 5.10(b)). Obviously, the parameter ξ (ζ) is a function of their distance d_1 (d_2), the OAM quantum number l, and the radial quantum number s. The Hamiltonian of the one-dimensional infinite nanodisk array therefore can be expressed as

$$\begin{pmatrix} \omega_1 & \xi_1 + \zeta_1 \mathrm{e}^{-ika} & 0 & 0 & \cdots \\ \xi_1 + \zeta_1 \mathrm{e}^{ika} & \omega_1 & 0 & 0 & \cdots \\ 0 & 0 & \omega_2 & \xi_2 + \zeta_2 \mathrm{e}^{-ika} & \cdots \\ 0 & 0 & \xi_2 + \zeta_2 \mathrm{e}^{ika} & \omega_2 & \cdots \\ \vdots & \vdots & \vdots & \vdots & \ddots \end{pmatrix}, \tag{5.43}$$

where a is the lattice constant, k is the wave vector, and ω_j ($j = 1, 2, 3. ..$) is the eigenfrequency of TMs. Here the size of the Hamiltonian is $2S \times 2S$ with S is the total number of TM modes. The spectra of this SSH model can be obtained as

$$E_j = \omega_j \pm \sqrt{\xi_j^2 + \zeta_j^2 + 2\xi_j\zeta_j \cos(ka)}, \quad \text{for } j = 1, 2, 3, \ldots. \tag{5.44}$$

From equation (5.43), we can see that for any choice of j, the system can support an analogous SSH band structure. The center frequencies of the band structures are eigenfrequencies of TMs ω_j. Therefore, when considering the finite nanodisk array as shown in figure 5.10(b), if $\xi_j/\zeta_j > 1$, the system can support the topologically protected edge state at each eigenfrequency of TMs. Here it is reasonable to assume that the conditions $\xi_j/\zeta_j > 1$ and $d_1/d_2 > 1$ are equivalent.

In this section, we considered the one-dimensional SSH model in nanodisk arrays. We show that the emerging topological edge states are localized at the end of the lattice and thus cannot propagate. However, if we consider a two-dimensional nanodisk lattice such as the Haldane [41] model, the system may support multi-channel chiral edge states, which is a very appealing topic. Very recently, the higher-order topological phases based on 'breathing lattice' have attracted a lot of attention [42–46]. If we construct a two- or three-dimensional 'breathing' nanodisk lattice, the so-called higher-order topological TMs may appear. In addition, the Weyl or Dirac semimetal phases based on TMs are also interesting topics to investigate.

References

[1] Allen L *et al* 1992 *Phys. Rev.* A **45** 8185
[2] Mair A *et al* 2001 *Nature* **412** 313–6

[3] Molina-Terriza G, Torres J P and Torner L 2007 *Nat. Phys.* **3** 305–10
[4] Padgett M J 2017 *Opt. Express* **25** 11265–74
[5] Ji Z *et al* 2020 *Science* **368** 763–7
[6] Uchida M and Tonomura A 2010 *Nature* **464** 737–9
[7] Verbeeck J, Tian H and Schattschneider P 2010 *Nature* **467** 301–4
[8] Mcmorran B J *et al* 2011 *Science* **331** 192–5
[9] Silenko A J, Zhang P and Zou L 2017 *Phys. Rev. Lett.* **119** 243903
[10] Lloyd S M *et al* 2017 *Rev. Mod. Phys.* **89** 035004
[11] Dashti P Z, Alhassen F and Lee H P 2006 *Phys. Rev. Lett.* **96** 043604
[12] Anhäuser A, Wunenburger R and Brasselet E 2012 *Phys. Rev. Lett.* **109** 034301
[13] Baresch D, Thomas J L and Marchiano R 2018 *Phys. Rev. Lett.* **121** 074301
[14] Marzo A, Caleap M and Drinkwater B W 2018 *Phys. Rev. Lett.* **120** 044301
[15] Bliokh K Y and Nori F 2019 *Phys. Rev.* B **99** 174310
[16] Clark C W *et al* 2015 *Nature* **525** 504–6
[17] Cappelletti R L, Jach T and Vinson J 2018 *Phys. Rev. Lett.* **120** 090402
[18] Larocque H *et al* 2018 *Nat. Phys.* **14** 1–2
[19] Jiang Y Y *et al* 2020 *Phys. Rev. Lett.* **124** 217204
[20] Jia C *et al* 2019 *Nat. Commun.* **10** 2077
[21] Chen M *et al* 2020 *Appl. Phys. Lett.* **116** 172403
[22] Jia C *et al* 2019 *J. Opt.* **21** 124001
[23] Neumann R R *et al* 2020 *Phys. Rev. Lett.* **125** 117209
[24] Schultheiss K *et al* 2019 *Phys. Rev. Lett.* **122** 097202
[25] Lee S and Kim S K 2022 *Front. Phys.* **10** 858614
[26] Gilbert T L 2004 *IEEE Trans. Magn.* **40** 3443–9
[27] Lim H S and Kuok M H 2011 *Handbook of Nanophysics: Nanotubes and Nanowires* ed K D Sattler (Boca Raton, FL: CRC Press) ch 31 https://doi.org/10.1201/9781420075434
[28] Arias R and Mills D L 2001 *Phys. Rev.* B **63** 134439
[29] Yan P *et al* 2013 *Phys. Rev.* B **88** 144413
[30] Vansteenkiste A *et al* 2014 *AIP Adv.* **4** 107133
[31] Wang X G *et al* 2012 *Phys. Rev.* B **86** 054445
[32] Li Z X *et al* 2022 *Phys. Rev.* B **105** 174433
[33] Gibson G *et al* 2005 *Opt. Express* **12** 5448–56
[34] Tamburini F *et al* 2012 *New J. Phys.* **14** 033001
[35] Wang J *et al* 2012 *Nat. Photon.* **6** 488–96
[36] Bozinovic N *et al* 2013 *Science* **340** 1545–8
[37] Yuan S S A *et al* 2021 *Phys. Rev. Appl.* **16** 064042
[38] Wang Z Y *et al* 2022 *Phys. Rev. Lett.* **129** 107203
[39] Su W P, Schrieffer J R and Heeger A J 1979 *Phys. Rev. Lett.* **42** 1698
[40] Sukhostavets O V, González J and Guslienko K Y 2013 *Phys. Rev.* B **87** 094402
[41] Haldane F D M 1988 *Phys. Rev. Lett.* **61** 2015
[42] Benalcazar W A, Bernevig B A and Hughes T L 2017 *Science* **357** 61–6
[43] Hassan A E *et al* 2019 *Nat. Photon.* **13** 697–700
[44] Xue H *et al* 2019 *Nat. Mater.* **18** 108–12
[45] Imhof S *et al* 2018 *Nat. Phys.* **14** 925–9
[46] Li Z X *et al* 2019 *npj Comput. Mater.* **5** 107

Chapter 6

Magnetic solitons

In this chapter we discuss another important excitation in the magnetic system, i.e. the magnetic soliton. A magnetic soliton represents a type of manifestation of topology in condensed matter physics which has a nano-scale topologically stable structure ranging from one dimension to three dimensions. Here we give a detailed introduction to both the static and dynamic properties of several representative magnetic solitons. We also address their applications in spintronic devices. It is noted that the collective dynamics of magnetic soliton arrays can also show the features of waves. In the last section, we introduce the band structure of artificial magnetic soliton lattices.

6.1 Structures, properties, and applications

Magnetic solitons are non-trivial topological structures which maintain essential properties invariant with a continuous deformation of the order parameter. For example, a magnetic vortex (or skyrmion) cannot be transformed to a ferromagnetic state without abrupt changes of the magnetization. Generally speaking, the two-dimensional magnetic solitons are characterized by the topological charge

$$Q = \frac{1}{4\pi} \iint q(\mathbf{r}) \mathrm{d}x \mathrm{d}y, \tag{6.1}$$

which counts the number of times the local normalized magnetic moment $\mathbf{m}(\mathbf{r})$ wraps the unit sphere, where $q(\mathbf{r}) = \mathbf{m}(\mathbf{r}) \cdot \left[\frac{\partial \mathbf{m}(\mathbf{r})}{\partial x} \times \frac{\partial \mathbf{m}(\mathbf{r})}{\partial y} \right]$ is the topological charge density.

The topological charge Q, as a mathematical invariant, can be reduced to a simple form after some algebra. In the polar coordinates, we have $\mathbf{r} = (r \cos \varphi, r \sin \varphi)$. By using the spherical coordinates and considering the spin configuration with the radial symmetry, the magnetization $\mathbf{m}(\mathbf{r})$ can be expressed as

$$\mathbf{m}(\mathbf{r}) = [\cos \phi(\varphi) \sin \theta(r), \sin \phi(\varphi) \sin \theta(r), \cos \theta(r)], \tag{6.2}$$

doi:10.1088/978-0-7503-5082-2ch6

where θ is polar angle and ϕ is azimuthal angle. Substituting equation (6.2) into equation (6.1), we obtain [1, 2]

$$Q = \frac{1}{4\pi}\int_0^\infty \mathrm{d}r \int_0^{2\pi} \mathrm{d}\varphi \frac{\mathrm{d}\theta(r)}{\mathrm{d}r}\frac{\mathrm{d}\phi(\varphi)}{\mathrm{d}\varphi} \sin\theta(r) = -\frac{1}{2}\cos\theta(r)\Big|_{r=0}^{r=\infty} \frac{1}{2\pi}\phi(\varphi)\Big|_{\varphi=0}^{\varphi=2\pi}. \tag{6.3}$$

In addition, we can define the vorticity

$$m = \frac{1}{2\pi}\phi(\varphi)\Big|_{\varphi=0}^{\varphi=2\pi}, \tag{6.4}$$

and helicity

$$\phi(\varphi) = m\varphi + \gamma_0, \tag{6.5}$$

with γ_0 being the phase to describe the angle between the selected magnetization and the coordinate axis.

6.1.1 Domain wall

In magnetic materials a domain wall (DW) exists in the transitional region between two magnetic domains with different magnetization orientations. The formation of DWs can be explained qualitatively as follows. The magnetic moment orientations of adjacent atoms are parallel due to the exchange interaction, which is called spontaneous magnetization, while the magnetocrystalline anisotropy forces the direction of spontaneous magnetization along the easy-axis. If the magnetic system only has these two interactions, the magnet will be magnetized along the easy-axis to minimize the total energy. However, a real magnet has a certain size and shape. To reduce the demagnetization energy, the crystal will be divided into different magnetic domains. The DW forms where the magnetic moment rotates gradually from one domain to the other. Moreover, the width of a DW is determined by the balance of different competing energies. The DW energy usually includes the exchange and magnetocrystalline and/or shape anisotropy energies. The smaller the exchange interaction (anisotropy) energy, the wider (narrower) the DW.

The most common type of DW is the 180° wall, which can be described well by a one-dimensional model, as shown in figure 6.1(a). The dark arrows denote the local

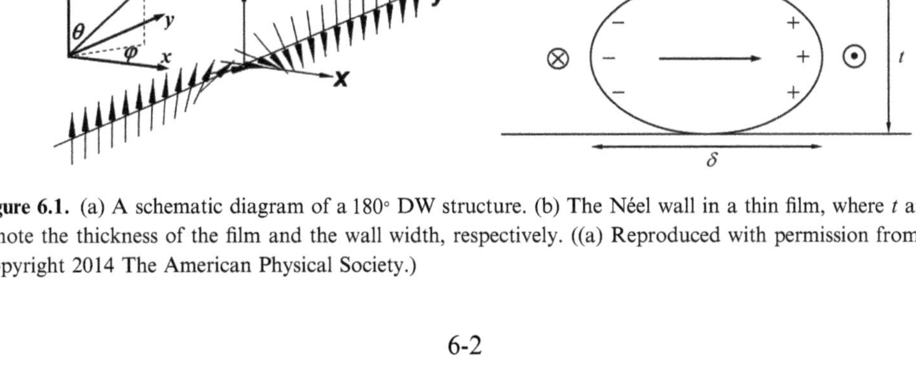

Figure 6.1. (a) A schematic diagram of a 180° DW structure. (b) The Néel wall in a thin film, where t and δ denote the thickness of the film and the wall width, respectively. ((a) Reproduced with permission from [3]. Copyright 2014 The American Physical Society.)

magnetization direction, and θ and φ are the polar and azimuthal angles in the spherical coordinate system, respectively. If $\varphi = 0$ the magnetic moments rotate in the plane perpendicular to the y-axis, such that this wall is called a Bloch DW. If $\varphi = \pi/2$, the magnetic moments rotate in the y–z-plane. In this case, the wall is called a Néel DW.

We consider a magnetic bulk material. Supposing the easy-axis is along z-axis, and the magnetocrystalline anisotropy constant $K > 0$, the total energy can be expressed as

$$E_{\text{tot}} = \int [A(\partial\theta/\partial y)^2 + K\sin^2\theta]\mathrm{d}y = \int F[\theta(y), \theta'(y)]\mathrm{d}y. \tag{6.6}$$

Here we have ignored the magnetostatic energy. A is the exchange stiffness. Minimizing the energy is equivalent to solving the Euler equation

$$\frac{\partial F}{\partial \theta} - \frac{\mathrm{d}}{\mathrm{d}y}\left(\frac{\partial F}{\partial \theta'}\right) = 0, \tag{6.7}$$

where $\theta' = \partial\theta/\partial y$. Then we obtain

$$K\sin\theta\cos\theta - A\partial^2\theta/\partial y^2 = 0. \tag{6.8}$$

Solving equation (6.8) and combining the boundary conditions ($\theta = \pi$ (0) when $y = +\infty$ ($-\infty$)) yields the Bloch DW equation

$$y = \sqrt{\frac{A}{K}}\ln\tan(\theta/2). \tag{6.9}$$

The Bloch DW width and the energy per unit domain wall area are

$$\delta_{\text{BW}} = \pi\sqrt{\frac{A}{K}}, \quad \gamma_{\text{BW}} = 4\sqrt{AK}. \tag{6.10}$$

The static DW structure can also be calculated directly by the LLG equation. The magnetization can be expressed as $\mathbf{M} = M_s(\sin\theta\cos\varphi, \sin\theta\sin\varphi, \cos\theta)$ in the spherical coordinate system. The effective field includes the exchange field and uniaxial magnetocrystalline anisotropy field

$$\mathbf{H}_{\text{eff}} = \frac{2A}{\mu_0 M_s^2}\frac{\partial^2\mathbf{M}}{\partial y^2} + \frac{2K}{\mu_0 M_s^2}M_z\hat{z}. \tag{6.11}$$

For the static DW, we have the equilibrium condition

$$-\gamma\mathbf{M}\times\mathbf{H}_{\text{eff}} = 0. \tag{6.12}$$

Then we obtain the wall equation which has the same form as equation (6.9).

If we consider a Néel DW, i.e. $\varphi = \pi/2$, the wall width and the wall energy density turn to

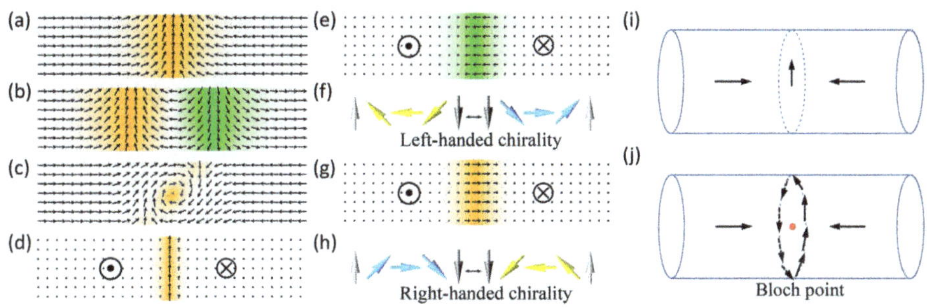

Figure 6.2. The micromagnetic structures of (a) 180° Néel-type, (b) 360° Néel-type, and (c) vortex-type DWs for the in-plane magnetized nanostrip. The (d) Bloch-type and (e), (g) chiral Néel-type DWs in the perpendicularly magnetized nanostrip. The schematic of chiral Néel-type DWs with (f) left-handed and (h) right-handed chiralities. The illustration of (i) transverse and (j) vortex DWs in the cylindrical nanowire, the red dot represents the Bloch point (singularity). ((f) and (h) Reproduced with permission from [19]. Copyright 2013 The American Physical Society.)

$$\delta_{\mathrm{NW}} = \pi\sqrt{\frac{A}{K + \mu_0 M_s^2/2}}, \quad \gamma_{\mathrm{NW}} = 4\sqrt{A\left(K + \frac{1}{2}\mu_0 M_s^2\right)}. \tag{6.13}$$

A detailed derivation of (6.13) can be found in [4]. One can clearly see that the expressions contain an extra term $\frac{1}{2}\mu_0 M_s^2$ (compared to equation (6.10)) which comes from the bulk magnetic charges emerging in the Néel DW. Therefore, for the bulk materials, the Néel DW always has a larger wall energy than the Bloch DW. In other words, the bulk magnetic system tends to form Bloch DWs. For a Bloch DW, the magnetic charge appears at the sample surface, which can be ignored if the system is large enough. However, as the system becomes thinner, the effect of magnetic charge for the Bloch DW begins to play a role. When the thickness (t) is less than a critical value (δ), with the sketch shown in figure 6.1(b), the demagnetization energy produced by surface magnetic charge will exceed the demagnetization energy produced by bulk magnetic charge in the Néel DW. In this case, the Néel DW has lower energy. We therefore have the conclusion that the Bloch DW is more stable in thick film, while the Néel DW exists in thin films.

The one-dimensional DWs discussed above are just ideal models, while real DWs generally have complex three-dimensional micromagnetic structures. For the in-plane magnetized nanostrip, there exist two types of DWs, i.e. the Néel-type [5, 6] and vortex-type [7, 8] DWs, as shown in figures 6.2(a) and (c), respectively. Here, it is difficult for the Bloch-type DWs to form because of the increasing demagnetization energy. Generally speaking, the narrower or thinner the nanostrips are, the easier it is to form a Néel-type DW. In contrast, it is easier to form a vortex-type DW in a wide and thick nanostrip. In addition, due to the mutual attraction between two π (180°) Néel-type DWs with opposite polarities, the 2π (360°) [9, 10] even $n\pi$ [11] Néel-type DWs also attract people's attention. The structure of a 2π Néel-type DW is shown in figure 6.2(b). For the perpendicularly magnetized nanostrip, Bloch-type and Néel-type DWs [12, 13] are the two most typical DWs, with the corresponding structures shown in figures 6.2(d), (e), and (g). Interestingly, the Dzyaloshinskii–Moriya interaction (DMI) [14, 15] can induce two different chiral Néel-type DWs [16–18] depending on the sign of D, i.e. left-handed (figures 6.2(e) and (f)) and right-handed (figures 6.2(g) and (h)) chiral Néel-type DWs. Moreover, the transverse and vortex DWs [20, 21] can also be found in cylindrical magnetic nanowires, with the structures shown in figures 6.2(i) and (j), respectively. Remarkably, the exotic singularity called the Bloch point appears at the center of a vortex DW [22, 23]. It is worth noting that there exist other type of DWs, for example, antivortex DWs [24], cross-tie DWs [25, 26], etc.

Due to the great number of potential applications of DW in magnetic storage devices, their dynamics stimulated by various external forces (magnetic field, electric current, spin-wave, etc) [27–38] have been studied extensively in recent years. For transverse DWs (Bloch-type and Néel-type) in the nanostrip, the velocity of the DWs increases steadily with the increase of external force strength when the external force is small. However, when the external force exceeds a critical value, the DW structure collapses and undergoes a series of complex cyclic transformations with a sharp reduction of the DW velocity, which is called Walker breakdown [27, 39–41]. We give an example to show this distinctive dynamics property. Figure 6.3(a) shows the steady velocity of transverse DWs in a nanostrip driven by spin-polarized currents [42]. One can see that when the non-adiabaticity coefficient β vanishes, the DWs cannot move only if the spin-polarized current density exceeds a critical value. If $\beta \neq 0$, the DW can always move no matter how small the current density is (the external pinning effect is not considered here). In addition, with the increase of β, the velocity of DW increases faster with the current density. The DW velocity will rapidly decrease when the current density goes beyond a critical value which corresponds to the Walker breakdown. For practical applications, the Walker breakdown is a major obstacle because it limits the development of DW based spintronic devices. Therefore, researchers in the community have made great efforts to overcome this limitation. For example, Yan *et al* [43] studied the dynamics of current-induced transverse DWs in cylindrical permalloy nanowires. They found that the DW can be regarded as a zero-mass micromagnetic object immune from the Walker breakdown. The main results are shown in figure 6.3(b). One can clearly see that the velocity of DW is a linear function of the current density for arbitrary β. In addition, they found that the Walker breakdown of transverse DW is completely suppressed in ferromagnetic nanotubes and the constant DW velocities above 1000 m s^{-1} can be achieved by small fields [44]. In addition, by using a comb-like nanostructure [45], transverse magnetic field [46, 47], thermal gradients [48], edge roughness [49], periodic holes [50], etc, the Walker breakdown can also be effectively suppressed.

Although the properties of the DW have been studied for a long time, it did not attract much attention until 2008, when Parkin *et al* [51] proposed the concept of a magnetic DW racetrack memory. The proposal made people realize the great application potential of DWs in high-density memory. Since then, the study of DWs has become a hot topic in the magnetism community. An illustration of a DW racetrack memory is shown in figure 6.4(a). The data are stored in the form of

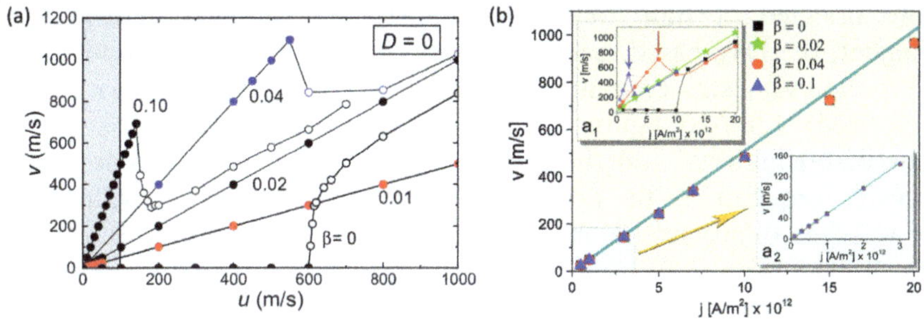

Figure 6.3. (a) The steady velocity of a transverse domain wall in a 120 × 5 nm^2 wire as a function of the velocity u representing the spin-polarized current density with a different non-adiabaticity coefficient β. Open symbols denote vortex nucleation. (b) Dependence of a simulated transverse domain wall velocity on the current density j for four different values of β in the permalloy (Py) cylindrical nanowire (diameter $d = 10$ nm). Here the Gilbert damping coefficient is $\alpha = 0.02$. ((a) Reproduced with permission from [42]. Copyright 2005 IOP Publishing. (b) Reproduced with permission from [43]. Copyright 2010 The American Physical Society.)

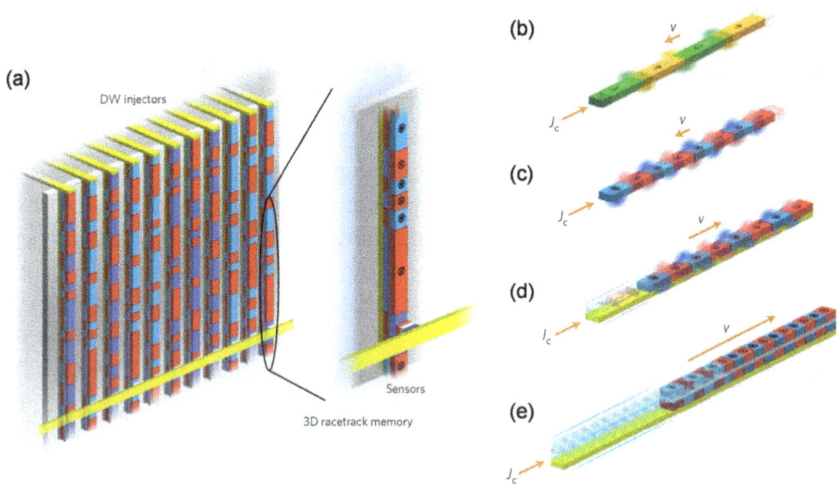

Figure 6.4. (a) Sketch of a magnetic DW based racetrack memory, with data encoded as a pattern of magnetic domains along a portion of the nanowire. The red and blue regions represent two domains with opposite directions. (b) Racetrack memory 1.0: in-plane magnetized racetracks. (c) Racetrack memory 2.0: perpendicularly magnetized racetracks. (d) Racetrack memory 3.0: DWs at high velocities driven by spin torque. (e) Racetrack memory 4.0: a giant exchange coupling torque driven DWs in synthetic antiferromagnet racetracks at extremely high velocity. (Reproduced with permission from [52]. Copyright 2015 Nature.)

magnetic domains, which have two opposite directions (denoted by red and blue regions) representing binary ‘0’ and ‘1’. The spin-polarized current can drive the entire pattern of DWs coherently along the wire over the reading and writing elements. The reading process is done by measuring the tunnel magnetoresistance of a magnetic tunnel junction connected to the racetrack, while the writing can be accomplished by the fringing fields of a DW moving in a second ferromagnetic nanowire oriented at right angles to the storage nanowire. During years of development, the DW based racetrack memory has experienced about four major steps [52] and every step contains some distinctive features, as follows. Racetrack memory (RM) 1.0 (see figure 6.4(b)): the racetrack is based on an in-plane magnetized soft material and the DWs are rather wide; the conventional volume spin transfer torque drives the DWs in the direction of electron flow; and DWs can achieve velocities of around 100 m s^{-1} for current densities in the order of 10^8 A cm^{-2} [31, 32, 53]. RM 2.0 (see figure 6.4(c)): the racetrack is perpendicularly magnetized with large magnetic anisotropy; the DWs are much narrower and more robust compared to those in RM 1.0; and the direction of DW propagation and the DW speed are roughly the same as for RM 1.0 [54, 55]. RM 3.0 (see figure 6.4(d)): the racetrack is perpendicularly magnetized; the chiral spin torque originating from spin–orbit coupling, DMI, and the spin Hall effect drives the DWs to propagate along the direction of current; and the DWs can achieve velocities of around 350 m s^{-1} for the same current densities ($\sim 10^8$ A cm^{-2}) with RM 1.0 and 2.0 [12, 33]. RM 4.0 (see figure 6.4(e)): the DWs move in racetracks that are formed from synthetic antiferromagnets; and DW velocities of about 1000 m s^{-1} can be achieved for the same current density ($\sim 10^8$ A cm^{-2}) [34].

In addition to the racetrack memory, DWs can also be used to realize logic circuits [56–59]. Compared to the traditional complementary metal-oxide semiconductor (CMOS) circuits, the DW based logic circuits have significant advantages, such as lower power, higher speed, higher density, nonvolatility, etc. In CMOS circuits the digital signals are represented by the absence or presence of electrical charge, while the directions of magnetization in the domain represent the binary digits in the DW logic circuit. In general, the practical logic circuit contains four basic function operations, i.e. fan-out, cross-over, NOT, and AND. The fan-out structure makes two identical copies of an input signal and the cross-over structure allows two signals to pass over each other without interference. Allwood *et al* [56] demonstrated these operations using a single rotating magnetic field in the DW logic circuit. In recent years the DW logic circuits have received extensive study [60–62]. However, most of the studies adopt magnetic fields to manipulate DWs, which limits their implementation in dense and large-scale chips. Luo *et al* [63] demonstrated a method for performing all-electric logic operations and cascading using DW racetracks. They realized a reconfigurable NAND/NOR gate with the structures shown in figures 6.5(a) and (b). It is well known that any Boolean function can be implemented using a combination of NAND or NOR gates. The main structure is composed of four out-of-plane (OOP) regions (the red circles in figure 6.5(b)) that form two logic inputs, one bias, and one logic output connected via in-plane (IP) regions (in blue). Figures 6.5(d) and (e) show devices with different logic-input configurations for NAND and NOR gates, respectively. In these devices, the two DW reservoirs are connected to the inputs via racetracks. The logic-input configurations are achieved by placing inverters after some of the DW reservoirs. The truth tables for the reconfigurable NAND and NOR logic operations are shown in figure 6.5(c).

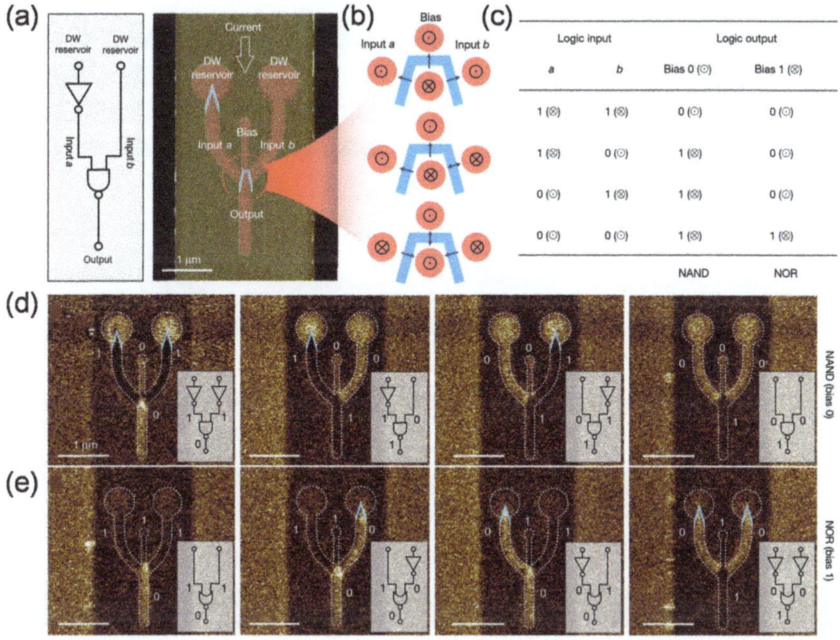

Figure 6.5. (a) Colored SEM image of a reconfigurable NAND/NOR logic gate (right) and corresponding logic circuit (left). (b) Schematics showing the relationship between the magnetization in the two logic inputs, the bias and the logic output for the NAND gate. (c) Truth table for the reconfigurable NAND and NOR logic operations. The MFM images of different logic operations for inputs of '11','10','01', and '00' with a bias of (d) '0' and (e) '1' and the corresponding logic-circuit diagrams. (Reproduced with permission from [63]. Copyright 2020 the Authors.)

Furthermore, when DWs are pinned by defects such as notches [64, 65], protrusions [66, 67], and kinks [68], the spin transfer torque can induce stable DW oscillations, acting as nano-oscillators [69–76]. In general, the DW based oscillators are more robust and flexible than conventional electronic devices. Figure 6.6(a) shows a typical model for a one-dimensional DW pinned by a notch, where the in-plane current flowing along x-axis stimulates the DW oscillation. The DW can maintain steady-state oscillation only if the current density j_e exceeds a threshold value [77]. Figure 6.6(b) plots the dependence of the DW position on time for different choices of j_e. Interestingly, the oscillation of DW is accompanied with continuous and repeated transitions between Bloch and Néel wall structures, as shown in figure 6.6(c). The spin current drift velocity u as a function of the current density j_e is defined by $u = j_e P\mu_B/eM_s$, where P is the spin polarization of the current, μ_B is the Bohr magneton, e is the (positive) electron charge, and M_s is the saturation magnetization. It has been shown that a moving DW can induce an electromotive force [78–80], which will generate a detectable voltage signal. From an experimental point of view, the voltage signals induced by the pinned DW oscillations are useful to design nano-scale microwave oscillators and generators. However, the signal generated by one DW oscillation is often too weak. Furthermore, Martinez *et al* [73]

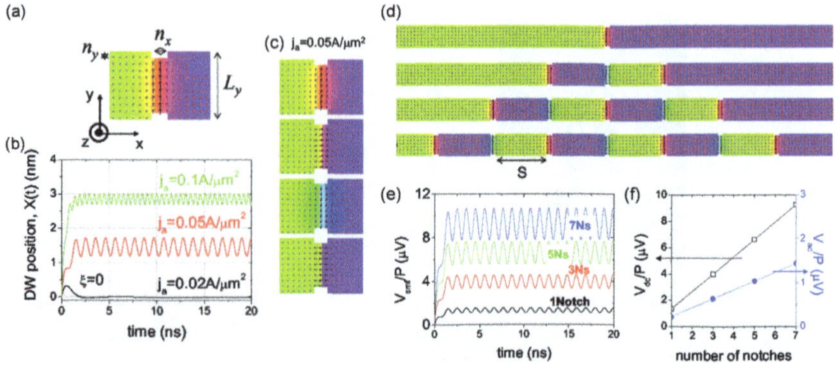

Figure 6.6. (a) Schematic illustration of the pinned DW structure. (b) DW position $X(t)$ as a function of time for different current densities. (c) The snapshots of DW structure with current density $j_e = 0.05$ A μm^{-2}. (d) Equilibrium states with a different number of Bloch DWs pinned at notches. (e) Voltage signal as a function of time for a strip with a different number of pinned DWs. (f) Static (V_{dc}) and oscillating (V_{ac}) contributions of the voltage signal V_{smf} (t) as a function of the number of pinned DWs. (Reproduced with permission from [73]. Copyright 2011 The American Physical Society.)

designed oscillator based multiple pinned DWs, as shown in figure 6.6(d). They found that with the increase of the number of DWs, the intensity of the voltage signal V_{smf} increases gradually which is conducive to experimental observation, as shown in figure 6.6(e). In addition, both dc voltage signal V_{dc} and oscillating contribution V_{ac} increase linearly with the number of pinned DWs (see figure 6.6(f)).

It is worth noting that DWs have other applications, for example, diode [81], memristor [82], etc, but we would not go much detail on these issues.

6.1.2 Vortex

A vortex is a fundamental topological structure in various systems, such as stirred fluids, cyclones, nebulas, etc. There are also vortex structures in magnetic systems due to the competition between exchange and magnetostatic couplings. A magnetic vortex has a spatial size typically ranging from hundreds of nanometers to a few microns [83–85]. Moreover, magnetic vortices usually exist in nanodisks and are characterized by the orientation of the magnetization circulating in-plane (chirality c) around a center spin (called vortex core) with an out-of-plane magnetization (polarity p) (see figure 6.7(a)). The chirality is defined by a clockwise ($c = +1$) or counterclockwise ($c = -1$) rotation of the in-plane magnetization. While the polarity is the orientation of the vortex core magnetization, either pointing up ($p = +1$) or down ($p = -1$). Naturally, one can obtain four different ground states of magnetic vortex for different combinations of c and p, as shown in figure 6.7(b).

For the static magnetic vortex configuration, the topological charge can be calculated through equation (6.3), with the boundary conditions $\mathbf{m}(\mathbf{r} \to \infty) = (\cos\phi, \sin\phi, 0)$, $\mathbf{m}(\mathbf{r} = 0) = (0, 0, \pm 1)$, and, $\phi(\varphi = 2\pi) - \phi(\varphi = 0) = 2\pi$. We therefore obtain the topological charge of magnetic vortex $Q = \pm 1/2$, indicating that the magnetic moments cover only half of the surface of the unit sphere in the order-parameter space.

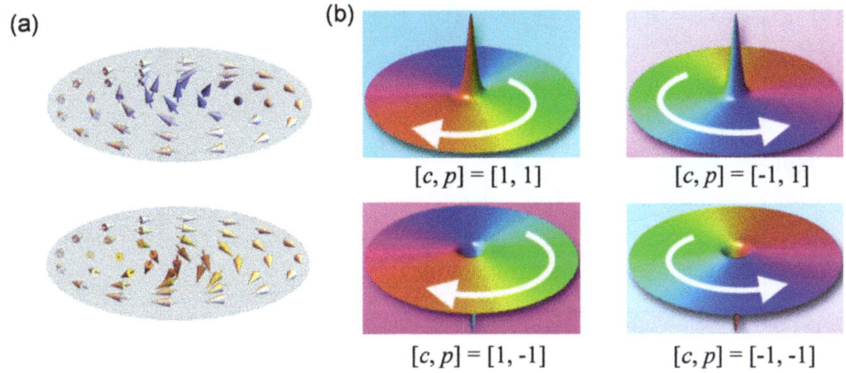

Figure 6.7. (a) Schematic diagram of a magnetic vortex. (b) Four different vortex ground states. (Reproduced with permission from [86]. Copyright 2017 The American Physical Society.)

Generally speaking, the nanodisk with a vortex state allows three different eigenmodes: the gyrotropic mode [87–89], the azimuthal spin-wave mode [90–93], and the radial spin-wave mode [94, 95]. The dynamic of the gyrotropic mode (low-energy excitation) can be characterized by the massless Thiele's equation [96]

$$G\hat{z} \times \frac{\mathrm{d}\mathbf{U}}{\mathrm{d}t} - \alpha D\frac{\mathrm{d}\mathbf{U}}{\mathrm{d}t} + \mathbf{F} = 0, \tag{6.14}$$

where $\mathbf{U} = \mathbf{R} - \mathbf{R}^0$ is the vortex core displacement $\mathbf{R}$ relative to its equilibrium position $\mathbf{R}^0$, $G = -4\pi QdM_s/\gamma$ is the gyroscopic constant with Q the topological charge, d the thickness of ferromagnetic layer, M_s the saturation magnetization, and γ the gyromagnetic ratio, and αD is the viscous coefficient with α being the Gilbert damping constant. The conservative force $\mathbf{F} = -\partial W/\partial\mathbf{U}$ with W denoting the total potential energy of the system. Considering a single vortex, the potential energy has a parabolic form (in the linear region): $W = W_0 + K\mathbf{U}^2/2$, where W_0 represents the potential energy when the vortex core locates at the disk center and K is the spring constant due to the confining potential from the boundary. Neglecting the damping term, we obtain the eigenfrequency for gyrotropic motion of a single vortex with $\omega_0 = K/|G|$. It has been proved that the gyration frequency is proportional to the thickness of the nanodisk (d) while being inversely proportional to the radius (r) of the nanodisk, i.e. $\omega_0 \propto d/r$ [87, 89].

For the gyrotropic mode, the trajectories of the vortex core with different polarities are plotted in figure 6.8(a). We see that the direction of vortex core precession is clockwise (counterclockwise) for $p = -1$ ($p = +1$), which can be explained by equation (6.14). At first, when the vortex core deviates from the equilibrium position, it will experience a restoring force $\mathbf{F} = -K\mathbf{U}$ pointing to the disk center (denoted by the red arrow in figure 6.8(b)). Because the gyrotropic force $G\hat{z} \times \mathrm{d}\mathbf{U}/\mathrm{d}t$ is always opposite to the direction of restoring force (green arrow in figure 6.8(b)), for the case of $Q = -1/2$ ($p = -1$), the vortex core gyration ($\mathrm{d}\mathbf{U}/\mathrm{d}t$) must be clockwise (denoted by purple arrow). Because of the existence of a damping

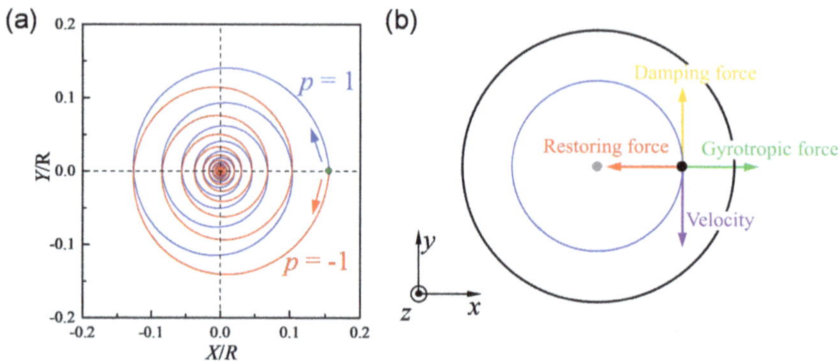

Figure 6.8. (a) Trajectories of vortex core precession around the equilibrium position (disk center) for different polarities. (b) Schematic plot of different forces acting on the vortex core.

force which is opposite to the velocity (represented by the orange arrow), the vortex core velocity will decrease, and the real trajectory exhibits a spiral line; see figure 6.8(a). However, if we can provide an external driving force (spin transfer torque, for instance) to balance the damping force, the vortex core can maintain a steady motion in a specific circular orbit.

In addition to the gyrotropic mode with a low frequency, the magnetic vortex can have higher frequency spin-wave modes, which are characterized by mode number (n, m) with n counting the number of nodes along the radial direction (radial spin-wave mode) and m counting the number of nodes along the azimuthal direction (azimuthal spin-wave mode) over half the nanodisk. By using the time-resolved magneto-optic Kerr microscopy and micromagnetic simulation, Buess *et al* [95, 97] identify the radial and azimuthal spin-wave mode with small n and m, with the results shown in figure 6.9(a). In addition, they find that the frequency of the radial spin-wave mode increases with the number of nodes (n), while the frequency of the azimuthal spin-wave modes decreases with the number of nodes (m). Furthermore, Schultheiss *et al* [98] derive the full dispersion curve of spin-wave modes in vortex for a Py disc with 50 nm thickness and 5.1 μm diameter, as shown in figure 6.9(b). They find that when n is small, the spin-wave frequency decreases at first and then increases with the increase of m. Figure 6.9(c) plots four different azimuthal spin-wave mode profiles when $n = 0$, which is also called the whispering gallery magnon.

In recent years, magnetic vortices have attracted significant attention since they exhibit great potential for future non-volatile data-storage devices [99, 100]. By using the unique characters of chirality and polarity, the magnetic vortices can realize multibit storage (remember, there are four different vortex states). Therefore, it becomes very important to control the polarity and chirality of the vortex independently. On the one hand, one of the most direct methods to reverse the vortex core is applying an external magnetic field which points along the opposite direction of the polarity of the vortex core. However, this method requires a large external magnetic field strength [101, 102]. Later, it was found that the vortex core

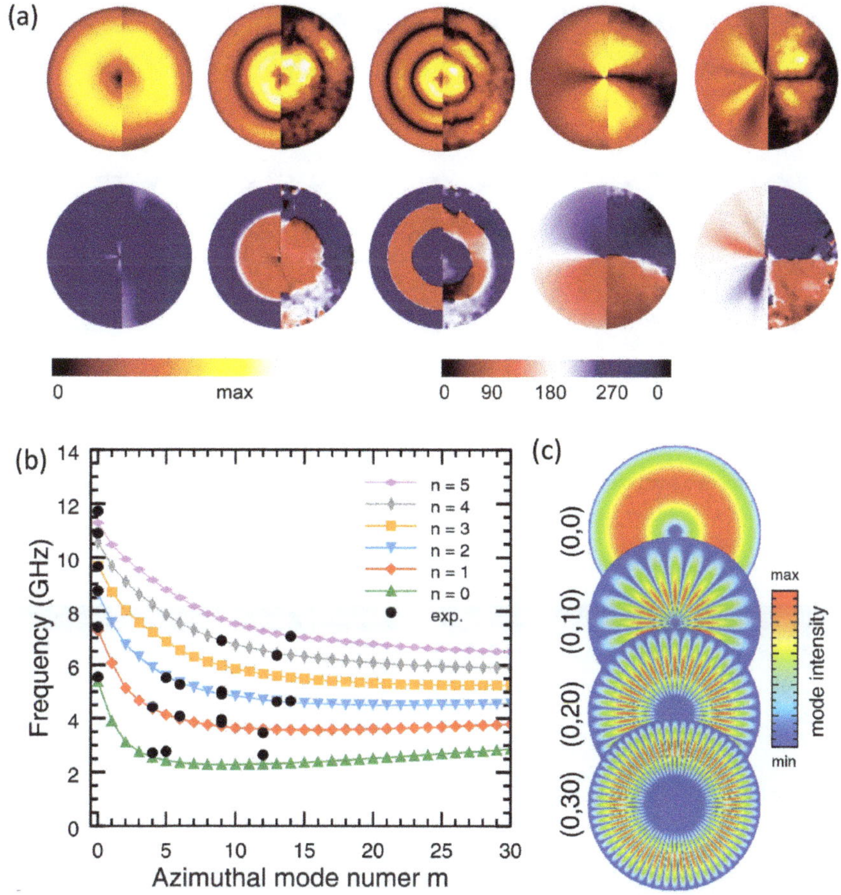

Figure 6.9. (a) Images of the eigenmodes of an isolated ferromagnetic Py disk with a radius $R = 3$ μm. The top row presents the distribution of the Fourier amplitude at resonance frequencies, the bottom row shows the phases of different eigenmodes. The images are composed of two parts: the left denote the micromagnetic simulation, while the right presents the experimental results. (b) Theoretical calculations of the magnon mode frequencies as a function of the radial and azimuthal mode numbers n and m. The black dots show the experimental results. (c) Four exemplary mode profiles resulting from analytical calculations. ((a) Reproduced with permission from [97]. Copyright 2005 The American Physical Society. (b), (c) Reproduced with permission from [98]. Copyright 2019 The American Physical Society.)

can resonantly reverse when the eigenmodes of the vortex are excited (including gyrotropic mode, radial and azimuthal spin-wave modes) [103–111]. In addition, it has been shown that the vortex core can be reversed by spin-polarized currents which can induce the vortex gyrotropic mode [112, 113]. It is more conducive to the integration of spintronic devices through current to reverse the vortex core. In general, the process of polarity switch of the vortex core is accompanied by the following process: when an external excitation is applied, the vortex core begins to accelerate. When the velocity of the vortex core is higher than a critical value [114–116], it will create a vortex–antivortex pair. Then the original vortex will

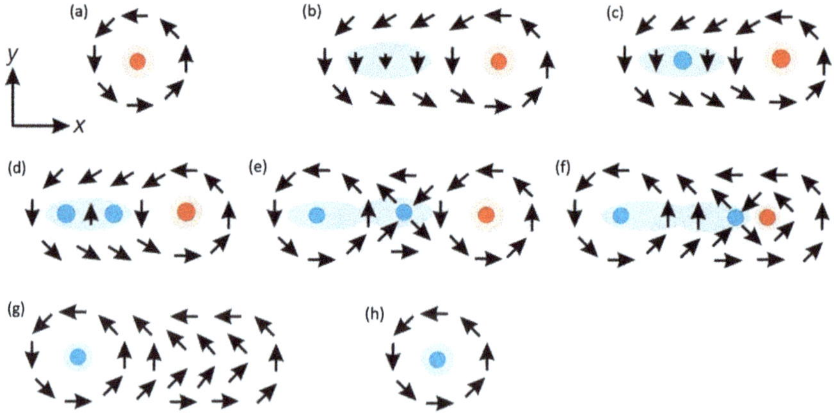

Figure 6.10. Schematic representation of different steps in the vortex core switching. The dark arrows represent the in-plane magnetization components, and the colored dots and rings represent the out-of-plane component (blue for up; red for down). (Reproduced with permission from [103]. Copyright 2006 Nature.)

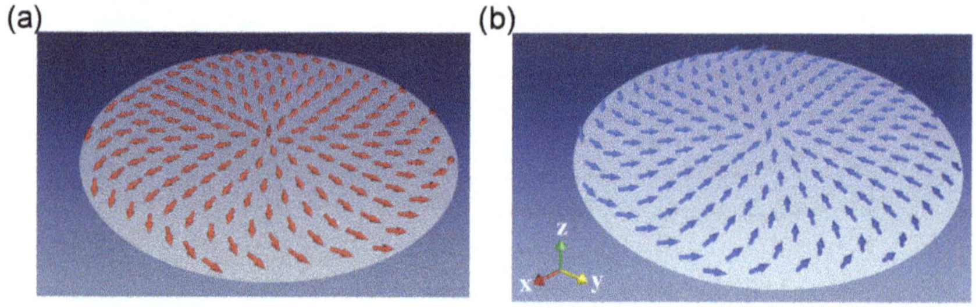

Figure 6.11. Schematic plot of the spin structure of a meron state with opposite chiralities. The arrows represent the direction of local magnetization. (Reproduced with permission from [130]. Copyright 2012 The American Physical Society.)

attract the antivortex due to their opposite topological charges and annihilate subsequently. So, only the vortex with opposite polarity is left and the process of polarity reverse is completed. An illustration of vortex polarity reverse is shown in figure 6.10. The vortex chirality can also be switched by nanosecond field pulses [117], electric fields [118], dipole-coupled nanomagnets [119], etc.

As discussed above, if we apply a proper spin-polarized current, the damping force can be balanced by the spin transfer torque and the vortex core may maintain a steady-state motion, which can be used to design magnetic nano-oscillators [120–124]. Importantly, the vortex based spin-torque nano-oscillators have many outstanding properties compared to conventional oscillators, such as low generation linewidth, high output power, and flexible frequency range. In addition, by manipulating the magnetic vortices one can realize the logic [125, 126] and transistor operations [127, 128].

There exists a special type of non-trivial topological spin state called meron, as shown in figure 6.11. The meron state exists only in pairs and originally emerges as a solution of the classical Yang–Mills theory [129]. Recently, Phatak *et al* [130] report the experimental observation of the magnetic meron in antiferromagnetically coupled NiFe disks at room temperature, in which the antiferromagnetic exchange interaction between the two disks plays a key role in the formation of the meron pair. Similarly, the topological charge of meron is $Q = \pm 1/2$ from equation (6.3). Compared to vortices, the magnetic merons have higher energies because of massive magnetic charges appearing at the boundary of the nanodisk. Therefore, to stabilize the merons in the nanodisk, extra energy, for example, the Ruderman–Kittel–Kasuya–Yosida (RKKY) interaction [131–133], interlayer coupling, etc, should be added. In chiral magnets such as $Co_8Zn_9Mn_3$, the paired merons can be replaced by an meron and antimeron pair which form a square lattice [134]. What's more, the meron and antimeron lattice can transform into a hexagonal lattice of skyrmions in the presence of a magnetic field.

6.1.3 Bubble

Due to the fundamental interest of potential applications in memory devices, the magnetic bubble has been studied extensively since its observation in ferromagnetic films with a strong perpendicular anisotropy [135–141]. The magnetic bubble state is a circular domain (inner region) surrounded by a domain with opposite magnetization orientation (outer region). Because of the high perpendicular anisotropy, the magnetizations in the inner and outer regions are out of plane. Between the two opposite magnetic domains, there exists a domain wall with the magnetization direction rotating gradually from one direction to the other. The micromagnetic structure of a magnetic bubble is shown in figure 6.12(a). It is worth noting that the perpendicular anisotropy is a necessary condition to stabilize the magnetic bubble, because the increased demagnetization energy when the magnetization is arranged vertically needs to be balanced by perpendicular anisotropic energy.

Similarly, the topological charge of the magnetic bubble can be calculated by using equation (6.3). For the static magnetization structure of figure 6.12(a) we have boundary conditions $\mathbf{m}(\mathbf{r} \to \infty) = (0, 0, -1)$, $\mathbf{m}(\mathbf{r} = 0) = (0, 0, 1)$, and $\phi(\varphi = 2\pi) - \phi(\varphi = 0) = 2\pi$. We therefore obtain the topological charge of magnetic bubble $Q = +1$. If the magnetization directions in the inner and outer regions are reversed, the topological charge will change from +1 to −1. It is worth noting that the value of vorticity $m = \frac{1}{2\pi}\phi(\varphi)|_{\varphi=0}^{\varphi=2\pi}$ can only be an integer $(0, \pm 1, \pm 2, \ldots)$, which means that the azimuthal angle ϕ can only wrap around in multiples of 2π. Therefore, if the azimuthal angle ϕ wraps around less than 2π, the vorticity m equal to 0, then the topological charge $Q = 0$, see figure 6.12(b), for the trivial magnetic bubble state. Interestingly, Moutafis *et al* [142] have shown that the strong field can induce switching of the topological bubble with $Q = 1$ to the trivial bubble with $Q = 0$, and the $Q = 0$ bubble can be switched back to a $Q = 1$ bubble by applying an external field gradient.

In general, the position of the magnetic bubble can be modeled by the so-called guiding center $\mathbf{R} = (R_x, R_y)$ defined by the following formulas [143]

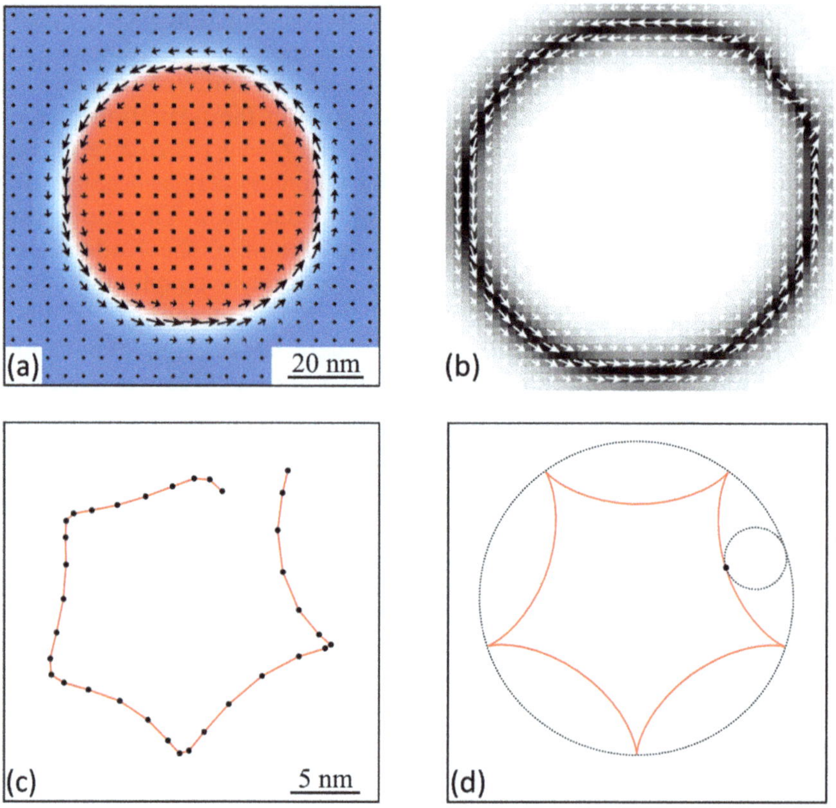

Figure 6.12. (a) Schematic of a circular magnetic bubble. The red and blue regions denote the out-of-plane magnetization with $M_z < 0$ and $M_z > 0$, respectively. The black arrows represent the in-plane magnetization. (b) The magnetic bubble with topological charge $Q = 0$. (c) The trajectory of the bubble's center observed in [142]. (d) A five-cusped hypocycloid with its directing and generating circles. ((a), (c), (d) Reproduced with permission from [144]. Copyright 2012 The American Physical Society. (b) Reproduced with permission from [142]. Copyright 2009 The American Physical Society.)

$$R_x = \frac{\int\int xq\mathrm{d}x\mathrm{d}y}{\int\int q\mathrm{d}x\mathrm{d}y}, \quad R_y = \frac{\int\int yq\mathrm{d}x\mathrm{d}y}{\int\int q\mathrm{d}x\mathrm{d}y}, \tag{6.15}$$

with q the topological charge density as defined in equation (6.1).

From previous discussions, we know that an isolated vortex exhibits gyrating motion at a circular orbit with frequency $\omega_0 = K/|G|$, which can be well characterized by the massless Thiele's equation (equation (6.14)). However, Moutafis *et al* [142] showed that this equation cannot reproduce the simulated motion of an isolated bubble. By considering the parabolic potential, they found that the magnetic bubble exhibits hypocycloidal gyrations and the trajectory of the bubble's guiding center is a pentagon instead of a circle, as shown in figure 6.12(c). To obtain the correct dynamical model of the magnetic bubble, Makhfudz *et al* [144] generalized the original Thiele's equation by adding an inertial term

$$-M\frac{\mathrm{d}^2\mathbf{U}}{\mathrm{d}t^2} + G\hat{z} \times \frac{\mathrm{d}\mathbf{U}}{\mathrm{d}t} - \alpha D\frac{\mathrm{d}\mathbf{U}}{\mathrm{d}t} + \mathbf{F} = 0, \tag{6.16}$$

where M is the effective mass of the magnetic bubble. Therefore, one can evaluate the gyration frequency of the magnetic bubble

$$\omega_\pm = -\frac{G}{2M} \pm \sqrt{\left(\frac{G}{2M}\right)^2 + \frac{K}{M}}. \tag{6.17}$$

The symbols '+' and '−' of the ω in equation (6.17) imply that there exist two different gyration modes along clockwise and counterclockwise directions, respectively. Makhfudz *et al* [144] further showed that if $\omega_-/\omega_+ = -4$, the hypocycloid motion of bubble has five cusps and resembles a pentagon, as shown in figure 6.12(d). Indeed, the numerical simulations showed that the bubble owns two eigenfrequencies with $\omega/2\pi = 0.97$ GHz and −4.27 GHz. Therefore, equation (6.17) exactly reproduces the bubble trajectory, which indicates that the mass effect should be considered for describing the magnetic bubble motion. Noticeably, the generalized Thiele's equation containing higher-order terms can be directly derived from the Landau–Lifshitz–Gilbert (LLG) equation (see section 7.1.1 for details).

Moon *et al* [145] have showed numerically that various regular polygon-like trajectories of magnetic bubble center positions can be obtained by applying an external magnetic field, with the results shown in figures 6.13(b) and (c). If an external field H_z is applied perpendicular to the magnetic bubble disk plane

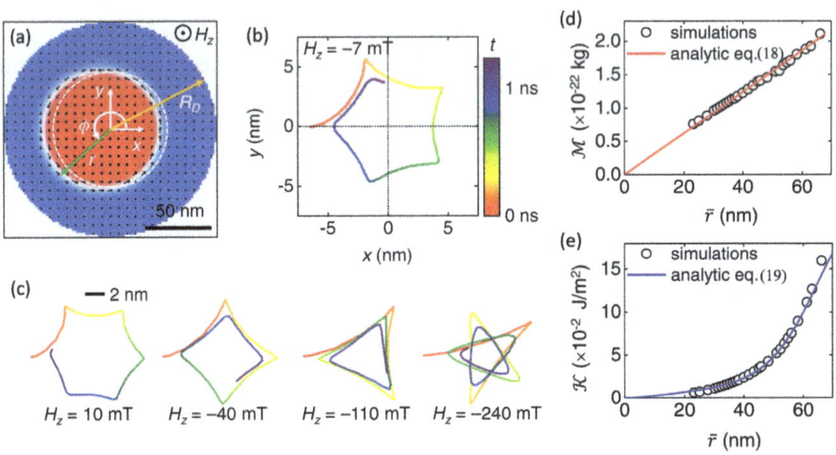

Figure 6.13. (a) Illustration of a magnetic bubble. R_D is the radius of the disk and r is the radius of the circular domain. The arrows represent the direction of local magnetization. (b) A regular pentagon-like trajectory of the bubble domain gyration with external field $H_z = -7$ mT. (c) Several examples of regular polygon-like trajectory with respect to the external field. The dependence of the (d) mass and (e) spring constants on the bubble radius. (Reproduced with permission from [145]. Copyright 2014 The American Physical Society.)

(see figure 6.13(a)), the bubble position has been shifted to the x-direction with a certain distance and then relaxes. For $H_z = -7$ mT the regular pentagon trajectory with five vertices is observed, as shown in figure 6.13(b). In addition, a hexagon under $H_z = 10$ mT, a square under -40 mT, and a triangle under -110 mT are also identified. In brief, when H_z decreases, the number of polygon sides and interior angles decrease. Interestingly, a pentagram trajectory was obtained when $H_z = -240$ mT; see figure 6.13(c).

In addition, Moon *et al* [145] derive the analytic expression for the mass constant M and spring constant K of magnetic bubble as

$$M \approx \frac{G^2}{\mu_0 M_s^2 \lambda}\left[4d\tan^{-1}\left(\frac{2d}{\lambda}\right) - \lambda\ln\left(1 + \left(\frac{2d}{\lambda}\right)^2\right)\right]^{-1}\bar{r}, \tag{6.18}$$

and

$$\begin{aligned} K \approx \frac{\mu_0\pi M_s^2 d^2}{12R_D}&\left[\frac{48}{\tau} - 24\left(\frac{1}{\rho_-} + \frac{1}{\rho_+}\right) + \pi^2\left(1 + \frac{2R_D^2}{d^2}\right)\left(\frac{1}{\rho_-} + \frac{13}{\rho_+} - \frac{14}{\tau}\right)\right. \\ &\left. + \frac{\pi^2 r}{R_D}\left\{\frac{1}{\rho_-^3} - \frac{13}{\rho_+^3} - \frac{2R_D^2}{d^2}\left(\frac{1}{\rho_-} - \frac{13}{\rho_+}\right)\right\}\right], \end{aligned} \tag{6.19}$$

where M_s is the saturation magnetization, d is the disk thickness, λ is the domain wall width, $\bar{r}$ is the averaged radii of the bubble domain, R_D is the radius of the nanodisk, $\tau = \sqrt{1 + (d/R_D)^2}$, and $\rho_\pm = \sqrt{(r/R_D \pm 1)^2 + (d/R_D)^2}$. The detailed derivation can be found in [145]. Figures 6.13(d) and (e) plot the dependence of parameters M and K on the magnetic bubble radius, respectively. The solid lines are obtained by equations (6.18) and (6.19), while the small circles denote simulation results. We can see that the analytical theory compares very well with numerical simulations. In addition, one can find that the mass constant is in exactly a linear relation with the bubble radius, which indicates that M is determined by the length of the domain wall instead of the area of the bubble domain. In contrast, the parameter K increases exponentially with the increase of $\bar{r}$.

An appealing application of a magnetic bubble is to design information storage devices. In magnetic materials, one can use the presence or absence of a bubble to represent the information bit [146].

6.1.4 Skyrmion

Skyrmions were originally introduced by the British particle physicist Tony Skyrme to describe localized and particle-like solutions of interacting pions in field theory about 60 years ago [147, 148]. After years of research, skyrmion structures have been found in broad fields of physics, such as optical systems [149–151], acoustic lattices [152], surface plasmonic systems [153, 154], liquid crystals [155, 156], string theory [157], Bose–Einstein condensate [158–160], and, recently, in magnetism [161, 162].

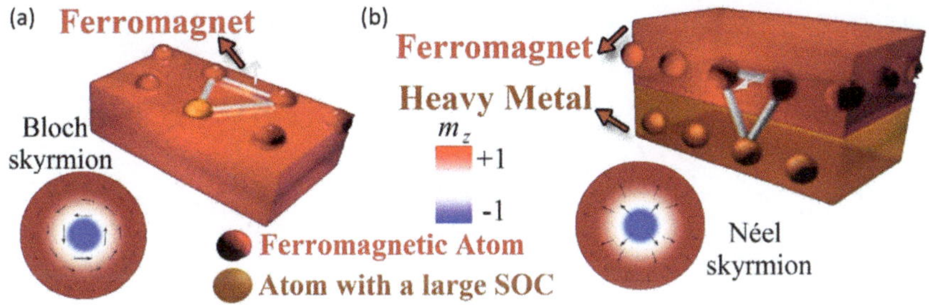

Figure 6.14. (a) The bulk DMI vector (white arrow) originating in a non-centrosymmetric crystal due to the interaction of the ferromagnetic atoms with an impurity with a large SOC. (b) The interfacial DMI vector (white arrow) in a ferromagnet/heavy metal bilayer. The magnetization distributions for Bloch and Néel skyrmions are plotted in the insets. (Reproduced with permission from [163]. Copyright 2016 IOP Publishing.)

In general, the magnetic skyrmion structure emerges in materials with out-of-plane magnetization and is stabilized by bulk or interfacial Dzyaloshinskii–Moriya interaction (DMI) [14, 15]. The bulk DMI originates from the broken crystal inversion symmetry in the ferromagnetic alloy which has high-spin–orbit coupling atoms (such as B20 materials including MnSi and $Fe_{1-x}Co_xSi$ etc). In this case, the skyrmion structure is of Bloch type, as shown in figure 6.14(a), while an interfacial DMI exists at the interface of multilayers with a thin perpendicular anisotropy film coupled to a large spin–orbit coupling (SOC) metal (such as Pt and Ir), in which the Néel-type skyrmions are more stable; see figure 6.14(b). In addition to the DMI, the magnetic skyrmion structure can also be stabilized by the magnetic dipolar interaction [164, 165], frustrated exchange interaction [166, 167], and higher-order exchange interaction [168, 169].

The topological charge of skyrmion can be directly calculated through equation (6.3) as $Q = \pm 1$. Noticeably, the topological charge of the skyrmion and non-trivial bubble are equal, which indicates that the two magnetic solitons are homotopic. Similar to a magnetic bubble, the skyrmion manifests an inertia in their gyrotropic motion and its dynamics should be described by the generalized Thiele's equation including a mass term (equation (6.16)). Therefore, there exist two different gyrotropic modes with clockwise and counterclockwise gyrations, as shown in figures 6.15(a) and (b). In addition, the skyrmion can exhibit an extra intrinsic mode with periodic variation of the size, i.e. a breathing mode; see figure 6.15(c).

Magnetic skyrmions have great potential applications in ultra-fast and ultra-high-density spintronic devices because of their outstanding features with small size, easy manipulation, and high stability. In recent years, a new discipline, skyrmionics [1, 2, 163, 171–178], has emerged, which focuses on the generation, propagation, manipulation, and detection of skyrmions.

Magnetic skyrmions can be created through various methods. For example, Romming *et al* [179] and Sampaio *et al* [180] showed that the individual skyrmion can be written and deleted by a local spin-polarized current, the skyrmion injection device is plotted in figure 6.16(a). After a local spin-polarized current being applied,

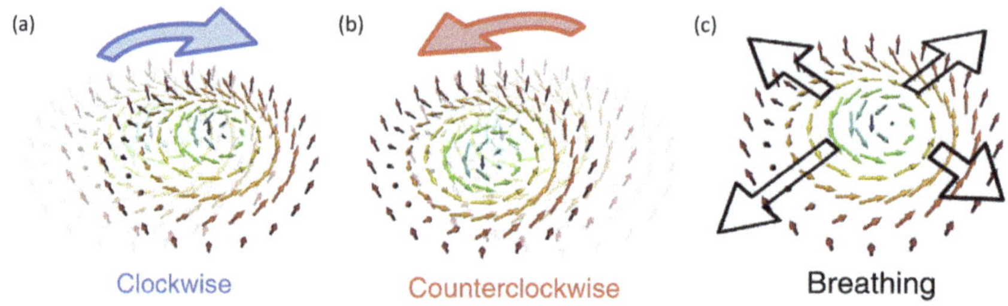

Figure 6.15. The intrinsic modes of a magnetic skyrmion. (a) The clockwise gyrotropic mode. (b) The counterclockwise gyrotropic mode. (c) The breathing mode. (Reproduced with permission from [170]. Copyright 2013 Nature.)

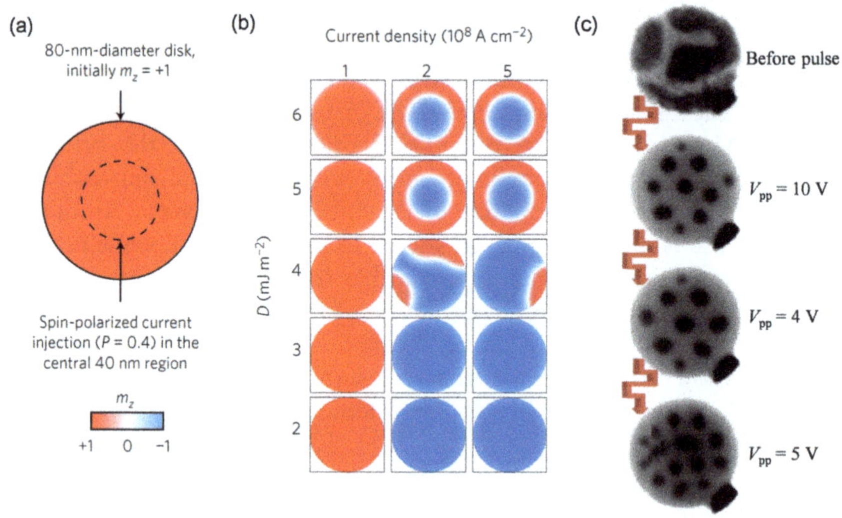

Figure 6.16. (a) Scheme of the injection device with the polarized current injected into the central region (dotted circle). (b) Magnetization in nanodisks (diameter d = 80 nm) of different DM parameters D after injecting spin-polarized currents of different densities for 2 ns. (c) The first image shows a labyrinth domain state induced by a static field. Then hexagonal skyrmion lattices (the rest images) are generated by applying a bipolar pulse train with different V_{pp}. ((a), (b) Reproduced with permission from [180]. Copyright 2013 Nature. (c) Reproduced with permission from [181]. Copyright 2016 Nature.)

the skyrmions can be generated for a proper current density and DMI strength (see figure 6.16(b)). Woo *et al* [181] have realized the creation and annihilation of skyrmions in a cylindrical nanodot at room temperature by applying a magnetic field pulse and the number of skyrmions can be controlled by changing the field amplitude, with the experimental results being shown in figure 6.16(c). Jiang *et al* [182] have observed that skyrmions can be generated by driving domain walls through a geometrical constriction. When the stripe domains pass through a constriction, the right-most parts of the stripe domain become unstable and are eventually converted into skyrmions, with the whole process being similar to blowing bubbles. In addition, magnetic skyrmions can also be created by local heating [183],

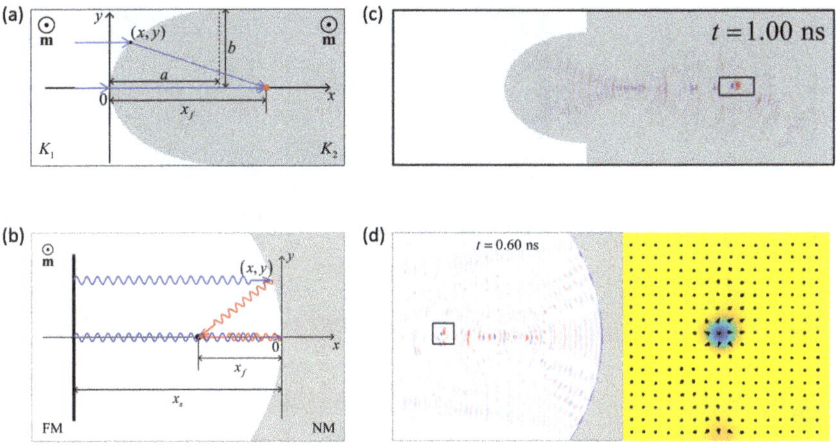

Figure 6.17. (a) Schematics of a spin-wave lens with an elliptical interface between two ferromagnetic films with different perpendicular magnetic anisotropies $K_{1,2}$. (b) Schematic of the film edge for total-reflection focusing of spin-waves. The red and black dots represent the focal points. After proper exciting fields are applied, the skyrmions are generated through (c) transmission and (d) total-reflection focusing spin-waves. ((a), (c) Reproduced with permission from [191]. Copyright 2020 AIP Publishing. (b), (d) Reproduced with permission from [192]. CC BY 4.0.)

an inhomogeneous spin Hall field [184], nanosecond electrical current pulse [185], constant applied field [186], resonance excitation field [187], spin-wave [188], etc.

Interestingly, it has been very recently demonstrated that a curved interface in two-dimensional ferromagnetic films can be used to construct a lens for spin-wave focusing [189, 190]. Further, by using this characteristic, Wang *et al* [191] and Yao *et al* [192] showed that the generation of skyrmions can be realized. Figures 6.17(a) and (b) plot the schematics for transmission and total-reflection focusing of spin-waves to generate skyrmions, respectively. By choosing the proper frequency and amplitude of the exciting magnetic field, skyrmions can be created around the focal point (see figures 6.17(c) and (d)).

Skyrmions are the promising candidate to design the next generation memory device. One important reason is that they can be manipulated by spin-polarized electrical current with an extremely low density (about 10^6 A m^{-2}) [193], in contrast to the very high current density for driving domain wall (about 10^{11} A m^{-2}) in conventional ferromagnetic systems. Jonietz *et al* [193] showed that when both the current (with density $j = 2.22 \times 10^6$ A m^{-2}) and temperature gradient are applied to the skyrmion lattice, the skyrmion lattice rotates counterclockwise. If the current direction reverses, the direction of motion also reverses. Here the temperature gradient induces inhomogeneous Magnus and drag forces which lead to a rotational torque. By utilizing Lorentz transmission electron microscopy (TEM), Yu *et al* [194] further showed that, for a current lower than a critical value, the skyrmion lattice exhibits random rotations. By increasing the current above a critical value, the continuous translational motion of the skyrmion lattice appears. These results are summarized in figure 6.18. In addition, a skyrmion can also be driven by other

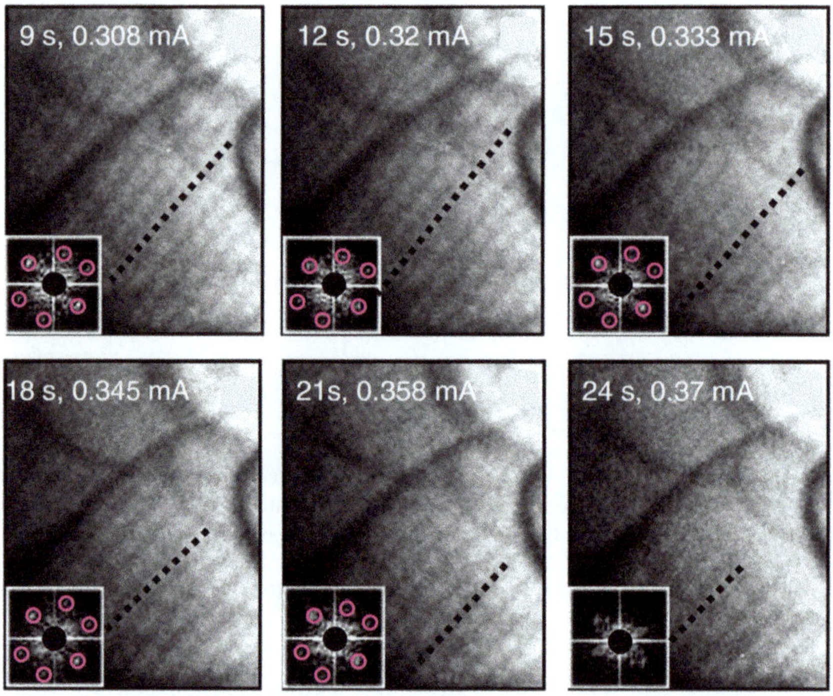

Figure 6.18. Lorentz transmission electron microscopy images of variations of the skyrmion crystal in a slowly increasing current. (Reproduced with permission from [194]. Copyright 2012 Nature.)

external forces, for example, spin-waves [195, 196], microwave fields [197, 198], temperature gradients [199–201], etc. Interestingly, very recently, it has been proved that twisted photons [202] and magnons [203] carrying orbital angular momentum (OAM) can act as 'optical tweezers' and 'magnetic tweezers' to induce the rotational motion of a skyrmion, the corresponding schematic diagrams are shown in figures 6.19(a) and (b), respectively. Theoretical calculations and numerical simulations consistently indicate that the radial and azimuthal quantum numbers of twisted photons (or magnons) can dominate both the magnitude and the direction of the rotational velocity of skyrmions.

Although skyrmions are believed to be ideal information carriers in spintronic devices, they cannot move in a straight line parallel to the direction of the driving current due to the Magnus force. The Magnus force can induce a shift of the skyrmion trajectory. Such a phenomenon is called a skyrmion Hall effect (SkHE) [204]. With the help of a polar magneto-optical Kerr effect (MOKE) microscope [205] and time-resolved x-ray microscopy [206], the SkHE has been observed in experiments. Figure 6.20 shows the experimental observations. When a current with density $j = +2.8 \times 10^6$ A cm^{-2} and a magnetic field $H_\perp = +5.4$ Oe are applied to a magnetic film, we can see that the skyrmion moves with a finite angle to the x-axis (see figures 6.20(a) and (b)), which is a typical feature of SkHE. In addition, if the topological charge of the skyrmion reverses, the transverse component of the

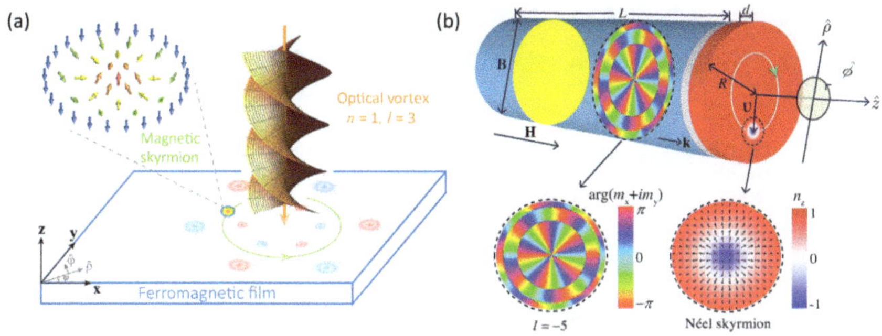

Figure 6.19. (a) A sketch map of the rotational motion of a Néel-type skyrmion in a thin ferromagnetic film driven by an optical vortex with radial quantum number $n = 1$ and OAM quantum number $l = 3$. (b) The schematic diagram of a heterostructured nanocylinder exchange-coupled to a chiral magnetic nanodisk hosting a Néel-type skyrmion. A spin-wave beam with the wavevector **k** and OAM quantum number $l = -5$ is excited, leading to a steady skyrmion gyrotropic motion around the disk center. ((a) Reproduced with permission from [202]. CC BY 4.0. (b) Reproduced with permission from [203]. Copyright 2020 The American Physical Society.)

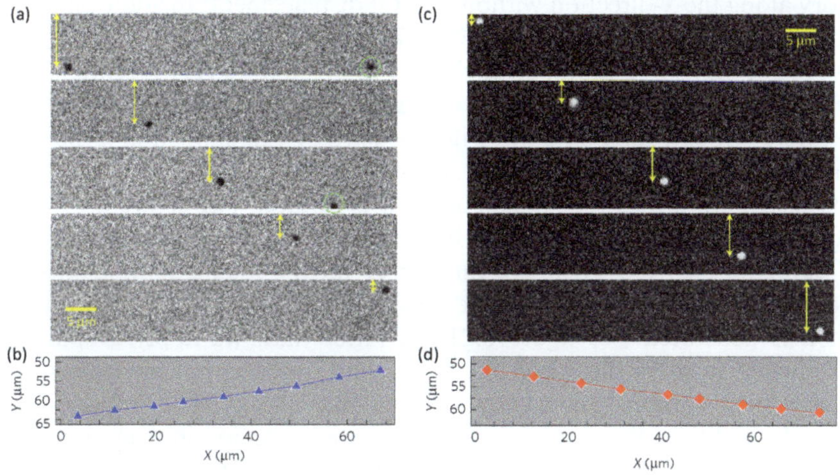

Figure 6.20. Snapshots of a skyrmion with topological charge (a) $Q = -1$ and (c) $Q = +1$ motion under the current and magnetic fields. Summary of the trajectory of the skyrmion with (b) $Q = -1$ and (d) $Q = +1$. (Reproduced with permission from [205]. Copyright 2017 Nature.)

skyrmion trajectory will be along the $-y$-direction because the reversed Magnus force acting on the skyrmion, as shown in figures 6.20(c) and (d). In real applications, skyrmions may be destroyed by the boundaries of the device because of the SkHE. To solve this problem, people have proposed various methods. For instance, Zhang *et al* [207] showed that the antiferromagnetically exchange-coupled bilayer containing two skyrmions with opposite polarities can suppress the SkHE. Figure 6.21(a) shows the result. When a current-perpendicular-to-plane (CPP) is

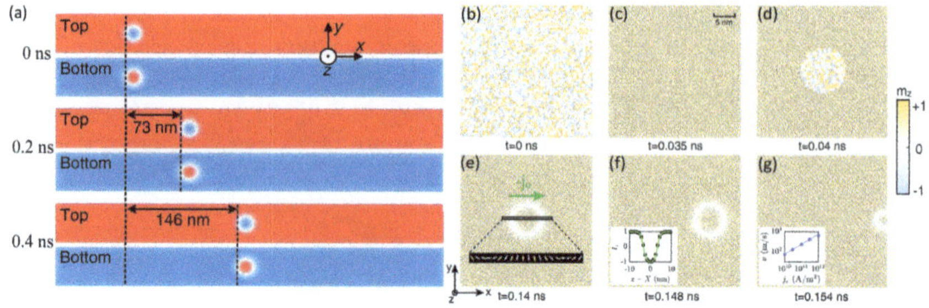

Figure 6.21. (a) Current-induced motion of skyrmions in the top and bottom ferromagnetic layers of an antiferromagnetic-coupled bilayer nanotrack. (b) Random spin configuration at $t = 0$ ns. (c) An AFM state evolved at $t = 0.035$ ns. (d) Randomizing spins inside the circle at $t = 0.04$ ns. (e) An AFM skyrmion stabilized at $t = 0.14$ ns. (f) Current-driven AFM skyrmion motion. (Inset) AFM skyrmion profile. (g) An AFM skyrmion annihilation at the film boundary. (Inset) Current dependence of the skyrmion velocity when it is far from the edge. ((a) Reproduced with permission from [207]. Copyright 2016 the Authors. (b)–(g) Reproduced with permission from [210]. Copyright 2018 The American Physical Society.)

applied, the two skyrmions with different polarities begin to move. Because the Magnus force acting on two skyrmions just offsets, it leads a perfectly straight trajectory along the x-direction without lateral displacement. In addition, it has been proved that the antiferromagnetic (AFM) skyrmion can also avoid SkHE [208, 209] because of the vanishing topological charge. Interestingly, very recently, Yang *et al* [210] showed that the AFM skyrmion can appear in a ferromagnetic film with gain, i.e. negative Gilbert damping α. Figures 6.21(b)–(g) show the process of the generation of an AFM skyrmion in ferromagnetic film, where α is set to -0.01 and the distribution of initial magnetization is random (see figure 6.21(b)). When $t = 0.035$ ns, the local magnetization has evolved to a stable antiparallelly aligned state (see figure 6.21(c)). Then, we randomize all the magnetic moments within a circle of radius 5 nm at the film center (see figure 6.21(d)). At $t = 0.14$ ns, an AFM skyrmion is generated (see figure 6.21(e)). After that, we apply a spin-polarized electric current to drive the AFM skyrmion, and observe that the trajectory of the AFM skyrmion exhibits an exactly straight line along the flowing direction of electrons. In addition, the current-induced skyrmion motion in magnetic nanotubes [211] can also make the skyrmion keep away from the boundary, because the tube is edgeless for the tangential skyrmion motion. In this case, even the current density is very large, the stable propagation of skyrmion can still survive without annihilation or accumulation. Here the nanotube can be viewed as a seamless, hollow tubular structure rolled from a planar nanostrip.

On the other side, the magnetic skyrmions can be used to realize logical operation with a great application potential [212–215]. Figure 6.22 shows a typical logic OR gate and AND gate based on skyrmions [213]. Here the skyrmion and ferromagnetic states represent bit ‘1’ and ‘0’, respectively. The logical OR operation can be realized through the following processes (see figure 6.22(a)). The left panel shows the logical operation of the OR gate ‘1 + 0 = 1’: there is a skyrmion in the input A and no skyrmion in the input B at the initial time, which represents input ‘1 + 0’; a current is

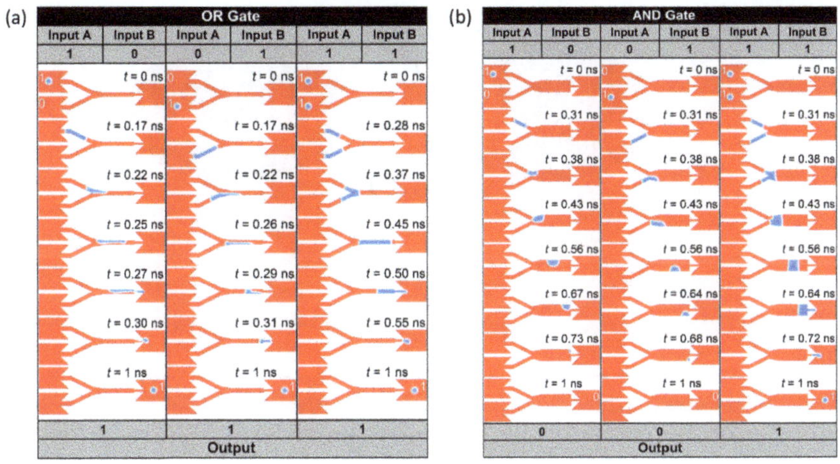

Figure 6.22. Illustration of the skyrmion based logical (a) OR and (b) AND operation. (Reproduced with permission from [213]. Copyright 2015 the Authors.)

applied along the $-x$-direction for $0 < t < 0.39$ ns then we relax the whole system; at $t = 1$ ns, a stable skyrmion is in the output side, which represents output '1'. The middle panel shows the logical operation of the OR gate '$0 + 1 = 1$', and the whole analysis process is similar to that of the left panel. The right panel shows the logical operation of the OR gate '$1 + 1 = 1$': there is a skyrmion in both the input A and the input B sides, which represents input '$1 + 1$'; a current is applied along the $-x$-direction for $0 < t < 0.64$ ns followed by a relaxation; at $t = 1$ ns, a stable skyrmion is in the output side, which represents output '1'. We can perform a similar analysis for AND gate operation; see figure 6.22(b).

Moreover, by using the gyrotropic mode of the skyrmion, one can design spin-torque nano-oscillators [216–218]. A typical model is shown in figure 6.23(a), in which the magnetic free layer is an ultrathin ferromagnet with a perpendicular magnetic anisotropy. So the skyrmion can be stabilized if a proper DMI is considered. When a current flows perpendicular to the reference layer electrons are polarized along the configuration of the reference layer, then the skyrmion in the free layer will be driven by the spin-polarized electrons through the spin transfer torque effect. The current dependence of the skyrmion gyration frequency for different external fields B_z is shown in figure 6.23(b). The role of applying a magnetic field is to change the size of the skyrmion. One can clearly see that the oscillation frequency of skyrmion can be tuned flexibly by the current density and external magnetic field (see figure 6.23(b)).

Skyrmions can also be used to design racetrack memory [219–221], neuromorphic devices [222–224], transistors [225–227], etc.

Notably, there also exist many other non-trivial spin textures, as shown in figure 6.24. The first three objects are different types of skyrmions which carry different helicities and vorticities: the antiskyrmion [228–230] with a vorticity $m = -1$ and topological charge $Q = -1$; the intermediate skyrmion [231] with

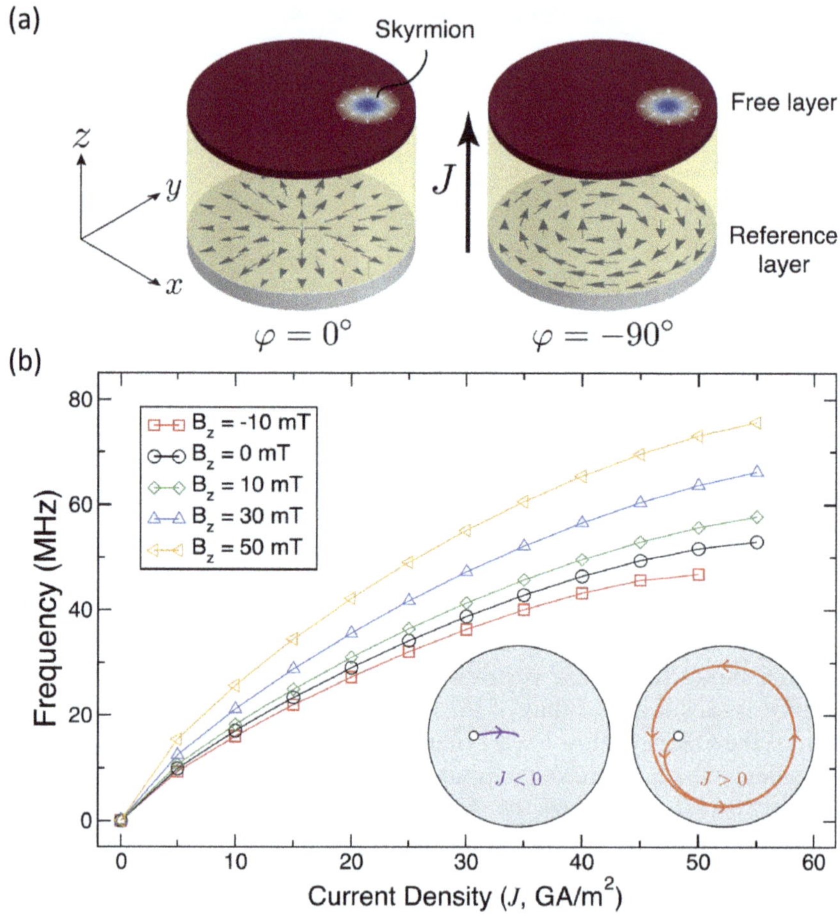

Figure 6.23. (a) Geometry and operating principle of a skyrmion based spin-torque nano-oscillator. (b) Current density dependence of the skyrmion gyration frequency for different perpendicular fields B_z. The inset shows the transient trajectories of the skyrmion core for opposite current polarities. (Reproduced with permission from [217]. Copyright 2016 IOP Publishing.)

helicity $\gamma = \pi/4$ between Bloch- and Néel-type skyrmions and $Q = +1$; and the higher-order skyrmion [232, 233] with $Q = +2$. Figure 6.24(d) shows a magnetic bimeron [234, 235] consisting of two merons which can be understood as a skyrmionic excitation in an in-plane magnetized medium, and its topological charge $Q = -1$. The middle row shows combinations of two skyrmions: the biskyrmion [236, 237] with $Q = +2$; the skyrmionium [238, 239] with $Q = 0$; the ferrimagnetic [240, 241] and synthetic antiferromagnetic [242, 243] skyrmions for which the topological charges of the two subskyrmions compensate each other. The bottom row shows the three-dimensional extension of skyrmions: the skyrmion tube [244, 245]; the chiral bobber [246, 247] as a discontinued skyrmion tube; a pair of Bloch and anti-Bloch points [248, 249] constituting the building block of a three-dimensional hedgehog lattice; and the magnetic hopfion [250–253].

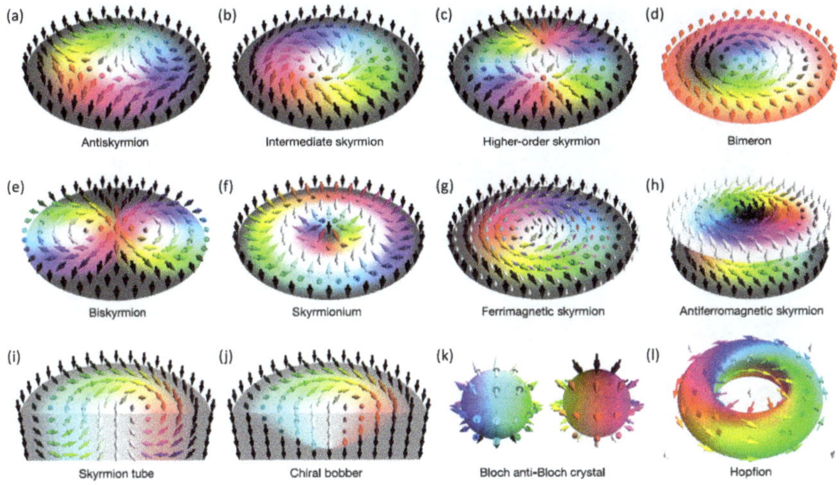

Figure 6.24. (a)–(l) Illustration of various non-trivial spin textures. The colored arrows denote the orientation of the magnetic moments.(Reproduced with permission from [2]. Copyright 2021 Elsevier.)

6.2 Collective dynamics of magnetic soliton crystals

As introduced in chapter 1, the collective dynamics of a magnetic soliton array manifests the feature of waves. In this section we discuss the collective oscillation of magnetic soliton crystals.

For the 1D case, figure 6.25(a) plots a scanning electron microscopy (SEM) image of the sample with five identical Py disks with even distance [254]. The out-of-plane and in-plane components of nanodisk magnetization can be measured by using scanning transmission x-ray microscopy (STXM), as shown in figures 6.25(b) and (c), respectively. One can identify the polarity and chirality of the vortex lattice as $P = [+1, -1, +1, -1, +1]$ and $C = [-1, -1, -1, -1, +1]$, as shown in figure 6.25(d). A current pulse with duration time 1.8 ns is applied into the electrode stripline to excite the gyrotropic motion of the first vortex. Because of the existence of dipolar interaction between vortices, the oscillation of the first vortex can spread to other vortices. The Fourier spectra of the position of vortex cores show that the system can support five different wave modes. The trajectories of vortex cores for different frequencies are plotted in figure 6.25(e). Here the red dots denote the positions of vortex cores. Below each mode, the physical quantity $C_n Y_n$ is plotted with the solid line (for $t = 0$) or dotted line ($t = T/2$). Here $T = 2\pi/\omega$ is the period of the gyrotropic mode, Y_n represents the displacement of the vortex core along the y-direction for the nth disk, and C_n denotes the chirality of the nth vortex. We observe that the collective gyrotropic motion of the vortex lattice exhibits various standing waves.

Moreover, if we consider the 1D array which contains more magnetic vortices, the feature of waves will be more feasible. Han *et al* [255] studied the 1D lattice composed of 25 disks based on pure NiMnSb or alternating NiMnSb and Py. Here the disks are in vortex states and have identical geometric parameters. For the case

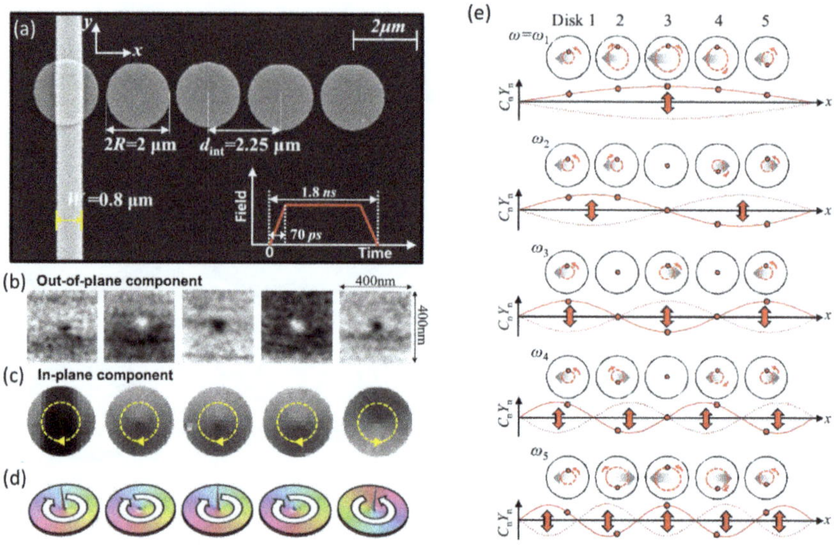

Figure 6.25. (a) An SEM image of a Py disk array with a vortex state. A stripline is placed to the left-end disk where a local magnetic field pulse is applied to excite the oscillation of the vortex. The inset plots a sketch map of the field pulse adopted in the experiment. (b) Out-of-plane and (c) in-plane magnetization components obtained from STXM measurements. (d) Schematic illustration of the initial states of the sample. (e) The position distributions of vortex cores for different modes. The red dotted lines in the disks represent the vortex core trajectories. (Reproduced with permission from [254]. Copyright 2013 the Authors.)

of the single-material array, two wavefronts are observed whereas for the alternative-material array, a weak and dispersive single wavefront appears. They further showed that the spectra of collective excitations of the vortex lattice exhibit the typical band structures of waves. In addition, they observed that the different $[p_n, C_n]$ orderings have the influence on the band structure.

Similarly, the 1D skyrmion lattice can also show the wave behavior [256–258]. Figure 6.26(a) plots a sketch map of a 1D skyrmion array in a nanostrip containing 25 skyrmions. The dispersion relations for different d_{int} and perpendicular magnetic field H_z are shown in figures 6.26(b) and (c), respectively. Here d_{int} denotes the distance between two adjacent skyrmions. One can clearly see that the interdistance and perpendicular magnetic field can modify the band structure. On the one hand, the bandwidth $\Delta\omega$ and the frequency ω_{BZ} at $k = k_{\text{BZ}}$ decrease with d_{int} increases. This result can be explained that with increasing d_{int}, the total energy density decreases, leading to the decrease of $\Delta\omega$ and ω_{BZ}. On the other hand, the bandwidth decreases with the increase of H_z, which can be understood as the application of the H_z field modifying the magnetization profiles of each skyrmion's cores, so that skyrmion shrinks with the positive external magnetic fields and expands with the negative external magnetic fields. When the skyrmion size changes, the corresponding eigenfrequencies ω_0 of the skyrmion gyration change. In addition, the 1D skyrmion lattice can also be composed of nanodisks, as plotted in figure 6.26(d). The

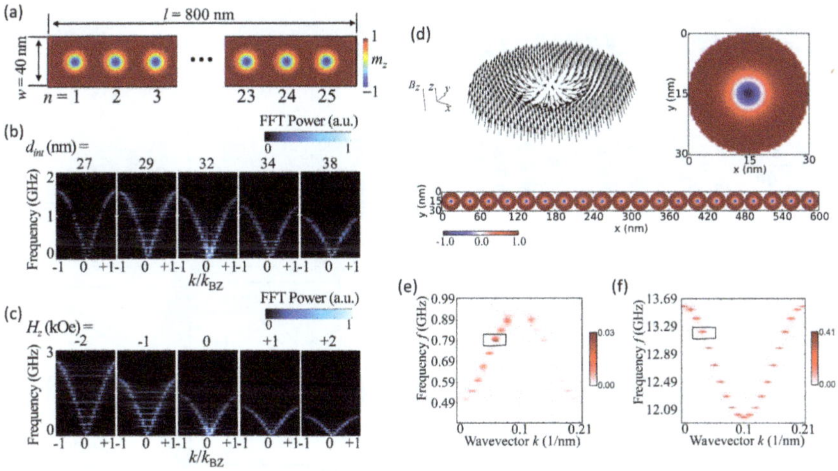

Figure 6.26. (a) A 1D skyrmion lattice in a nanostrip containing 25 skyrmions. The band structures of a 1D skyrmion array with different (b) interdistances and (c) perpendicular magnetic fields. (d) Schematic diagrams of a single skyrmion disk with a diameter of 30 nm and the array of 20 disks. Dispersion relations of the (e) clockwise gyrotropic and (f) breathing mode excitation in a 1D skyrmion lattice with an out-of-plane magnetic field applied locally to the central disk. ((a)-(c) Reproduced with permission from [256]. Copyright 2017 the Authors. (d)-(f) Reproduced with permission from [257]. Copyright 2016 The American Physical Society.)

collective excitation of gyrotropic and breathing modes in a skyrmion array shows obvious dispersion curves, as shown in figures 6.26(e) and (f), respectively.

The collective modes of 2D magnetic soliton crystals have also been studied [259–264]. Here we choose the 2D vortex array [263] as an example. The diagrammatic sketch and scanning electron micrograph of the vortex lattice are plotted in figures 6.27(a) and (b), respectively. One can see that there exist three striplines of copper on top of the nanodisks (figure 6.27(b)). Here all three striplines are used to tune the polarization pattern of the vortices by self-organized state formation [265, 266], while the two thin striplines are used for band structure measurements. The harmonic excitations $\mathbf{H}_1 = [H_0 \sin(\omega t), 0, 0]$ and $\mathbf{H}_2 = [H_0 \sin(\omega t + \Delta\varphi), 0, 0]$ are applied to the first and second column of the vortices with different phase relations $\Delta\varphi$ which can be used to create waves with different wavelengths. Figure 6.27(c) plots the absorption spectra varying with the phase difference for homogeneous (the polarities of the vortices are identical) and horizontal (the polarities of vortices alternate along the y-axis) vortex polarization patterns. The red and blue lines denote the maxima of the absorption calculated by Lorentzian fits. One can see that the resonance frequencies increase with the increase of the phase difference. The dispersion relations with different vortex polarization patterns [267] are plotted in figure 6.27(d). Here the markers denote the measured absorption, while the solid and dashed lines represent the calculations using the extended Thiele's model and an infinite lattice without damping, respectively. These results strongly indicate that the dispersion relation of a 2D vortex array can be tuned by the vortex polarization pattern.

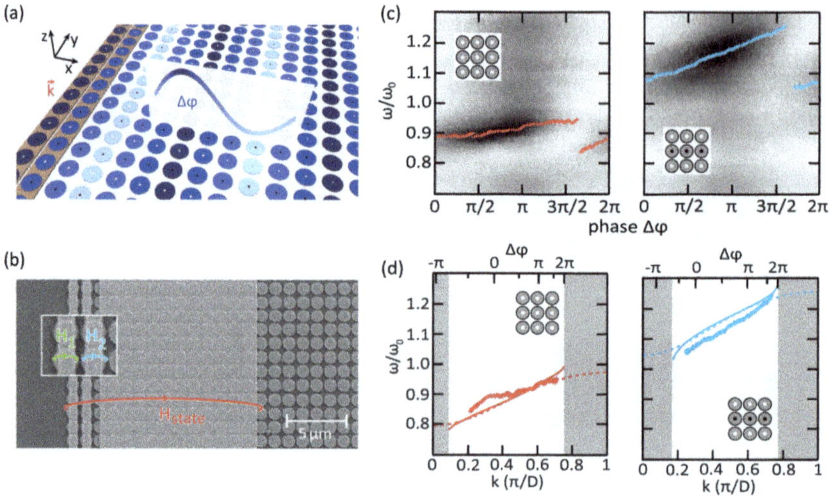

Figure 6.27. (a) Schematic diagram of gyration waves in a two-dimensional vortex lattice. (b) A scanning electron micrograph of a vortex lattice where the shaded areas denote copper striplines. (c) The experimentally measured absorption intensity for different combinations of frequency and phase difference with homogeneous (left image) and horizontal (right image) vortex polarization patterns. (d) The corresponding dispersion relation measurements (markers) and calculations using the extended Thiele's equation without damping (dashed lines) and with damping (solid lines). The insets in (c) and (d) show illustrations of a vortex lattice, the white and black dots denote different vortex polarities. (Reproduced with permission from [263]. Copyright 2015 The American Physical Society.)

Very recently, Hänze *et al* [268] studied the collective dynamics of 3D vortex lattices by utilizing ferromagnetic resonance spectroscopy and scanning transmission x-ray microscopy. They showed that the collective vortex spectrum is determined by both vertical and horizontal couplings which are closely related to the polarity and chirality of the vortices. Specifically, for the lateral arrangements, the in-plane dipoles strongly influence the interaction between vortices and lead to the polarity-dependent resonance frequencies. In addition, for the vertical coupling, because the chirality can affect the out-of-plane dipoles of the vortex core region, the resonance frequencies of the system thus additionally depend on the chirality.

References

[1] Nagaosa N and Tokura Y 2013 *Nat. Nanotechnol.* **8** 899–911

[2] Göbel B, Mertig I and Tretiakov O A 2021 *Phys. Rep.* **895** 1–28

[3] Wang X G *et al* 2014 *Phys. Rev.* B **89** 144418

[4] Kronmüller H 2007 General micromagnetic theory *Handbook of Magnetism and Advanced Magnetic Materials* ed H Kronmüller and S Parkin (Chichester: Wiley) pp 1–30 https://doi.org/10.1002/9780470022184.hmm201

[5] McMichael R D and Donahue M J 1997 *IEEE Trans. Magn.* **33** 4167–9

[6] Catalan G *et al* 2012 *Rev. Mod. Phys.* **84** 119

[7] He J, Li Z and Zhang S 2006 *Phys. Rev.* B **73** 184408

[8] Clarke D J *et al* 2008 *Phys. Rev.* B **78** 134412

[9] Mascaro M D and Ross C A 2010 *Phys. Rev.* B **82** 214411
[10] Muratov C B and Osipov V V 2008 *J. Appl. Phys.* **104** 053908
[11] Li Z X *et al* 2015 *J. Magn. Magn. Mater.* **388** 10–5
[12] Miron I M *et al* 2011 *Nat. Mater.* **10** 419–23
[13] Haazen P P J *et al* 2013 *Nat. Mater.* **12** 299–303
[14] Dzyaloshinsky I 1958 *J. Phys. Chem. Solids* **4** 241–55
[15] Moriya T 1960 *Phys. Rev. Lett.* **4** 228
[16] Heide M, Bihlmayer G and Blügel S 2008 *Phys. Rev.* B **78** 140403(R)
[17] Thiaville A *et al* 2012 *Europhys. Lett.* **100** 57002
[18] Emori S *et al* 2013 *Nat. Mater.* **12** 611–6
[19] Chen G *et al* 2013 *Phys. Rev. Lett.* **110** 177204
[20] Wieser R, Nowak U and Usadel K D 2004 *Phys. Rev.* B **69** 064401
[21] Wieser R *et al* 2010 *Phys. Rev.* B **82** 144430
[22] Col S D *et al* 2014 *Phys. Rev.* B **89** 180405(R)
[23] Schöbitz M *et al* 2019 *Phys. Rev. Lett.* **123** 217201
[24] Roy P E *et al* 2009 *Phys. Rev.* B **79** 060407(R)
[25] Jr E E H, Smith D O and Goodenough J B 1958 *J. Appl. Phys.* **29** 294–5
[26] Metlov K L 2001 *Appl. Phys. Lett.* **79** 2609–11
[27] Beach G S D *et al* 2005 *Nat. Mater.* **4** 741–4
[28] Mougin A *et al* 2007 *Europhys. Lett.* **78** 57007
[29] Wang X R *et al* 2009 *Ann. Phys.* **324** 1815–20
[30] Goussev A, Robbins J M and Slastikov V 2010 *Phys. Rev. Lett.* **104** 147202
[31] Hayashi M *et al* 2008 *Science* **320** 209–11
[32] Vlaminck V and Bailleul M 2008 *Science* **322** 410–3
[33] Ryu K S *et al* 2013 *Nat. Nanotechnol.* **8** 527–33
[34] Yang S H, Ryu K S and Parkin S 2015 *Nat. Nanotechnol.* **10** 221–6
[35] Hinzke D and Nowak U 2011 *Phys. Rev. Lett.* **107** 027205
[36] Yan P, Wang X S and Wang X R 2011 *Phys. Rev. Lett.* **107** 177207
[37] Wang X G *et al* 2012 *Phys. Rev.* B **86** 054445
[38] Han J *et al* 2019 *Science* **366** 1121–5
[39] Schryer N L and Walker L R 1974 *J. Appl. Phys.* **45** 5406
[40] Kläui M *et al* 2005 *Phys. Rev. Lett.* **95** 026601
[41] Hayashi M *et al* 2007 *Nat. Phys.* **3** 21–5
[42] Thiaville A *et al* 2005 *Europhys. Lett.* **69** 990–6
[43] Yan M *et al* 2010 *Phys. Rev. Lett.* **104** 057201
[44] Yan M *et al* 2011 *Appl. Phys. Lett.* **99** 122505
[45] Lewis E R *et al* 2010 *Nat. Mater.* **9** 980–3
[46] Lee J Y, Lee K S and Kim S K 2007 *Appl. Phys. Lett.* **91** 122513
[47] Richter K *et al* 2010 *Appl. Phys. Lett.* **96** 182507
[48] Fan L *et al* 2016 *J. Magn. Magn. Mater.* **401** 484–7
[49] Nakatani Y, Thiaville A and Miltat J 2003 *Nat. Mater.* **2** 521–3
[50] Xu M *et al* 2021 *J. Magn. Magn. Mater.* **532** 167988
[51] Parkin S S P, Hayashi M and Thomas L 2008 *Science* **320** 190–4
[52] Parkin S and Yang S H 2015 *Nat. Nanotechnol.* **10** 195–8
[53] Hayashi M *et al* 2007 *Phys. Rev. Lett.* **98** 037204
[54] Chiba D *et al* 2010 *Appl. Phys. Express* **3** 073004

[55] Thomas L *et al* 2011 *Int. Electron Devices Meeting (IEDM)* pp 24.2.1–4
[56] Allwood D A *et al* 2005 *Science* **309** 1688–92
[57] Xu P *et al* 2008 *Nat. Nanotechnol.* **3** 97–100
[58] Allwood D A, Xiong G and Cowburn R P 2006 *J. Appl. Phys.* **100** 123908
[59] Omari K A and Hayward T J 2014 *Phys. Rev. Appl.* **2** 044001
[60] Currivan J A *et al* 2012 *IEEE Magn. Lett.* **3** 3000104
[61] Roohi A, Zand R and DeMara R F 2016 *IEEE Trans. Magn.* **52** 1–7
[62] Omari K A *et al* 2019 *Adv. Funct. Mater.* **29** 1807282
[63] Luo Z *et al* 2020 *Nature* **579** 214–8
[64] Hayashi M *et al* 2006 *Phys. Rev. Lett.* **97** 207205
[65] Lepadatu S *et al* 2009 *Phys. Rev. Lett.* **102** 127203
[66] Lewis E R *et al* 2011 *Appl. Phys. Lett.* **98** 042502
[67] Franchin M *et al* 2011 *Phys. Rev.* B **84** 094409
[68] Glathe S and Mattheis R 2012 *Phys. Rev.* B **85** 024405
[69] He J and Zhang S 2007 *Appl. Phys. Lett.* **90** 142508
[70] Franchin M *et al* 2008 *Phys. Rev.* B **78** 054447
[71] Ono T and Nakatani Y 2008 *Appl. Phys. Express* **1** 061301
[72] Bisig A *et al* 2009 *Appl. Phys. Lett.* **95** 162504
[73] Martinez E, Torres L and Lopez-Diaz L 2011 *Phys. Rev.* B **83** 174444
[74] Franken J H *et al* 2011 *Appl. Phys. Lett.* **98** 102512
[75] Sharma S, Muralidharan B and Tulapurkar A 2015 *Sci. Rep.* **5** 14647
[76] Voto M, Lopez-Diaz L and Martinez E 2017 *Sci. Rep.* **7** 13559
[77] Tatara G and Kohno H 2004 *Phys. Rev. Lett.* **92** 086601
[78] Berger L 1986 *Phys. Rev.* B **33** 1572
[79] Barnes S E and Maekawa S 2007 *Phys. Rev. Lett.* **98** 246601
[80] Yang S A *et al* 2009 *Phys. Rev. Lett.* **102** 067201
[81] Whyte J R and Gregg J M 2015 *Nat. Commun.* **6** 7361
[82] McConville J P V *et al* 2020 *Adv. Fun. Mater.* **30** 2000109
[83] Shinjo T *et al* 2000 *Science* **289** 930–2
[84] Miltat J and Thiaville A 2002 *Science* **298** 555
[85] Wachowiak A *et al* 2002 *Science* **298** 577–80
[86] Kim S K and Tserkovnyak Y 2017 *Phys. Rev. Lett.* **119** 077204
[87] Guslienko K Y *et al* 2002 *J. Appl. Phys.* **91** 8037
[88] Park J P *et al* 2003 *Phys. Rev.* B **67** 020403(R)
[89] Guslienko K Y *et al* 2006 *Phys. Rev. Lett.* **96** 067205
[90] Park J P and Crowell P A 2005 *Phys. Rev. Lett.* **95** 167201
[91] Zaspel C E *et al* 2005 *Phys. Rev.* B **72** 024427
[92] Ivanov B A and Zaspel C E 2005 *Phys. Rev. Lett.* **94** 027205
[93] Guslienko K Y *et al* 2008 *Phys. Rev. Lett.* **101** 247203
[94] Vogt K *et al* 2011 *Phys. Rev.* B **84** 174401
[95] Buess M *et al* 2004 *Phys. Rev. Lett.* **93** 077207
[96] Thiele A A 1973 *Phys. Rev. Lett.* **30** 230
[97] Buess M *et al* 2005 *Phys. Rev.* B **71** 104415
[98] Schultheiss K *et al* 2019 *Phys. Rev. Lett.* **122** 097202
[99] Bussmann K *et al* 1999 *Appl. Phys. Lett.* **75** 2476
[100] Choe S B *et al* 2004 *Science* **304** 420–2

[101] Höllinger R, Killinger A and Krey U 2003 *J. Magn. Magn. Mater.* **261** 178–89
[102] Thiaville A *et al* 2003 *Phys. Rev.* B **67** 094410
[103] Waeyenberge B V *et al* 2006 *Nature* **444** 461–4
[104] Lee K S *et al* 2007 *Phys. Rev.* B **76** 174410
[105] Hertel R *et al* 2007 *Phys. Rev. Lett.* **98** 117201
[106] Xiao Q F *et al* 2006 *Appl. Phys. Lett.* **89** 262507
[107] Kravchuk V P *et al* 2007 *J. Appl. Phys.* **102** 043908
[108] Curcic M *et al* 2008 *Phys. Rev. Lett.* **101** 197204
[109] Yoo M W, Lee J and Kim S K 2012 *Appl. Phys. Lett.* **100** 172413
[110] Bauer H G *et al* 2014 *Phys. Rev. Lett.* **112** 077201
[111] Yoo M W and Kim S K 2015 *J. Appl. Phys.* **117** 023904
[112] Yamada K *et al* 2007 *Nat. Mater.* **6** 270–3
[113] Kim S K *et al* 2007 *Appl. Phys. Lett.* **91** 082506
[114] Lee K S *et al* 2008 *Phys. Rev. Lett.* **101** 267206
[115] Khvalkovskiy A V *et al* 2010 *Appl. Phys. Lett.* **96** 022504
[116] Yoo M W *et al* 2010 *Phys. Rev.* B **82** 174437
[117] Uhlíř V *et al* 2013 *Nat. Nanotechnol.* **8** 341–6
[118] Zhang Y *et al* 2020 *Sci. Bull.* **65** 1260–7
[119] Haldar A and Adeyeye A O 2015 *Appl. Phys. Lett.* **106** 032404
[120] Pribiag V S *et al* 2007 *Nat. Phys.* **3** 498–503
[121] Manfrini M *et al* 2009 *Appl. Phys. Lett.* **95** 192507
[122] Guslienko K Y, Aranda G R and Gonzalez J 2011 *J. Phys.: Conf. Ser.* **292** 012006
[123] Sanches F *et al* 2014 *Phys. Rev.* B **89** 140410(R)
[124] Hrkac G *et al* 2015 *J. Phys. D: Appl. Phys.* **48** 453001
[125] Jung H *et al* 2012 *ACS Nano* **6** 3712–7
[126] Vigo-Cotrina H and Guimarães A P 2017 *J. Magn. Magn. Mater.* **441** 14–20
[127] Kumar D, Barman S and Barman A 2014 *Sci. Rep.* **4** 4108
[128] Mondal S, Barman S and Barman A 2020 *J. Magn. Magn. Mater.* **502** 166520
[129] Alfaro V D, Fubini S and Furlan G 1976 *Phys. Lett.* B **65** 163–6
[130] Phatak C, Petford-Long A K and Heinonen O 2012 *Phys. Rev. Lett.* **108** 067205
[131] Ruderman M A and Kittel C 1954 *Phys. Rev.* **96** 99
[132] Kasuya T 1956 *Prog. Theor. Phys.* **16** 45–57
[133] Yosida K 1957 *Phys. Rev.* **106** 893
[134] Yu X Z *et al* 2018 *Nature* **564** 95–8
[135] Bobeck A H and Scovil H E D 1971 *Sci. Am.* **224** 78–91 https://www.jstor.org/stable/24922754
[136] Leeuw F H D, Doel R V D and Enz U 1980 *Rep. Prog. Phys.* **43** 689
[137] Bobeck A H, Bonyhard P I and Geusic J E 1975 *Proc. IEEE* **63** 1176–95
[138] Eschenfelder A H 2012 *Magnetic Bubble Technology* vol 14 (Berlin: Springer)
[139] O'Dell T H 1981 *Ferromagnetodynamics: The Dynamics of Magnetic Bubbles, Domains, and Domain Walls* (New York: Wiley)
[140] Fukumura T *et al* 1999 *Science* **284** 1969–71
[141] Hehn M *et al* 1996 *Science* **272** 1782–5
[142] Moutafis C, Komineas S and Bland J A C 2009 *Phys. Rev.* B **79** 224429
[143] Papanicolaou N and Tomaras T N 1991 *Nucl. Phys.* B **360** 425–62
[144] Makhfudz I, Krüger B and Tchernyshyov O 2012 *Phys. Rev. Lett.* **109** 217201
[145] Moon K W *et al* 2014 *Phys. Rev.* B **89** 064413

[146] Malozemoff A P and Slonczewski J C 2016 *Magnetic Domain Walls in Bubble Materials: Advances in Materials and Device Research* vol 1 (New York: Academic)
[147] Skyrme T H R 1961 *Proc. R. Soc. Lond.* A **260** 127–38
[148] Skyrme T H R 1962 *Nucl. Phys.* **31** 556–69
[149] Tsesses S *et al* 2018 *Science* **361** 993–6
[150] Du L *et al* 2019 *Nat. Phys.* **15** 650–4
[151] Shen Y, Martínez E C and Rosales-Guzmán C 2022 *ACS Photon.* **9** 296–303
[152] Ge H *et al* 2021 *Phys. Rev. Lett.* **127** 144502
[153] Deng Z L *et al* 2022 *Nat. Commun.* **13** 8
[154] Yang J *et al* 2022 *Laser Photon. Rev.* **16** 2200007
[155] Fukuda J I and Žumer S 2011 *Nat. Commun.* **2** 246
[156] Leonov A O *et al* 2014 *Phys. Rev.* E **90** 042502
[157] Vilenkin A and Shellard E P S 1994 *Cosmic Strings and Other Topological Defects* (Cambridge: Cambridge University Press)
[158] Khawaja U A and Stoof H 2001 *Nature* **411** 918–20
[159] Battye R A, Cooper N R and Sutcliffe P M 2002 *Phys. Rev. Lett.* **88** 080401
[160] Leslie L S *et al* 2009 *Phys. Rev. Lett.* **103** 250401
[161] Mühlbauer S *et al* 2009 *Science* **323** 915–9
[162] Yu X Z *et al* 2010 *Nature* **465** 901–4
[163] Finocchio G *et al* 2016 *J. Phys. D: Appl. Phys.* **49** 423001
[164] Ezawa M 2010 *Phys. Rev. Lett.* **105** 197202
[165] Dai Y Y *et al* 2013 *Phys. Rev.* B **88** 054403
[166] Okubo T, Chung S and Kawamura H 2012 *Phys. Rev. Lett.* **108** 017206
[167] Hayami S, Lin S Z and Batista C D 2016 *Phys. Rev.* B **93** 184413
[168] Heinze S *et al* 2011 *Nat. Phys.* **7** 713–8
[169] Paul S *et al* 2020 *Nat. Commun.* **11** 4756
[170] Okamura Y *et al* 2013 *Nat. Commun.* **4** 2391
[171] Kang W *et al* 2016 *Proc. IEEE* **104** 2040–61
[172] Wiesendanger R 2016 *Nat. Rev. Mater.* **1** 16044
[173] Garst M, Waizner J and Grundler D 2017 *J. Phys. D: Appl. Phys.* **50** 293002
[174] Fert A, Reyren N and Cros V 2017 *Nat. Rev. Mater.* **2** 17031
[175] Jiang W *et al* 2017 *Phys. Rep.* **704** 1–49
[176] Everschor-Sitte K *et al* 2018 *J. Appl. Phys.* **124** 240901
[177] Zhou Y 2019 *Natl. Sci. Rev.* **6** 210–2
[178] Zhang X *et al* 2020 *J. Phys.: Condens. Matter* **32** 143001
[179] Romming N *et al* 2013 *Science* **341** 636–9
[180] Sampaio J *et al* 2013 *Nat. Nanotechnol.* **8** 839–44
[181] Woo S *et al* 2016 *Nat. Mater.* **15** 501–6
[182] Jiang W J *et al* 2015 *Science* **349** 283–6
[183] Koshibae W and Nagaosa N 2014 *Nat. Commun.* **5** 5148
[184] Heinonen O *et al* 2016 *Phys. Rev.* B **93** 094407
[185] Yuan H Y and Wang X R 2016 *Sci. Rep.* **6** 22638
[186] Han J H *et al* 2010 *Phys. Rev.* B **82** 094429
[187] Li Z X *et al* 2017 *J. Magn. Magn. Mater.* **433** 216–21
[188] Liu Y *et al* 2015 *Appl. Phys. Lett.* **107** 152411
[189] Toedt J N *et al* 2016 *Sci. Rep.* **6** 33169

[190] Bao W W *et al* 2020 *Phys. Rev.* B **102** 014423
[191] Wang Z Y *et al* 2020 *Appl. Phys. Lett.* **117** 222406
[192] Yao X L *et al* 2021 *Front. Phys.* **9** 729967
[193] Jonietz F *et al* 2010 *Science* **330** 1648–51
[194] Yu X Z *et al* 2012 *Nat. Commun.* **3** 988
[195] Zhang X *et al* 2015 *Nanotechnology* **26** 225701
[196] Psaroudaki C and Loss D *Phys. Rev. Lett.* **120** 237203
[197] Wang W *et al* 2015 *Phys. Rev.* B **92** 020403(R)
[198] Zhang B *et al* 2015 *Appl. Phys. Lett.* **106** 102401
[199] Everschor K *et al* 2012 *Phys. Rev.* B **86** 054432
[200] Kong L and Zang J 2013 *Phys. Rev. Lett.* **111** 067203
[201] Lin S Z *et al* 2014 *Phys. Rev. Lett.* **112** 187203
[202] Yang W R *et al* 2018 *Opt. Express* **26** 8778–90
[203] Jiang Y Y *et al* 2020 *Phys. Rev. Lett.* **124** 217204
[204] Chen G 2017 *Nat. Phys.* **13** 112–3
[205] Jiang W *et al* 2017 *Nat. Phys.* **13** 162–9
[206] Litzius K *et al* 2017 *Nat. Phys.* **13** 170–5
[207] Zhang X, Zhou Y and Ezawa M 2016 *Nat. Commun.* **7** 10293
[208] Zhang X, Zhou Y and Ezawa M 2016 *Sci. Rep.* **6** 24795
[209] Barker J and Tretiakov O A 2016 *Phys. Rev. Lett.* **116** 147203
[210] Yang H H *et al* 2018 *Phys. Rev. Lett.* **121** 197201
[211] Wang X F *et al* 2019 *J. Phys. D: Appl. Phys.* **52** 225001
[212] Luo S *et al* 2018 *Nano Lett.* **18** 1180–4
[213] Zhang X, Ezawa M and Zhou Y 2015 *Sci. Rep.* **5** 9400
[214] Chauwin M *et al* 2019 *Phys. Rev. Appl.* **12** 064053
[215] Yang H H *et al* 2018 *Phys. Rev.* B **98** 014433
[216] Zhang S *et al* 2015 *New J. Phys.* **17** 023061
[217] Garcia-Sanchez F *et al* 2016 *New J. Phys.* **18** 075011
[218] Luo S *et al* 2017 *Appl. Phys. Lett.* **110** 112402
[219] Tomasello R *et al* 2014 *Sci. Rep.* **4** 6784
[220] Fert A, Cros V and Sampaio J 2013 *Nat. Nanotechnol.* **8** 152–6
[221] Zhang X *et al* 2015 *Sci. Rep.* **5** 7643
[222] Song K M *et al* 2020 *Nat. Electron.* **3** 148–55
[223] Li S *et al* 2017 *Nanotechnology* **28** 31LT01
[224] Huang Y *et al* 2017 *Nanotechnology* **28** 08LT02
[225] Zhang X *et al* 2015 *Sci. Rep.* **5** 11369
[226] Jung D H *et al* 2021 *Phys. Rev.* B **104** L060408
[227] Hong I S and Lee K J 2019 *Appl. Phys. Lett.* **115** 072406
[228] Nayak A K *et al* 2017 *Nature* **548** 561–6
[229] Koshibae W and Nagaosa N 2016 *Nat. Commun.* **7** 10542
[230] Hoffmann M *et al* 2017 *Nat. Commun.* **8** 308
[231] Garlow J A *et al* 2019 *Phys. Rev. Lett.* **122** 237201
[232] Ozawa R, Hayami S and Motome Y 2017 *Phys. Rev. Lett.* **118** 147205
[233] Rózsa L *et al* 2017 *Phys. Rev.* B **95** 094423
[234] Göbel B *et al* 2019 *Phys. Rev.* B **99** 060407(R)
[235] Shen L *et al* 2020 *Phys. Rev.* B **102** 104427

[236] Yu X Z *et al* 2014 *Nat. Commun.* **5** 3198
[237] Peng L *et al* 2017 *Nano Lett.* **17** 7075–9
[238] Zhang X *et al* 2016 *Phys. Rev.* B **94** 094420
[239] Zhang S *et al* 2018 *Nano Lett.* **18** 1057–63
[240] Hirata Y *et al* 2019 *Nat. Nanotechnol.* **14** 232–6
[241] Woo S *et al* 2018 *Nat. Commun.* **9** 959
[242] Legrand W *et al* 2020 *Nat. Mater.* **19** 34–42
[243] Dohi T *et al* 2019 *Nat. Commun.* **10** 5153
[244] Birch M T *et al* 2020 *Nat. Commun.* **11** 1726
[245] Wolf D *et al* 2022 *Nat. Nanotechnol.* **17** 250–5
[246] Zheng F *et al* 2018 *Nat. Nanotechnol.* **13** 451–5
[247] Ahmed A S *et al* 2018 *Phys. Rev. Mater.* **2** 041401(R)
[248] Fujishiro Y *et al* 2019 *Nat. Commun.* **10** 1059
[249] Okumura S *et al* 2020 *Phys. Rev.* B **101** 144416
[250] Wang X S, Qaiumzadeh A and Brataas A 2019 *Phys. Rev. Lett.* **123** 147203
[251] Liu Y *et al* 2020 *Phys. Rev. Lett.* **124** 127204
[252] Raftrey D and Fischer P 2021 *Phys. Rev. Lett.* **127** 257201
[253] Kent N *et al* 2021 *Nat. Commun.* **12** 1562
[254] Han D S *et al* 2013 *Sci. Rep.* **3** 2262
[255] Han D S, Jeong H B and Kim S K 2013 *Appl. Phys. Lett.* **103** 112406
[256] Kim J *et al* 2017 *Sci. Rep.* **7** 45185
[257] Mruczkiewicz M *et al* 2016 *Phys. Rev.* B **93** 174429
[258] Kim J *et al* 2018 *J. Appl. Phys.* **123** 053903
[259] Galkin A Y, Ivanov B A and Zaspel C E 2006 *Phys. Rev.* B **74** 144419
[260] Vogel A *et al* 2010 *Phys. Rev. Lett.* **105** 037201
[261] Sukhostavets O V, González J and Guslienko K Y 2013 *Phys. Rev.* B **87** 094402
[262] Vogel A *et al* 2014 *Phys. Rev.* B **89** 104403
[263] Behncke C *et al* 2015 *Phys. Rev.* B **91** 224417
[264] Behncke C *et al* 2018 *Sci. Rep.* **8** 186
[265] Hänze M *et al* 2015 *Phys. Rev.* B **91** 104428
[266] Adolff C F *et al* 2013 *Phys. Rev.* B **88** 224425
[267] Shibata J and Otani Y 2004 *Phys. Rev.* B **70** 012404
[268] Hänze M *et al* 2016 *Sci. Rep.* **6** 22402

IOP Publishing

Topology in Collective Magnetization Dynamics

Zhixiong Li, Yunshan Cao and Peng Yan

Chapter 7

Topological phases in soliton crystals

7.1 Topological solitonic insulators

As introduced in chapter 6, the collective dynamics or wave behaviors of magnetic soliton crystals have received much attention in recent years, while the possible topological phases are rarely discussed. In 2017, by solving the massless Thiele's equation and mapping it to the Haldane model [1], Kim *et al* [2] predicted that the magnetic solitons (such as vortex and bubble) in a two-dimensional honeycomb lattice can realize chiral edge states and the propagation direction is related to the sign of the topological charge of the constituted magnetic solitons. Subsequently, Li *et al* [3] generalized this method by adding a second-order inertial term and a third-order non-Newtonian gyroscopic term to explain the multiband chiral edge states in a two-dimensional honeycomb lattice composed of magnetic skyrmions. In addition, the Su–Schrieffer–Heeger (SSH) state [4] in a one-dimensional magnetic soliton array was also reported [5, 6]. Interestingly, the higher-order topological solitonic insulators have been realized by constructing the so-called 'breathing' lattice in both two- and three-dimensional magnetic soliton crystals [7–10]. In this section, we aim to develop the theory and implement micromagnetic simulation to study the topological solitonic insulators.

7.1.1 First-order topological insulating phases

The SSH [4] and Haldane [1] models are two typical models being studied in various systems. For magnetic solitons, we can construct the proper lattice structure to realize the first-order topological insulating phases predicted by these models.

7.1.1.1 One-dimensional lattice

Similar to the diatomic chain in the original SSH model, magnetic solitons can also form a one-dimensional 'breathing' lattice, with an illustration shown in figure 7.1(a). The magnetization dynamics is described by the generalized LLG equation [11, 12]

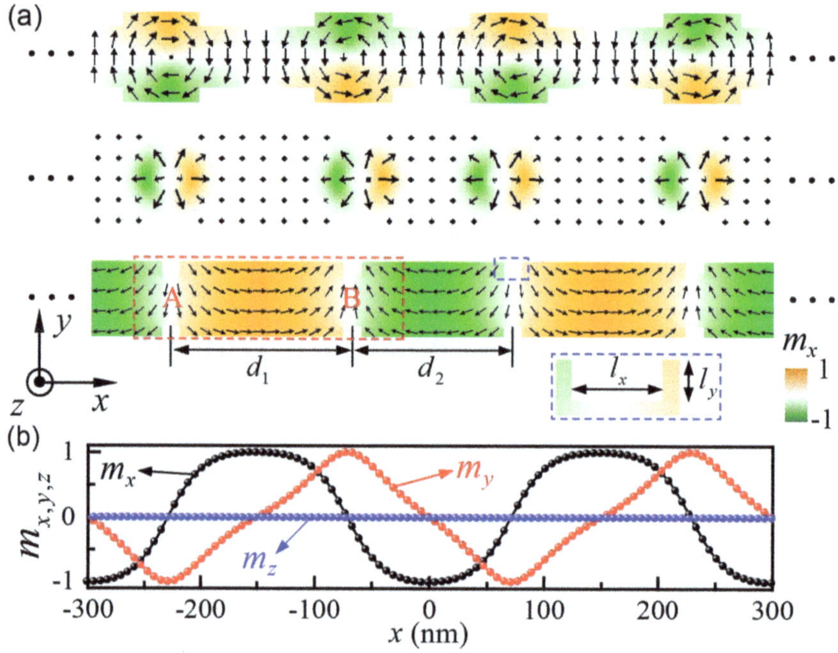

Figure 7.1. (a) A schematic diagram of vortex, skyrmion, and DW racetracks, where the magnetic solitons are pinned by periodic notches. The micromagnetic structure of a one-dimensional Néel-type DW lattice is highlighted. Here the cuboid notches with $l_x = 20$ nm, $l_y = 9$ nm, and $l_z = 5$ nm are considered. The red dotted box denotes the unit cell containing two DWs. d_1 and d_2 are the alternate distances between two nearest notches. (b) The magnetization components as a function of position in a DW racetrack with parameters $d_1 = 160$ nm and $d_2 = 140$ nm. (Reproduced with permission from [5]. Copyright 2021 The American Physical Society.)

$$\frac{\partial \mathbf{m}}{\partial t} = -\gamma \mathbf{m} \times \mathbf{H}_{\text{eff}} + \alpha \mathbf{m} \times \frac{\partial \mathbf{m}}{\partial t} + \Gamma_{\text{st}}, \tag{7.1}$$

where $\mathbf{m} = \mathbf{M}/M_s$ is the unit vector along the local magnetic moment with M_s the saturation magnetization, α is the Gilbert damping coefficient, and γ is the gyromagnetic ratio. $\mathbf{H}_{\text{eff}}$ is the effective field including the exchange field, magneto-crystalline anisotropic field, external magnetic field, and dipolar field. Γ_{st} denotes the spin–orbit or spin-transfer torque. For spin-transfer torque (STT), we have $\Gamma_{\text{st}} = b_J(\hat{J} \cdot \nabla)\mathbf{m} - \beta b_J \mathbf{m} \times (\hat{J} \cdot \nabla)\mathbf{m}$ with $b_J = JPg\mu_B/2|e|M_s$. $\hat{J}$ is the flow direction of the spin-polarized current, β is a non-adiabaticity coefficient, J is the current density, P is the spin polarization, g is the Lande g-factor, μ_B is the Bohr magneton, and e is the electron charge.

We choose the domain wall (DW) as the representative magnetic solitons to study the topological insulating phases. The well-known collective-coordinate method is used to describe the motion of DWs [13, 14]. In spherical coordinates, we have

$$\mathbf{m} = (\cos\theta, \sin\theta\cos\phi, \sin\theta\sin\phi), \tag{7.2}$$

where $\phi(x, t)$ and $\theta(x, t)$ are the azimuthal angle and polar angle, respectively. Here the azimuthal angle is defined as the angle between the y-axis and the projection of magnetization in the y–z-plane. Assuming the electric current, external magnetic field, and magnetocrystalline anisotropy field are all along the x-axis, the total effective field can be expressed as

$$\mathbf{H}_{\text{eff}} = \frac{2A}{\mu_0 M_s}\nabla^2\mathbf{m} + \frac{2K_u}{\mu_0 M_s}m_x\hat{x} + H_0\hat{x} + \mathbf{H}_{\text{dmg}}, \tag{7.3}$$

where A is exchange stiffness, K_u is the uniaxial magnetocrystalline anisotropy constant, H_0 is the external magnetic field strength, μ_0 is the vacuum permeability, and $\mathbf{H}_{\text{dmg}} = -\mu_0\mathbf{N}\cdot\mathbf{M}$ is the demagnetizing field with $\mathbf{N}$ a diagonal tensor. In general, we can express $\mathbf{H}_{\text{dmg}} = -\mu_0(N_x M_x\hat{x} + N_y M_y\hat{y} + N_z M_z\hat{z})$. Here N_x, N_y, and N_z are the demagnetizing factors in different directions and satisfy the relation $N_x + N_y + N_z = 1$. Substituting equations (7.2) and (7.3) into equation (7.1), we obtain

$$\begin{aligned}\frac{\partial\theta}{\partial t} + \alpha\sin\theta\frac{\partial\phi}{\partial t} = \gamma\frac{2A}{\mu_0 M_s}&\left(2\cos\theta\frac{\partial\phi}{\partial x}\frac{\partial\theta}{\partial x} + \sin\theta\frac{\partial^2\phi}{\partial x^2}\right)\\ &- \gamma\mu_0 M_s(N_z - N_y)\sin\theta\sin\phi\cos\phi\\ &+ b_J\frac{\partial\theta}{\partial x} + \beta b_J\sin\theta\frac{\partial\phi}{\partial x},\end{aligned} \tag{7.4}$$

$$\begin{aligned}\sin\theta\frac{\partial\phi}{\partial t} - \alpha\frac{\partial\theta}{\partial t} &= -\gamma\frac{2A}{\mu_0 M_s}\left[\frac{\partial^2\theta}{\partial x^2} - \sin\theta\cos\theta\left(\frac{\partial\phi}{\partial x}\right)^2\right]\\ &\quad + \gamma\left(H_0 + \frac{2K_u}{\mu_0 M_s}\cos\theta\right)\sin\theta\\ &\quad + \gamma\mu_0 M_s\sin\theta\cos\theta[(N_y - N_x) + (N_z - N_y)\sin^2\phi]\\ &\quad + b_J\sin\theta\frac{\partial\phi}{\partial x} - \beta b_J\frac{\partial\theta}{\partial x}.\end{aligned} \tag{7.5}$$

Considering the DW profile satisfying the Walker's trial solution [13, 14]

$$\begin{aligned}\phi(x, t) &= \phi(t),\\ \tan\frac{\theta(x, t)}{2} &= \mathrm{e}^{\frac{x-q(t)}{\Delta}},\end{aligned} \tag{7.6}$$

with q the position of DW and Δ being the DW width, we have

$$\frac{\partial\theta}{\partial x} = \frac{\sin\theta}{\Delta},\quad \frac{\partial\theta}{\partial t} = -\frac{1}{\Delta}\sin\theta\frac{\partial q}{\partial t},\quad \text{and}\quad \frac{\partial\phi}{\partial x} = 0. \tag{7.7}$$

Then we obtain the dynamic equations of DW

$$\frac{1}{\Delta}\frac{\partial q}{\partial t} - \alpha\frac{\partial \phi}{\partial t} = \frac{1}{2}\gamma\mu_0 M_s(N_z - N_y)\sin 2\phi - \frac{b_J}{\Delta}, \tag{7.8}$$

and

$$\frac{\partial \phi}{\partial t} + \frac{\alpha}{\Delta}\frac{\partial q}{\partial t} = \gamma H_0 - \frac{\beta b_J}{\Delta} + c, \tag{7.9}$$

with

$$c = \cos\theta\left\{-\gamma\frac{2A}{\mu_0 M_s \Delta^2} + \gamma\frac{2K_u}{\mu_0 M_s} + \gamma\mu_0 M_s\left[(N_y - N_x) + (N_z - N_y)\sin^2\phi\right]\right\}. \tag{7.10}$$

The left side of equation (7.9) is x independent and thus the prefactor in front of $\cos\theta$ on the right side of equation (7.9) must be zero [14], i.e. $c = 0$. Then the dynamic equations of DW can be re-written as

$$\begin{aligned}(1+\alpha^2)\frac{\mathrm{d}q}{\mathrm{d}t} &= \alpha\gamma\Delta H_0 + \frac{1}{2}\gamma\mu_0 M_s \Delta(N_z - N_y)\sin 2\phi - (b_J + \alpha c_J),\\ (1+\alpha^2)\frac{\mathrm{d}\phi}{\mathrm{d}t} &= \gamma H_0 - \frac{1}{2}\alpha\gamma\mu_0 M_s(N_z - N_y)\sin 2\phi + \frac{\alpha}{\Delta}b_J - \frac{c_J}{\Delta}.\end{aligned} \tag{7.11}$$

Since we focus on the collective dynamics of DWs around the pinning notches, the STT terms in equation (7.11) therefore can be dropped temporarily. In addition, the external magnetic field H_0 should be replaced by H_{pin} which includes the pinning field due to the existence of notch and DW–DW interactions. Then we have

$$\begin{aligned}(1+\alpha^2)\frac{\mathrm{d}q_j}{\mathrm{d}t} &= \gamma\alpha H_{\text{pin},\,j}\Delta_j + \frac{1}{2}\gamma\mu_0(N_z - N_y)\Delta_j M_s\sin 2\phi_j,\\ (1+\alpha^2)\frac{\mathrm{d}\phi_j}{\mathrm{d}t} &= \gamma H_{\text{pin},\,j} - \frac{1}{2}\gamma\alpha\mu_0(N_z - N_y)M_s\sin 2\phi_j,\end{aligned} \tag{7.12}$$

where the subscript j represents the jth DW and DW width Δ_j is expressed as

$$\Delta_j = \sqrt{2A\Big/\left\{2K_u + \mu_0 M_s^2\left[(N_y - N_x) + (N_z - N_y)\sin^2\phi_j\right]\right\}}\,. \tag{7.13}$$

Generally speaking, $H_{\text{pin},j}$ can be calculated with the spatial derivative to the total potential

$$H_{\text{pin},j} = -\frac{1}{2\mu_0 M_s L_y L_z}\frac{\partial U}{\partial q_j}, \tag{7.14}$$

where L_y and L_z are the width and thickness of the nanostrip, respectively, and U is the total energy of the system: $U = \sum_j K q_j^2/2 + \sum_{j\neq k} I(d_{jk})q_j q_k/2$. Here K is the spring constant determined by the shape of the notch and $I(d_{jk})$ is the coupling constant depending on the distance d_{jk} between DWs. Assuming a small ϕ and neglecting the damping terms, we obtain the linear form of equation (7.12)

$$M\frac{\mathrm{d}^2 q_j}{\mathrm{d}t^2} + Kq_j + \sum_{k\in\langle j\rangle} I(d_{jk})q_k = 0, \tag{7.15}$$

where $M = 2\mu_0 L_y L_z/\gamma^2(N_z - N_y)\Delta$ is the DW mass, $\Delta = \sqrt{2A/\left[2K_u+\mu_0 M_s^2(N_y - N_x)\right]}$, $\langle j\rangle$ denotes the nearest neighbor of j. In addition, $I(d_{jk}) = I_1$ or I_2 when indices j and k have an intracellular and intercellular connection, respectively. Here $I_{1,2} = I(d_{1,2})$ with d_1 and d_2 are the alternating distances between DWs.

The d-dependence of I in equation (7.15) is the key for calculating the dispersion relations of the DW lattice. The explicit analytic function of $I(d)$ can be determined by micromagnetic simulations. We consider the systems of both single DW and DW–DW pair pinned by notches in nanostrip, the distance between DWs is set to be 144 nm. The material parameters of Ni are used as in [15]: the exchange stiffness $A = 1.05 \times 10^{-11}$ J m^{-1}, the saturation magnetization $M_s = 0.477 \times 10^6$ A m^{-1}, the magnetocrystalline anisotropy constant $K_u = 0$, and the Gilbert damping constant $\alpha = 0.001$. The geometric size of the nanostrip has length $L_x = 7000$ nm, width $L_y = 60$ nm, and thickness $L_z = 5$ nm. The cellsize is set to be $2 \times 3 \times 5$ nm^3. To excite the oscillation of the DW lattice, we apply a sinc-function magnetic field $H(t) = H_1 \sin[2\pi f_0(t - t_0)]/[2\pi f_0(t - t_0)]$ for 1 μs along the x-axis direction with $H_1 = 10$ mT, $f_0 = 20$ GHz, and $t_0 = 1$ ns. By performing a Fourier analysis for single DW and DW–DW pair oscillations, we obtain the eigenfrequencies for a single DW $\omega_0/2\pi = 1.794$ GHz, the acoustic in-phase mode $\omega_1/2\pi = 1.71$ GHz and optical out-of-phase mode $\omega_2/2\pi = 2.033$ GHz for the DW–DW pair. In addition, from equation (7.15) we have the following identity

$$K = M\omega_0^2, \quad I = M(\omega_2^2 - \omega_1^2)/2. \tag{7.16}$$

The effective mass of a single DW is evaluated as $M = 2\mu_0 L_y L_z/\gamma^2(N_z - N_y)\Delta = 5.2097 \times 10^{-25}$ kg, where the demagnetizing factors can be analytically determined [16]: $N_x = 0$, $N_y = 0.106$, and $N_z = 0.894$. We therefore obtain $K = 6.6194 \times 10^{-5}$ J m^{-2} and $I = 1.2433 \times 10^{-5}$ J m^{-2} for $d = 144$ nm. In these simulations, the extrinsic disorder and defects are not considered. Therefore, the obtained K reflects purely the spring constant due to the pinning potential.

By systematically changing the DW–DW distance, we can obtain the function $I(d)$. It has been proved that the DW–DW interaction can be divided into three parts: the monopole–monopole ($\propto 1/d_{jk}$), the exchange ($\propto 1/d_{jk}^2$), and the dipole–dipole ($\propto 1/d_{jk}^3$) [15]. Figure 7.2(a) plots the numerical results (symbols) and theoretical fitting curve (solid line) with the formula $I(d) = c_1/d + c_2/d^2 + c_3/d^3$, where $c_1 = -9.2635 \times 10^{-12}$ J m^{-1}, $c_2 = 2.294 \times 10^{-18}$ J, and $c_3 = -1.0111 \times 10^{-25}$ J m. The d-dependence of ω_1 and ω_2 are plotted in figure 7.2(b), one can see that with the increase of d, ω_1 increases while ω_2 decreases. One naturally predicts that $\omega_1 = \omega_2 = \omega_0$ when $d \to \infty$, where $\omega_0 = \sqrt{K/M}$ is the eigenfrequency of a single pinned DW. Ideally, we can calculate the pinning-potential stiffness K by measuring ω_0 in experiments. However, this method suffers from an inevitable issue that the

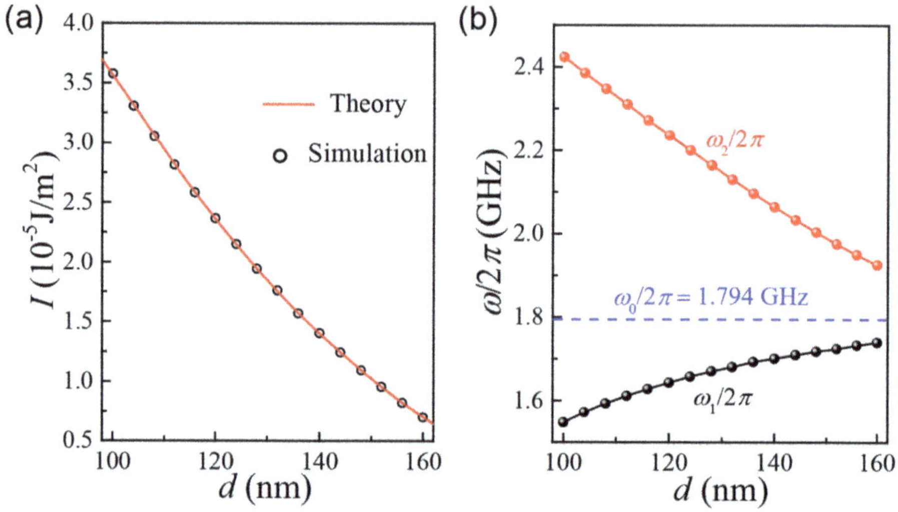

Figure 7.2. (a) Dependence of the coupling parameter I on d. The black circles and red solid line represent the simulation results and analytical formula, respectively. (b) The eigenfrequencies of a DW–DW pair as a function of d. (Reproduced with permission from [5]. Copyright 2021 The American Physical Society.)

dynamics of the DW can be easily modified by material defects and disorder and thus the genuine profile of the pinning potential cannot be determined precisely. In what follows, we introduce a topological method to overcome this issue.

We consider a one-dimensional DW array, as shown in figure 7.1(a), where the dashed red rectangle denotes the unit cell with basis vector $\mathbf{a} = a\hat{x}$, with $a = d_1 + d_2$. The band structure can be calculated by a plane-wave expansion $q_j = q_j \exp[\mathrm{i}(\omega t + nka)]$ with $j = A, B$ for different sublattices, n an integer, and k the wave vector. Then the Hamiltonian of the one-dimensional infinite DW lattice can be expressed in the momentum space as below:

$$H = \begin{pmatrix} K & I_1 + I_2 e^{-ika} \\ I_1 + I_2 e^{ika} & K \end{pmatrix} = [I_1 + I_2 \cos(ka)]\sigma_x + I_2 \sin(ka)\sigma_y + K\sigma_0, \quad (7.17)$$

where $\sigma_x = \begin{pmatrix} 0 & 1 \\ 1 & 0 \end{pmatrix}$, $\sigma_y = \begin{pmatrix} 0 & -\mathrm{i} \\ \mathrm{i} & 0 \end{pmatrix}$ are the Pauli matrices and $\sigma_0 = \begin{pmatrix} 1 & 0 \\ 0 & 1 \end{pmatrix}$ is the identity matrix. It can be seen clearly that our model contains an extra constant diagonal term $K\sigma_0$ when compared equation (7.17) to the Hamiltonian of the standard SSH model [4, 17]. Here the constant term describes the on-site energy and is irrelevant to the wave vector. It therefore only modifies the position of the band structure and does not affect the topological properties of the system. I_1 and I_2 in equation (7.17) denote the alternating hopping strengths between sites 'A' and 'B', the Hamiltonian (7.17) thus can be mapped to the SSH model. By solving (7.17) we obtain the dispersion relation

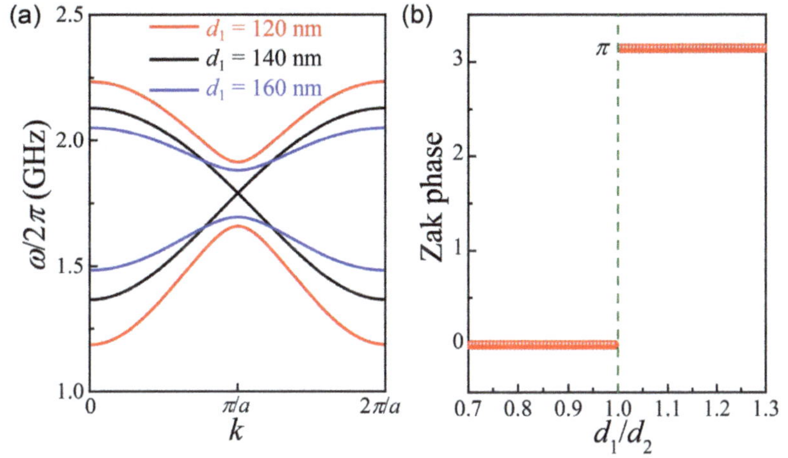

Figure 7.3. (a) The dispersion relations of a one-dimensional DW lattice with different intracellular distances: d_1= 120, 140, and 160 nm, here d_2 is fixed at 140 nm. (b) The topological invariant Zak phase varying with the ratio d_1/d_2. (Reproduced with permission from [5]. Copyright 2021 The American Physical Society.)

$$\omega_{\pm}(k) = \sqrt{\frac{K \pm \sqrt{I_1^2 + I_2^2 + 2I_1 I_2 \cos ka}}{M}}, \tag{7.18}$$

where the symbols + and − denote the optical and acoustic branch, respectively. The band structures with different geometric parameters are shown in figure 7.3(a). Here we fix d_2 to 140 nm and the parameters of Ni [15] are used. When $d_1 = d_2$, the two bands merge together forming a gapless structure (black curve in figure 7.3(a)), while if we let $d_1 \neq d_2$, a gap will open at $k = \pi/a$ (red and blue curves in figure 7.3(a)) and an insulating phase emerges. The topological invariant Zak phase [18] is used to identify whether these insulating phases are topologically protected. The dependence of the Zak phase Z on the different values of d_1/d_2 is plotted in figure 7.3(b). We can clearly see that Z is quantized to 0 and π when $d_1/d_2 < 1$ and $d_1/d_2 > 1$, respectively, indicating that the two regions support two topologically distinct phases. Noticeably, this conclusion always holds no matter what value d_2 takes.

Then we consider a finite system containing 39 DWs to check the bulk–boundary correspondence. Figure 7.4(a) plots the spectrum of numerical results, and one can see that the in-gap state appears as long as $d_1/d_2 \neq 1$. We choose the parameter $d_1/d_2 = 8/7$ (>1) as an example; the eigenfrequencies of the system are plotted in figure 7.4(b), from which we see that there exists one in-gap state (marked by a red dot). In addition, the oscillation distribution of this in-gap state is highly confined to the left end of the array (see figure 7.4(c)), in sharp contrast to the bulk state shown in figure 7.4(d). By using the *Ansätze* to describe the localized mode as $q_j = q_j \exp(\mathrm{i}\omega_0 t) z^n$ where $|z| < 1$, the profile of the edge state can be obtained by solving the following equations:

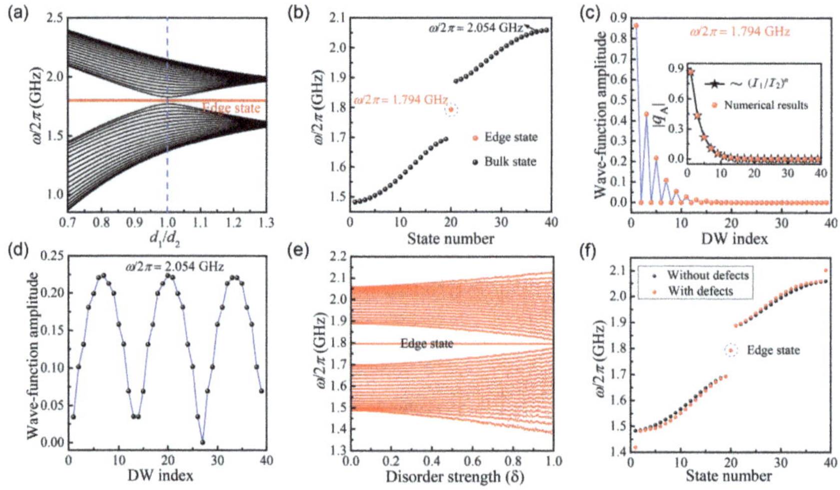

Figure 7.4. (a) The eigenfrequencies of the finite DW array with different ratios d_1/d_2. The dotted blue line represents the boundary separating two different topological phases, and the red line indicates the in-gap modes. (b) The eigenfrequencies of the finite DW array with $d_1/d_2 = 8/7$. The oscillation distribution of the DW lattice for the (c) edge and (d) bulk states. The inset in (c) shows the comparison between numerical and analytical results. (e) The spectrum of the DW array with different disorder strengths. (f) The eigenfrequencies of the DW array without (black dots) and with (red dots) defects. (Reproduced with permission from [5]. Copyright 2021 The American Physical Society.)

$$\begin{aligned} (I_1 + I_2 z)q_A(n) &= 0, \quad \text{for } n = 1, 2, 3, \ldots, \\ (I_1 + I_2 z^{-1})q_B(n) &= 0, \quad \text{for } n = 2, 3, \ldots, \end{aligned} \tag{7.19}$$

where the boundary condition is $I_1 q_B(1) = 0$ (the A-site DW is in the outmost left boundary). In addition, because $I_{1,2} \neq 0$, we obtain $q_B(n) = 0 \;\forall\, n$ and $z = -I_1/I_2$. As a result, the wave function of the DWs at the A-site follow an exponentially decaying formula $|q_A| = |q_0|(I_1/I_2)^n$ for $n = 1, 2, 3, \ldots$. The analytical formula compares excellently with numerical results, as shown in the inset of figure 7.4(c). Remarkably, one can naturally predict that if we consider the parameter $d_1/d_2 < 1$, the edge state will be localized at the right end. In addition, the localized edge states emerge in both ends if the lattice contains an even number of DWs.

The topological stability of the edge states emerging in figure 7.4(c) can be identified by analyzing the band structures when the disorder and defects are considered, with the corresponding results being shown in figures 7.4(e) and (f). To introduce the disorder, we assume that the coupling parameters I_1 and I_2 have a random shift, i.e. $I_1 \to I_1(1 + \delta W)$, $I_2 \to I_2(1 + \delta W)$, where δ represents the strength of disorder and W is a random number which uniformly distributed between -1 and 1. For the defects, we assume the parameters I_1 and I_2 experiencing a variation, i.e. $I_1 \to 10I_1$, $I_2 \to 0.1I_2$, only for the second and fourth DWs. From figures 7.4(e) and (f), we can clearly see that the frequency of the edge state can be immune from disorder and defects, while the bulk states are sensitive to them. It is noted that, in the SSH model, the chiral symmetry protects the topological robustness of the edge

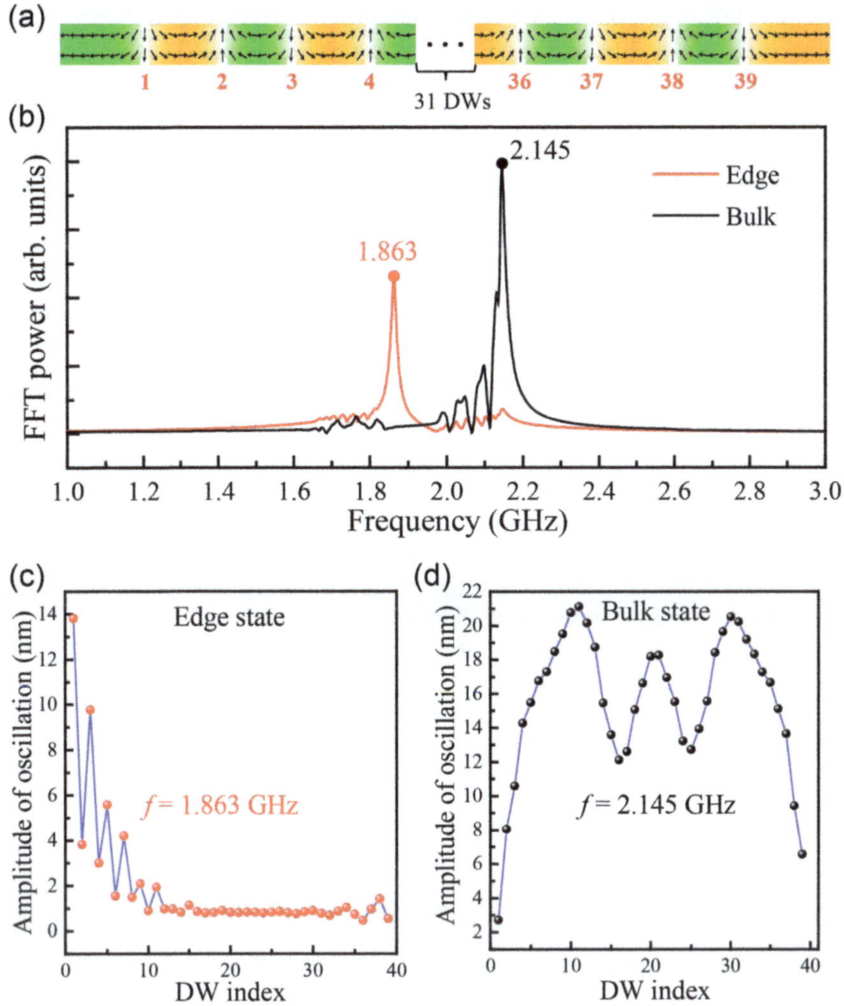

Figure 7.5. (a) The schematic diagram of a finite DW array containing 39 Néel-type DWs. The geometric parameters $d_1 = 160$ nm and $d_2 = 140$ nm are used. (b) The spectra of the oscillation of first and twentieth DWs for the edge and bulk bands, respectively. The oscillation distribution of the DW array for (c) edge and (d) bulk modes. (Reproduced with permission from [5]. Copyright 2021 The American Physical Society.)

modes. A detailed discussion has been presented in chapter 2. In our calculations, we introduce the disorder and defects into the off diagonal elements of the Hamiltonian, as a result the chiral symmetry of the system still is respected and the 'zero-energy' edge states can therefore survive. However, if we introduce the disorder to parameter K, the chiral symmetry of the system is broken and the topological robustness of the edge states will disappear.

Micromagnetic simulations are used to verify the analytical results. We consider a finite Ni nanostrip containing 39 DWs with length $L_x = 7000$ nm, as plotted in figure 7.5(a). Then we apply a sinc-function magnetic field along x-axis to excite the collective oscillation of DWs. All DW positions q_j are recorded every 0.1 ns with $q_j = \int x|m_y|^2 dx / \int |m_y|^2 dx$ and the integration region confined within jth DW. Here $|m_y|$ can act as the weight function because the closer to the DW center, the greater the $|m_y|$. By analyzing the spectra of the DW oscillation at two different positions (DW 1 and DW 20), we can find the frequency range for supporting the edge and bulk states. Figure 7.5(b) plots the results, where the black and red lines represent the bulk and edge bands, respectively. To further show the difference between the edge and bulk modes, we apply a excitation field $\mathbf{h}(t) = h_0 \sin(2\pi ft)\hat{x}$ over the whole system with frequencies $f = 1.863$ and 2.145 GHz as marked in figure 7.5(b). The oscillation distributions of the DW array for edge and bulk modes are shown in figures 7.5(c) and (d), respectively. We can clearly identify the localized (extended) characteristic for the edge (bulk) state. Full simulations therefore agree well with the theoretical predictions.

It is noted that the value of non-adiabaticity coefficient β in STT is still highly controversial in different experiments [19–22]. One important reason is the lack of a reliable method for determining the pinning potential. Because the extrinsic pinning is often inevitably mixed up with the intrinsic pinnings originating from material defects and disorder [23, 24], leading to a seemingly insuperable difficulty in distinguishing the two contributions. The magnetic soliton based SSH model may provide a solution to solve this issue. By linearizing equation (7.11) and considering the pinning field (the dissipation terms are also neglected), we have

$$\frac{d^2 q}{dt^2} + \frac{K}{M} q = -\beta\gamma(N_z - N_y)M_s b_J. \tag{7.20}$$

The solution of (7.20) thus can be expressed as

$$q(t) = q_0 \exp(i\omega_0 t) - \frac{\beta\gamma(N_z - N_y)M_s b_J}{\omega_0^2}. \tag{7.21}$$

From equation (7.21) one can clearly see that the STT term does not change the oscillation frequency of DW but leads to a shift to the equilibrium position:

$$X = |\langle q(t)\rangle| = \frac{\beta\gamma(N_z - N_y)M_s b_J}{\omega_0^2}. \tag{7.22}$$

By using the topological robust edge state mentioned above, we can accurately determine the eigenfrequency ω_0. Then the parameter β can be accurately calculated when the slope of the X–b_J curve is measured experimentally, where the slope $\lambda = \beta\gamma(N_z - N_y)M_s/\omega_0^2$. For the typical parameters $\beta = 0.01$, $J = 3 \times 10^{12}$A m^{-2}, and $P = 0.7$, we can estimate the average displacement $|\langle q(t)\rangle| \approx 10.5$ nm.

We conclude that the topologically robust edge state of the DW array can be used as the standard for measuring the DW frequency, which is helpful to measure the pinning profile and to finally resolve the controversy about the β parameter. It is worth noting these results are also applicable to other magnetic solitons, such as magnetic vortices and skyrmions.

Recently, Go *et al* [6] investigated a one-dimensional bipartite chain composed of magnetic vortex (or skyrmion) disks. They showed that there are no bound states existing in the system if the defects are not introduced. However, when the domain-wall and anti-domain-wall defects are considered, they proved that a midgap state will appear confined at a domain wall connecting two topologically distinct configurations, which is similar to the SSH model.

7.1.1.2 Two-dimensional lattice

In addition to the one-dimensional SSH model, the two-dimensional Haldane model can also support the first-order topological phase. To describe the collective dynamics of magnetic vortices arranged as a two-dimensional honeycomb network, Kim *et al* [2] considered the massless Thiele's equation

$$G\hat{z} \times \frac{\mathrm{d}\mathbf{U}_j}{\mathrm{d}t} - \alpha D\frac{\mathrm{d}\mathbf{U}_j}{\mathrm{d}t} + \mathbf{F}_j = 0. \tag{7.23}$$

Unlike the single vortex, here the potential energy W should include the contributions from both the confinement from the disk boundary and the interaction between vortices, i.e. $W = \sum_j K\mathbf{U}_j^2/2 + \sum_{j\neq k} U_{jk}/2$ with $U_{jk} = I_\parallel U_j^\parallel U_k^\parallel - I_\perp U_j^\perp U_k^\perp$ [25, 26], where $\mathbf{U}_j = \mathbf{R}_j - \mathbf{R}_j^0$ is the displacement $\mathbf{R}_j$ of the vortex from the equilibrium position $\mathbf{R}_j^0$, $U_j^\parallel = \hat{e}_{jk} \cdot \mathbf{U}_j$ is the projection of the displacement $\mathbf{U}_j$ onto the line connecting two vortices. Here $\hat{e}_{jk} = (\mathbf{R}_k^0 - \mathbf{R}_j^0)/|\mathbf{R}_k^0 - \mathbf{R}_j^0|$ is a unit vector, $U_j^\perp = (\hat{z} \times \hat{e}_{jk}) \cdot \mathbf{U}_j$ is the projection of $\mathbf{U}_j$ onto the line perpendicular to $\hat{e}_{jk}$. $I_\perp$ and $I_\parallel$ denote the transverse and longitudinal coupling coefficients, respectively, because the dipole–dipole interactions between vortices are highly anisotropic.

Assuming $\mathbf{U}_j = (u_j, v_j)$, equation (7.23) can be re-written as the following coupled equations for u_j and v_j

$$\mathrm{sgn}(Q)\begin{pmatrix} \dot{v}_j \\ -\dot{u}_j \end{pmatrix} = \omega_K\begin{pmatrix} u_j \\ v_j \end{pmatrix} + \sum_{k\in\langle j\rangle}\begin{pmatrix} \zeta + \xi\cos 2\theta_{jk} & \xi\sin 2\theta_{jk} \\ \xi\sin 2\theta_{jk} & \zeta - \xi\cos 2\theta_{jk} \end{pmatrix}\begin{pmatrix} u_k \\ v_k \end{pmatrix}, \tag{7.24}$$

where $\omega_K = K/|G|$, $\zeta = (I_\parallel - I_\perp)/2|G|$, $\xi = (I_\parallel + I_\perp)/2|G|$, θ_{jk} is the angle describing the direction of $\hat{e}_{jk}$ from the x-axis, and $\langle j\rangle$ denotes the nearest neighbor of j. Here the damping term is neglected. As an example, we consider the case of the negative topological charge, i.e. $Q = -1/2$. Defining $\psi_j = u_j + \mathrm{i}v_j$, equation (7.24) becomes the following form

$$\mathrm{i}\frac{d\psi_j}{\mathrm{d}t} = \omega_K\psi_j + \sum_{k\in\langle j\rangle}(\zeta\psi_k + \xi e^{\mathrm{i}2\theta_{jk}}\psi_k^*). \tag{7.25}$$

The complex variable ψ_j can be expanded as

$$\psi_j = \chi_j(t)\exp(-\mathrm{i}\omega_0 t) + \eta_j(t)\exp(\mathrm{i}\omega_0 t). \tag{7.26}$$

By substituting equation (7.26) into equation (7.25) and matching the coefficients of $\exp(\mathrm{i}\omega_0 t)$ and $\exp(-\mathrm{i}\omega_0 t)$, we obtain

$$\begin{aligned} \mathrm{i}\dot{\chi}_j &= \sum_{k\in\langle j\rangle} (\zeta\chi_k + \xi e^{2\mathrm{i}\theta_{jk}}\eta_k^*), \\ \mathrm{i}\dot{\eta}_j &= 2\omega_0\eta_j + \sum_{k\in\langle j\rangle} (\zeta\eta_k + \xi e^{2\mathrm{i}\theta_{jk}}\chi_k^*). \end{aligned} \tag{7.27}$$

Here we can use the method of rotating wave approximation to solve equations (7.27) for the reason that the rotation direction (clockwise or counterclockwise) of the vortex core is fixed for a given polarity of the core, so that the oscillation with another direction of rotation can be dropped safely. Assuming that the interactions are weak, i.e. $\zeta, \xi \ll \omega_0$, the soliton dynamics is mostly associated with the resonance frequency ω_0. For vortex gyrations with $Q = -1/2$, the rotation direction of the vortex core is clockwise, one can therefore justify $|\chi_j| \gg |\eta_j|$. The second equation of (7.27) can be simplified as

$$2\omega_0\eta_j \approx -\sum_{k\in\langle j\rangle} \xi e^{2\mathrm{i}\theta_{jk}}\chi_k^*. \tag{7.28}$$

By using the above relation, the first equation of (7.27) becomes

$$\begin{aligned} \mathrm{i}\frac{d\psi_j}{dt} = \left(\omega_K - \frac{3\xi^2}{2\omega_K}\right)\psi_j + \zeta\sum_{k\in\langle j\rangle}\psi_k - \frac{\xi^2}{2\omega_K}\sum_{l\in\langle\langle j\rangle\rangle}\cos(2\bar{\theta}_{jl})\psi_l - \mathrm{i}\frac{\xi^2}{2\omega_K} \\ \sum_{l\in\langle\langle j\rangle\rangle}\sin(2\bar{\theta}_{jl})\psi_l, \end{aligned} \tag{7.29}$$

where $\bar{\theta}_{jl} = \theta_{jk} - \theta_{kl}$ is the relative angle from the bond $k \to l$ to the bond $j \to k$ with k between j and l, $\langle\langle j\rangle\rangle$ represents the second-nearest neighbor of j. By solving equation (7.29) we can obtain the dispersion relations of the system.

Equation (7.29) is therefore mapped to the Haldane model which originally describes the topological properties of electrons in a honeycomb lattice [1, 27]. Here the last term on the right-hand side of equation (7.29) denotes the next-to-nearest hopping that breaks the time-reversal symmetry, leading to the emergence of chiral edge state. Illustrations of a vortex based finite honeycomb lattice (zigzag edges) for $Q = 1/2$ and $Q = -1/2$ are plotted in figures 7.6(a) and (c), respectively. Figures 7.6(b) and (d) show the corresponding dispersion relations for vortex gyration modes, from which we can easily identify the chiral edge modes. In addition, one can see that when the sign of the topological charge of vortices switches, the chirality of the edge modes also reverses. Further, Kim *et al* [2] showed that when the vortices are replaced by magnetic bubbles, the chiral edge modes still exist. However, if we set $\bar{\theta}_{jl} = \pm\pi/2$, the last term in on the right-hand side of equation (7.29) will vanish, as a result, the time-reversal symmetry is maintained and the band structure becomes gapless, then the chiral edge modes will disappear.

It has been shown that the magnetic bubbles and skyrmions can exhibit an inertia in their gyrotropic motion [28–31]. So, if we want to build an accurate theory to

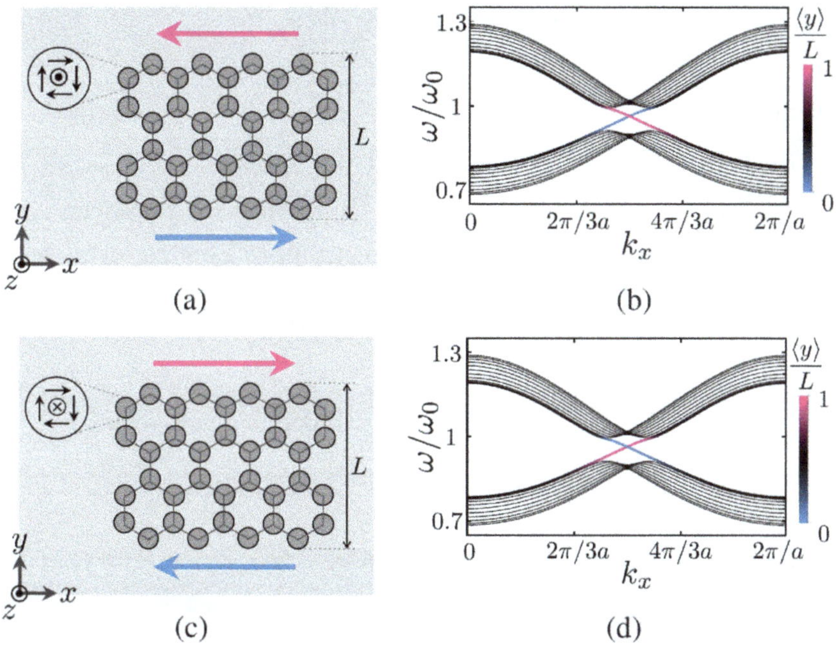

Figure 7.6. Schematic diagrams of a finite honeycomb lattice (zigzag edges) based on ferromagnetic vortices with topological charge (a) $Q = 1/2$ and (c) $Q = -1/2$. The colored arrows denote the propagation directions of the chiral edge states. Panels (b) and (d) show the corresponding band structures considered in (a) and (c), respectively. (Reproduced with permission from [2]. Copyright 2017 The American Physical Society.)

describe the collective oscillations of magnetic soliton, the second-order inertial term or even higher-order terms should be considered. Below, we derive the generalized form of Thiele's equation by using the LLG equation. With the help of the tensor notation, the LLG equation can be re-written as

$$\frac{1}{\gamma}\dot{m}_i + \epsilon_{jki}m_jH_k^{\text{eff}} - \frac{\alpha}{\gamma}\epsilon_{jki}m_j\dot{m}_k = 0. \tag{7.30}$$

The LLG equation also has another alternative form

$$\mathbf{m} \times \mathbf{H}^t = 0, \tag{7.31}$$

where $\mathbf{H}^t$ denotes the total effective field $H_i^t = H_i^1 + H_i^2 + H_i^3$. Here $H_i^1 = -\frac{1}{\gamma}\epsilon_{jki}m_j\dot{m}_k$, $H_i^2 = -\frac{\alpha}{\gamma}\dot{m}_i$, and $H_i^3 = H_i^{\text{eff}}$ represent the gyroscopic equivalent field, the dissipative equivalent field, and the conventional effective field, respectively. Moreover, we can also define the local force density $f_i^\nu = -M_sH_j^\nu\frac{\partial m_j}{\partial r_i}$, with $\nu = 1, 2, 3$. For every r_i, we need the balance of forces, i.e. $f_i^1 + f_i^2 + f_i^3 = 0$. After that, we assume that the magnetization depends on the position, velocity, and acceleration of the magnetic solitons simultaneously. We therefore obtain $m_j = m_j[\mathbf{r} - \mathbf{U}(t), \dot{\mathbf{U}}(t), \ddot{\mathbf{U}}(t)]$. Then we have

$$\dot{m}_j = \frac{\partial m_j}{\partial U_i}\dot{U}_i + \frac{\partial m_j}{\partial \dot{U}_i}\ddot{U}_i + \frac{\partial m_j}{\partial \ddot{U}_i}\dddot{U}_i. \tag{7.32}$$

Subsequently, the various forces can be determined. First, $f_i^1 = g_{ij}^1\dot{U}_j + g_{ij}^2\ddot{U}_j + g_{ij}^3\dddot{U}_j$, with $g_{ij}^1 = -\frac{M_s}{\gamma}\epsilon_{kpq}m_k\frac{\partial m_p}{\partial r_i}\frac{\partial m_q}{\partial U_j}$, $g_{ij}^2 = -\frac{M_s}{\gamma}\epsilon_{kpq}m_k\frac{\partial m_p}{\partial r_i}\frac{\partial m_q}{\partial \dot{U}_j}$, and $g_{ij}^3 = -\frac{M_s}{\gamma}\epsilon_{kpq}m_k\frac{\partial m_p}{\partial r_i}\frac{\partial m_q}{\partial \ddot{U}_j}$. Second, f_i^2 can be dropped for considering the small damping. At last, since we assume all magnetizations oscillate in a steady-state manner, only the externally applied fields have the contribution to the reversible energy force. We therefore have $f_i^3 = f_i^{\mathrm{ex}} = -M_s H_j^{\mathrm{ex}}\frac{\partial m_j}{\partial r_i}$.

For the gyroscopic term, we define vector $g_t^\nu = -\frac{1}{2}\epsilon_{tij}g_{ij}$, where $\nu = 1, 3$ such that $g_{ij}^1\dot{U}_j = \epsilon_{ijk}g_i^1\dot{U}_j$ and $g_{ij}^3\dddot{U}_j = \epsilon_{ijk}g_i^3\dddot{U}_j$. Then we can define a new vector

$$\mathbf{G} = \int \mathrm{d}V\mathbf{g}^1. \tag{7.33}$$

Obviously, if we consider a two-dimensional system, gyroscopic term $\mathbf{G} = (0, 0, G)$. Then we obtain

$$G = -\frac{M_s}{\gamma}\int \mathrm{d}V[\mathbf{m}\cdot(\partial_{r_x}\mathbf{m}\times\partial_{r_y}\mathbf{m})] = -\frac{4\pi \mathrm{d}QM_s}{\gamma}. \tag{7.34}$$

Similarly, a third-order gyroscopic term can be defined as

$$\mathbf{G}_3 = (0, 0, G_3) = \int \mathrm{d}V\mathbf{g}^3, \tag{7.35}$$

where

$$G_3 = \frac{\mathrm{d}M_s}{2\gamma}\int \mathrm{d}S[\mathbf{m}\cdot(\partial_{r_x}\mathbf{m}\times\partial_{\ddot{U}_y}\mathbf{m}) - \mathbf{m}\cdot(\partial_{r_y}\mathbf{m}\times\partial_{\ddot{U}_x}\mathbf{m})]. \tag{7.36}$$

In addition, one can define a mass tensor

$$M_{ij} = \int \mathrm{d}Vg_{ij}^2 = \frac{\mathrm{d}M_s}{\gamma}\int \mathrm{d}S[\mathbf{m}\cdot(\partial_{r_i}\mathbf{m}\times\partial_{\dot{U}_j}\mathbf{m})]. \tag{7.37}$$

Here we assume that $M_{xx} = M_{yy} = -M$ and $M_{xy} = M_{yx} = 0$, where

$$M = \frac{\mathrm{d}M_s}{\gamma}\int \mathrm{d}S[\mathbf{m}\cdot(\partial_{r_x}\mathbf{m}\times\partial_{\dot{U}_x}\mathbf{m})]. \tag{7.38}$$

Finally, the LLG equation is converted into the generalized Thiele's equation

$$G_3\hat{z}\times\frac{\mathrm{d}^3\mathbf{U}_j}{\mathrm{d}t^3} - M\frac{\mathrm{d}^2\mathbf{U}_j}{\mathrm{d}t^2} + G\hat{z}\times\frac{\mathrm{d}\mathbf{U}_j}{\mathrm{d}t} + \mathbf{F}_j = 0, \tag{7.39}$$

where M is the effective mass [28–31], G_3 is the third-order non-Newtonian gyroscopic coefficient [32–34], and $\mathbf{F}_j = \int \mathrm{d}V\mathbf{f}_j^{\mathrm{ex}}$ represents the external force.

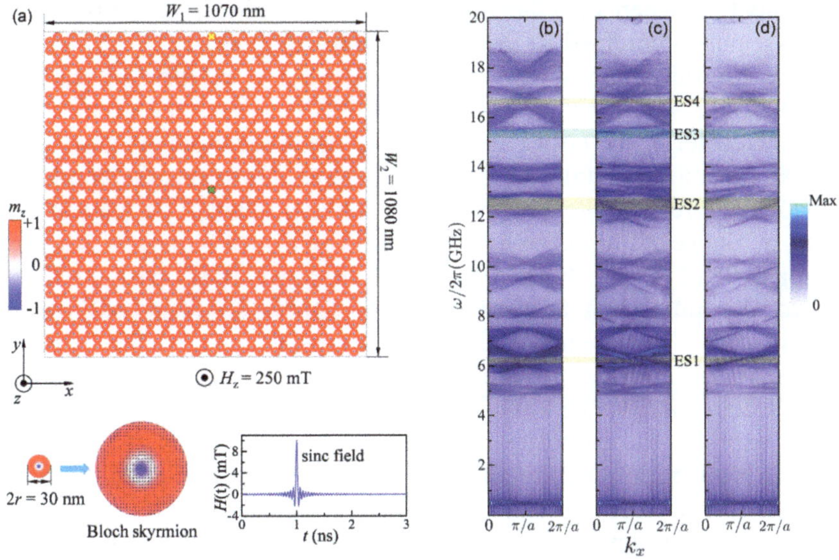

Figure 7.7. (a) A schematic diagram of a two-dimensional honeycomb lattice based on Bloch skyrmions. The green and yellow crosses indicate the positions where the excitation fields are applied. The dispersion relations of the skyrmion lattice when the exciting field is located at (b) the center or at the edge by calculating the spectra over (c) the upper and (d) the lower parts of the lattice. (Reproduced with permission from [3]. Copyright 2018 The American Physical Society.)

Noticeably, the above derivation is purely phenomenological and the microscopic mechanisms are still needed to explain the origin of the effective mass and non-Newtonian gyroscopic behavior of the magnetic solitons.

In addition to the vortices and magnetic bubbles mentioned above, the massive skyrmions arranged as a two-dimensional honeycomb lattice can also support chiral edge states. We first present the results of micromagnetic simulations. Figure 7.7(a) shows a plot of a honeycomb lattice containing 984 identical nanodisks with skyrmion states. Here we consider the Bloch-type skyrmion made of MnSi [35] which provides the bulk Dzyakoshinskii–Moriya interaction (DMI) [36, 37]. It is worth noting that the distance between two nearest skyrmions is equal to the diameter of the disk, which indicates that skyrmions can strongly couple with each other through the exchange interaction. The material parameters are used as follows [35]: the exchange stiffness $A = 3.2 \times 10^{-13}$ J m^{-1}, the saturation magnetization $M_s = 1.52 \times 10^5$ A m^{-1}, and the strength of the bulk DMI $D = 0.15$ mJ m^{-2}. To stabilize the skyrmions, we apply an external magnetic field with strength $H_z = 250$ mT perpendicular to the disk plane. A sinc-function magnetic field is applied confined to a nanodisk either at the edge or in the center to excite the oscillation of skyrmion lattice, as marked in figure 7.7(a). Then we record the guiding centers $\mathbf{R}_j = (R_{j,x}, R_{j,y})$ of all skyrmions every 5 ps. Here, $R_{j,x}$ and $R_{j,y}$ are defined as $R_{j,x} = \frac{\iint xq\mathrm{d}x\mathrm{d}y}{\iint q\mathrm{d}x\mathrm{d}y}$ and $R_{j,y} = \frac{\iint yq\mathrm{d}x\mathrm{d}y}{\iint q\mathrm{d}x\mathrm{d}y}$. Here $q = \mathbf{m} \cdot (\frac{\partial \mathbf{m}}{\partial x} \times \frac{\partial \mathbf{m}}{\partial y})$ denotes the skyrmion charge density [38], where $\mathbf{m}$ is the unit vector along the local magnetization and the integration region is confined to the jth nanodisk.

After the Fourier analysis, we plot the dispersion relation of the collective oscillations of the skyrmion lattice in figure 7.7(b), where the excitation field is located at the center of the lattice as marked by the green cross in figure 7.7(a). One can see that there are no states existing in the gaps. However, the band structures show obvious differences when the excitation field is applied at the edge of the array, as marked by the yellow cross in figure 7.7(a). The Fourier analyses are performed for the upper ($W_2/2 < y < W_2$) and the lower ($0 < y < W_2/2$) parts of the system, with the corresponding results being plotted in figures 7.7(c) and (d), respectively. We can easily find four edge modes in the band gaps, which are labeled as ES1–ES4. In addition, we can also identify the chiralities of these edge modes by determining the group velocity $\mathrm{d}\omega/\mathrm{d}k_x$. ES1 and ES2 propagate in a counterclockwise direction, while ES4 exhibits the opposite characteristics. In addition, ES3 manifests a bidirectional propagation feature and thus belongs to the non-chiral mode.

To further identify the chirality of these emerging edge modes, we can plot the propagation of skyrmion oscillations for different modes in real space. Here the excitation field $\mathbf{h}(t) = h_0 \sin(2\pi f t)\hat{x}$ is merely applied to one disk at the top edge, as marked in figures 7.8(a)–(d) with blue arrows. We choose four representative frequencies to visualize the propagation characteristics for different edge modes, with the results being shown in figures 7.8(a)–(d). One can clearly see that the edge states ES1 and ES2 propagate in a counterclockwise manner, while ES4 exhibits in the opposite way. In contrast, the edge state ES3 shows a feature of bidirectional propagation (see figure 7.8(c)). This non-chiral edge state can be easily understood by the Tamm–Shockley mechanism [39, 40] which predicts that the conducting surface/edge state will appear when the crystal periodicity is broken by the boundary.

One important feature of topological edge states is their robust transport against structure defects, while it is not so for the Tamm–Shockley ones. To verify that it is indeed the case for these skyrmionic edge states, we introduce defects by arbitrarily removing skyrmions at the film boundary as shown in figures 7.9(a)–(c). We then apply a sinusoidal field at the edge with the amplitude $h_0 = 10$ mT and the frequency $f = 12.62$, 16.65, and 15.3 GHz to excite the skyrmionic edge modes ES2, ES4, and ES3, respectively. We can clearly see that the chiral edge states (ES2 and ES4) are robust against the artificial defects, while the Tamm–Shockley state (ES3) cannot pass them.

We can use the generalized Thiele's equation (equation (7.39)) to explain the emerging multiband chiral edge states in skyrmion lattice. After some algebra, we obtain the simplified form of equation (7.39):

$$\begin{aligned}\hat{D}\psi_j = (\omega_K - 3\xi^2/2\bar{\omega}_K)\psi_j + \zeta \sum_{k\in\langle j\rangle} \psi_k - (\xi^2/2\bar{\omega}_K) \sum_{l\in\langle\langle j\rangle\rangle} \cos(2\bar{\theta}_{jl})\psi_l \\ - \mathrm{i}(\xi^2/2\bar{\omega}_K) \sum_{l\in\langle\langle j\rangle\rangle} \sin(2\bar{\theta}_{jl})\psi_l,\end{aligned} \tag{7.40}$$

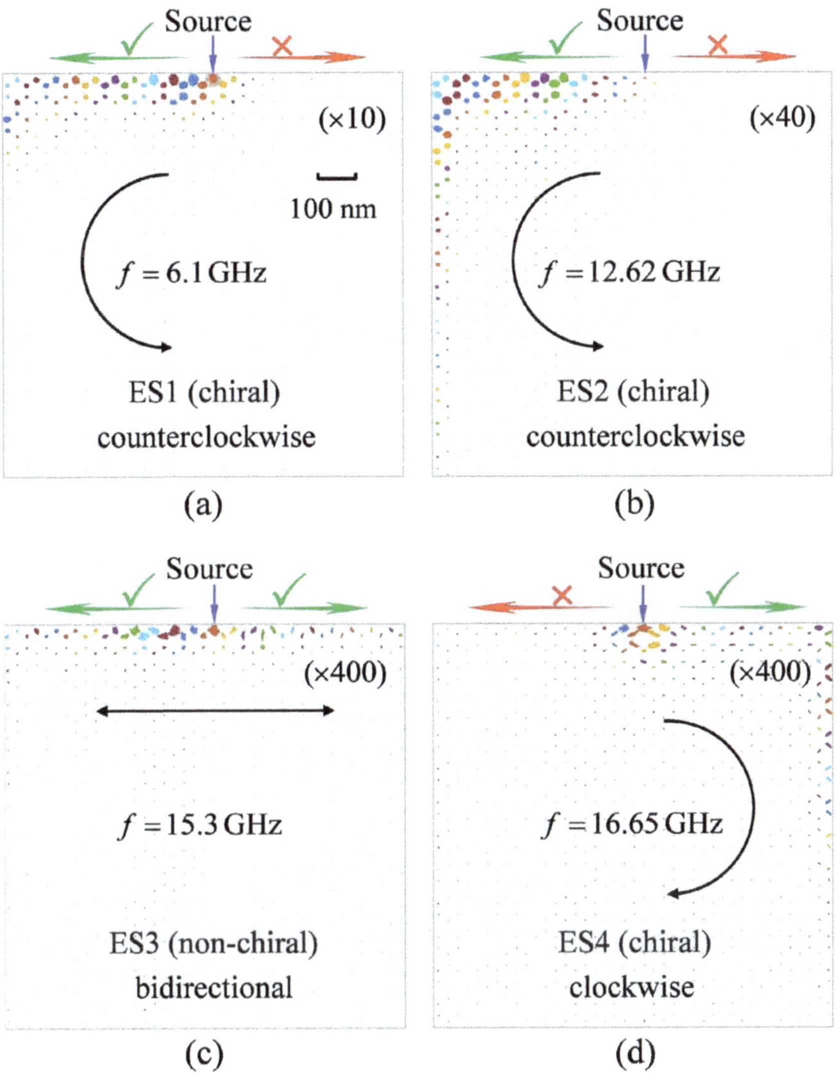

Figure 7.8. Snapshots of the oscillation propagation of different edge modes with frequency (a) $f = 6.1$, (b) 12.62, (c) 15.3, and (d) 16.65 GHz at $t = 40$ ns. The oscillation amplitudes of the skyrmions have been magnified for better observation, as marked in the panels. (Reproduced with permission from [3]. Copyright 2018 The American Physical Society.)

where the operator $\hat{D} = \mathrm{i}\omega_3 \frac{\mathrm{d}^3}{\mathrm{d}t^3} - \omega_M \frac{\mathrm{d}^2}{\mathrm{d}t^2} + \mathrm{i}\frac{\mathrm{d}}{\mathrm{d}t}$, $\omega_M = M/|G|$, $\omega_3 = G_3/|G|$, and $\bar{\omega}_K = \omega_K - \omega_0^2\omega_M$ with ω_0 meeting the condition

$$\omega_K = \omega_M\omega_0^2 \pm (\omega_3\omega_0^3 - \omega_0), \tag{7.41}$$

for clockwise and counterclockwise gyrotropic motions, respectively. In addition, the parameters ω_K, ζ, and ξ have the same definition as those in equation (7.24).

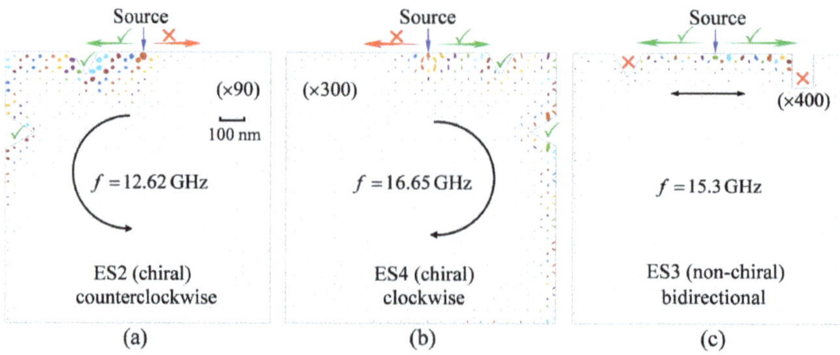

Figure 7.9. Snapshots of the propagation of both chiral and non-chiral edge states in the presence of artificial defects at the lattice boundary, with the frequency (a) $f = 12.62$ GHz, (b) 16.65 GHz, and (c) 15.3 GHz at $t = 50$ ns. Blue arrows represent the location of the exciting field. (Reproduced with permission from [3]. Copyright 2018 The American Physical Society.)

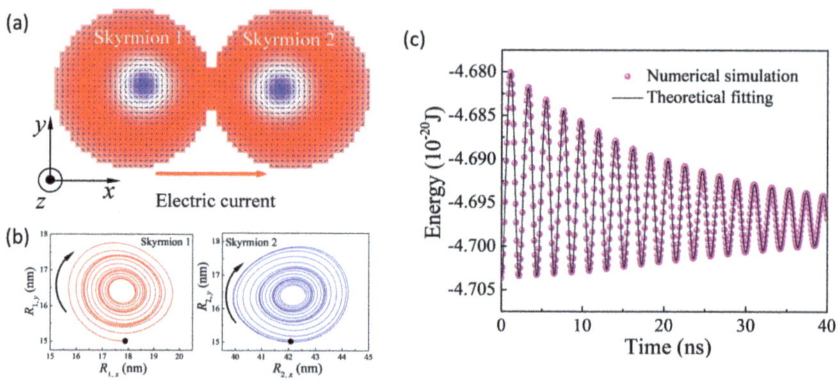

Figure 7.10. (a) Illustration of the double-skyrmion system. The red arrow represents the direction of the current. (b) The trajectories of skyrmion guiding centers. (c) Simulation and fitting results of the system energy. (Reproduced with permission from [3]. Copyright 2018 The American Physical Society.)

To calculate the band structures of the system, we need to identify the parameters $I_\parallel$, $I_\perp$, G_3, M, and K. First, the coupling coefficients $I_\parallel$ and $I_\perp$ can be calculated by considering a simplified system containing two nanodisks, as shown in figure 7.10(a). By applying a spin-polarized electrical current along the x-axis, with the current density $J = 2 \times 10^{10}$ A m^{-2} and the spin polarization $P = 0.7$, the two skyrmions are driven to oscillate/gyrate. Figure 7.10(b) shows the trajectories of the guiding centers of the two skyrmions. The simulation lasts for 40 ns. Figure 7.10(c) shows the time dependence of the system energy, with the magenta dots representing the numerical results while the black curve is the best fitting of $W = \sum_j K\mathbf{U}_j^2/2 + \sum_{j\neq k} U_{jk}/2$ with $U_{jk} = I_\parallel U_j^\parallel U_k^\parallel - I_\perp U_j^\perp U_k^\perp$. We find an excellent agreement between theory and

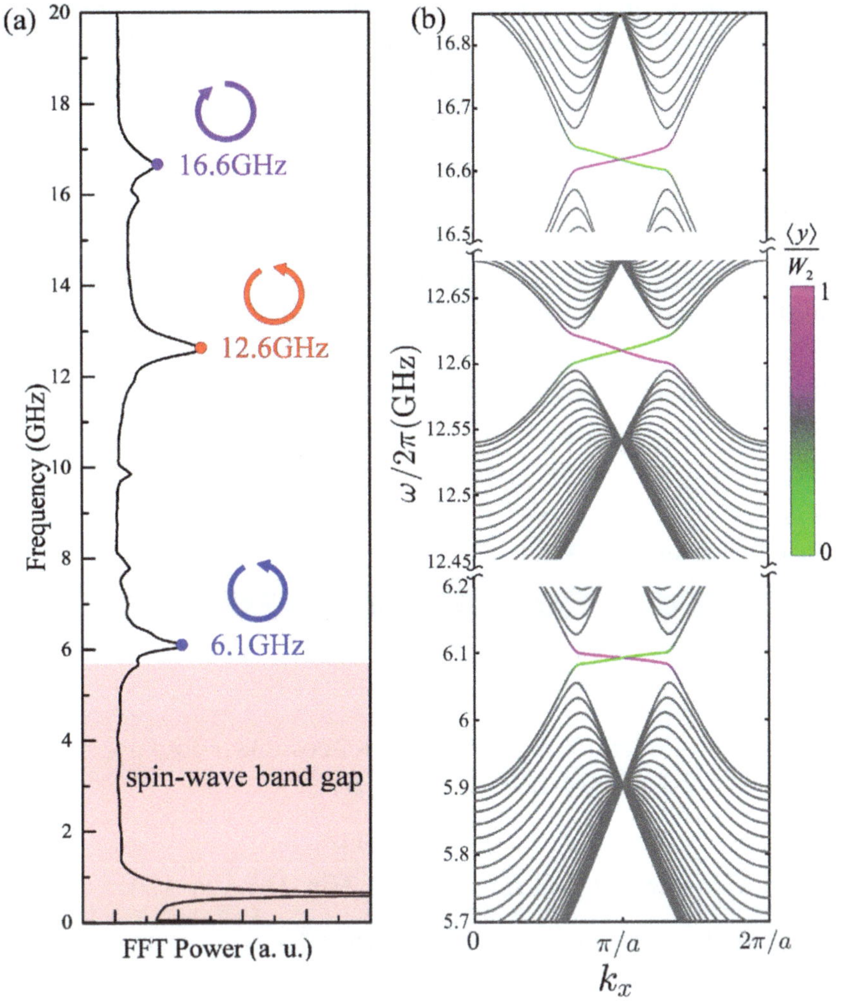

Figure 7.11. (a) The spectrum of a skyrmion lattice when the excitation field is applied to the whole system. (b) The numerically calculated dispersion relations of the skyrmion lattice. (Reproduced with permission from [3]. Copyright 2018 The American Physical Society.)

simulation. We thus obtain the values of the parameters $I_{\parallel} = -2.902 \times 10^{-6}$ J m^{-2} and $I_{\perp} = -1.456 \times 10^{-5}$ J m^{-2}. Similarly, the parameters G_3, M, and K can also be calculated from micromagnetic simulations. The spectrum for the collective oscillation of the skyrmion lattice is plotted in figure 7.11(a). Above the band gap of spin-wave, we can read three resonance peaks with $\omega_{0,1}/2\pi = 6.1$ GHz, $\omega_{0,2}/2\pi = 12.6$ GHz, and $\omega_{0,3}/2\pi = 16.6$ GHz. The chiralities of these modes for skyrmion gyrations are also marked. By using equation (7.41), we obtain the spring constant $K = 3.744\,6 \times 10^{-4}$ J m^{-2}, effective mass $M = 1.561\,2 \times 10^{-26}$ kg, and the third-order gyroscopic parameter $G_3 = 1.183\,2 \times 10^{-36}$J s^3 m^{-2}.

Considering the periodic boundary condition along the x-axis direction and the open boundary condition along the y-axis direction, i.e. zigzag edge, we obtain the band structure near the resonance frequencies of the skyrmion lattice by solving equation (7.40), as shown in figure 7.11(b). The average positions of different modes $\langle y\rangle \equiv \sum_j R^0_{j,y}|\mathbf{U}_j|^2/\sum_j|\mathbf{U}_j|^2$ are also calculated, where $R^0_{j,y}$ represents the projection of the skyrmion equilibrium position onto the y-axis and the colors are closer to magenta (green) indicating that the modes are more localized at the upper (lower) edge.

Interestingly, one can easily note that the chirality of ES1 (or ES2) is opposite to ES4. This feature can be explained by the sign switch of the topological invariant—the Chern number. First, by mapping the equation (7.40) into the Haldane model, we obtain the following Hamiltonian:

$$H = \sum_{\langle jk\rangle} t_1\psi_j^*\psi_k + \sum_{\langle\langle jk\rangle\rangle} t_2 \exp(\mathrm{i}2\pi\phi)\psi_j^*\psi_k + \mathrm{c.c.}, \tag{7.42}$$

where $t_1 = \zeta$, $t_2 = -\xi^2/2\bar{\omega}_K$, and $\phi = 1/3$. By utilizing the Bloch's theorem, we diagonalize the equation (7.42) as

$$H = \sum_{i=0,x,y,z} h_i(\mathbf{k})\sigma_i, \tag{7.43}$$

with

$$\begin{aligned}
h_0(\mathbf{k}) &= -2t_2\cos(2\pi\phi)(\cos 2\sqrt{3}k_x r + 2\cos\sqrt{3}k_x r\cos k_y r),\\
h_x(\mathbf{k}) &= -t_1(\cos 2k_y r + 2\cos k_x r\cos k_y r),\\
h_y(\mathbf{k}) &= -t_1(\sin 2k_y r - 2\cos\sqrt{3}k_x r\sin k_y r),\\
h_z(\mathbf{k}) &= -2t_2\sin(2\pi\phi)(\sin 2\sqrt{3}k_x r - 2\sin\sqrt{3}k_x r\cos k_y r).
\end{aligned} \tag{7.44}$$

Equation (7.43) describes a two-band model which is analogous to the dynamic of a single spin when a $\mathbf{k}$-dependent magnetic field is applied. In the first Brillouin zone, the variation of $\mathbf{k}$ forms a closed surface $\mathbf{S}$. Then the Berry curvature can be expressed as

$$\Omega_\eta = \eta\frac{\mathbf{e}_h}{2h^2}, \tag{7.45}$$

where $h = \sqrt{h_x^2 + h_y^2 + h_z^2}$ is the amplitude of the 'effective magnetic field' $\mathbf{h} = (h_x, h_y, h_z)$ with the unit direction along $\mathbf{e}_h = \mathbf{h}/h$. Here $\eta = \pm$ denotes the index of the band. Therefore, the Chern number is given by

$$C_\eta = \eta\frac{1}{2\pi}\oint_{\mathbf{S}}\Omega_\eta(h)\cdot d\mathbf{S}. \tag{7.46}$$

It is straightforward to obtain that the Chern number $C_\eta = -\eta$ when the closed surface $\mathbf{S}$ contains the degeneracy point $h = 0$, otherwise it is zero. In addition, if we consider the gyrotropic motion of skyrmions have an opposite chirality, i.e. $\omega_0 \rightarrow -\omega_0$, then the Haldane Hamiltonian becomes

$$\begin{aligned} H &= \sum_{\langle jk \rangle} t_1 \psi_j^* \psi_k - \sum_{\langle\langle jk \rangle\rangle} t_2 \exp(\mathrm{i}2\pi\phi) \psi_j^* \psi_k + \text{c.c.} \\ &= \sum_{\langle jk \rangle} t_1 \psi_j^* \psi_k + \sum_{\langle\langle jk \rangle\rangle} t_2 \exp[\mathrm{i}(2\pi\phi + \pi)] \psi_j^* \psi_k + \text{c.c.}. \end{aligned} \tag{7.47}$$

One can see that the only change of the Hamiltonian is the extra π phase introduced in the factor $2\pi\phi \rightarrow 2\pi\phi + \pi$. It is equivalent to perform a time-reversal operation to equation (7.44): $H(\mathbf{k}) \rightarrow H^*(-\mathbf{k})$, then we obtain equation (7.47). After this operation, the sign of the Berry curvature naturally changes. As a result, the edge mode also switches its chirality.

7.1.2 Second-order topological insulating phases

In above sections, we have discussed the first-order TIs based on a magnetic soliton lattice. In this section, we show that the second-order topological insulating phase can also be realized in a magnetic soliton system. Here we use the two-dimensional breathing honeycomb lattices of magnetic vortices [8] as an example.

It has been proved that the perfect graphene structure has a gapless dispersion relation with Dirac cones emerging in the momentum space [41]. However, when the magnetic flux [1] or spin–orbit coupling [42] is considered, a gap will open at the Dirac point, then the FOTI appears. Very recently, it has been shown that one can realize the gap opening or closing by adjusting the intercellular and intracellular couplings in photonic [43] and elastic [44] honeycomb lattices. In this case, the HOTI can be created. In what follows, based on the breathing honeycomb lattice of vortices, we show the existence of HOTI in a magnetic system.

Figure 7.12 plots a two-dimensional breathing honeycomb lattice based on vortices, where the topological charges of vortices are $Q = 1/2$, and d_1 and d_2 denote the alternate distance between two nearest vortices. The generalized Thiele's equation (equation (7.39)) is used to describe the collective dynamics of the vortex lattice. Noticeably, because there exist two different coupling distances in this breathing lattice, one should modify the equation as

$$\begin{aligned} \left(\mathrm{i}\omega_3 \frac{\mathrm{d}^3}{\mathrm{d}t^3} - \omega_M \frac{\mathrm{d}^2}{\mathrm{d}t^2} - \mathrm{i}\frac{\mathrm{d}}{\mathrm{d}t}\right)\psi_j &= \left(\omega_K - \frac{\xi_1^2 + 2\xi_2^2}{2\bar{\omega}_K}\right)\psi_j + \zeta_1 \sum_{k\in\langle j_1 \rangle} \psi_k + \zeta_2 \sum_{k\in\langle j_2 \rangle} \psi_k \\ &\quad - \frac{\xi_1 \xi_2}{2\bar{\omega}_K} \sum_{s\in\langle\langle j_1 \rangle\rangle} \mathrm{e}^{\mathrm{i}2\bar{\theta}_{js}} \psi_s - \frac{\xi_2^2}{2\bar{\omega}_K} \sum_{s\in\langle\langle j_2 \rangle\rangle} \mathrm{e}^{\mathrm{i}2\bar{\theta}_{js}} \psi_s, \end{aligned} \tag{7.48}$$

where $\xi_l = (I_{\parallel, l} + I_{\perp, l})/2|G|$ and $\zeta_l = (I_{\parallel, l} - I_{\perp, l})/2|G|$, with $l = 1$ ($l = 2$) denoting that the distance between two nearest vortices is d_1 (d_2).

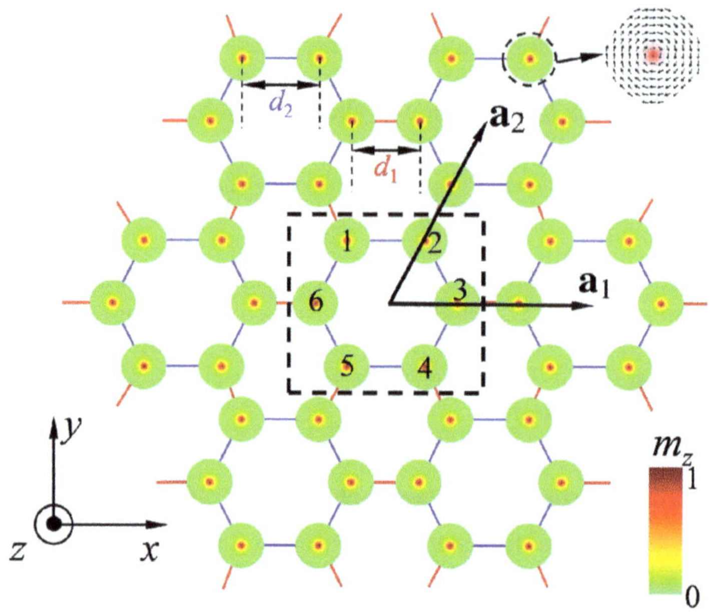

Figure 7.12. A schematic diagram of a breathing honeycomb lattice based on magnetic vortices. Here parameters d_1 and d_2 represent the alternating distances of intercellular and intracellular bonds, respectively. The radius of the nanodisk is $r = 50$ nm and the thickness is $w = 10$ nm. (Reproduced with permission from [8]. Copyright 2020 The American Physical Society.)

For an infinite system, the unit cell can be chosen with the dashed black rectangle, as plotted in figure 7.12, where $\mathbf{a}_1 = a\hat{x}$ and $\mathbf{a}_2 = \frac{1}{2}a\hat{x} + \frac{\sqrt{3}}{2}a\hat{y}$ denote two basis vectors with $a = d_1 + 2d_2$. We consider a plane-wave expansion of $\psi_j = \phi_j \exp(\mathrm{i}\omega t)\exp[\mathrm{i}(n\mathbf{k}\cdot\mathbf{a}_1 + m\mathbf{k}\cdot\mathbf{a}_2)]$, where $j = 1, 2, 3, 4, 5, 6$ represent different vortices in the unit cell, n and m are integers, and $\mathbf{k}$ denotes the wave vector. Substituting the expression of ψ_j into equation (7.48), we obtain the Hamiltonian in a matrix form

$$H = \begin{pmatrix} Q_0 & \zeta_2 & Q_1 & Q_2 & Q_3 & \zeta_2 \\ \zeta_2 & Q_0 & \zeta_2 & Q_4 & \zeta_1\exp(\mathrm{i}\mathbf{k}\cdot\mathbf{a}_2) & Q_5 \\ Q_1^* & \zeta_2 & Q_0 & \zeta_2 & Q_6 & \zeta_1\exp(\mathrm{i}\mathbf{k}\cdot\mathbf{a}_1) \\ Q_2^* & Q_4^* & \zeta_2 & Q_0 & \zeta_2 & Q_7 \\ Q_3^* & \zeta_1\exp(-\mathrm{i}\mathbf{k}\cdot\mathbf{a}_2) & Q_6^* & \zeta_2 & Q_0 & \zeta_2 \\ \zeta_2 & Q_5^* & \zeta_1\exp(-\mathrm{i}\mathbf{k}\cdot\mathbf{a}_1) & Q_7^* & \zeta_2 & Q_0 \end{pmatrix}, \tag{7.49}$$

where the elements read

$$
\begin{aligned}
Q_0 &= \omega_K - \frac{\xi_1^2 + 2\xi_2^2}{2\bar{\omega}_K},\\
Q_1 &= -\frac{\xi_1\xi_2}{2\bar{\omega}_K}\exp\left(\mathrm{i}\frac{2\pi}{3}\right)\{\exp[\mathrm{i}\mathbf{k}\cdot(\mathbf{a}_2-\mathbf{a}_1)] + \exp(-\mathrm{i}\mathbf{k}\cdot\mathbf{a}_1)\}\\
&\qquad - \frac{\xi_2^2}{2\bar{\omega}_K}\exp(\mathrm{i}\frac{2\pi}{3}),\\
Q_2 &= \zeta_1\exp[\mathrm{i}\mathbf{k}\cdot(\mathbf{a}_2-\mathbf{a}_1)],\\
Q_3 &= -\frac{\xi_1\xi_2}{2\bar{\omega}_K}\exp\left(-\mathrm{i}\frac{2\pi}{3}\right)\{\exp[\mathrm{i}\mathbf{k}\cdot(\mathbf{a}_2-\mathbf{a}_1)] + \exp(\mathrm{i}\mathbf{k}\cdot\mathbf{a}_2)\}\\
&\qquad - \frac{\xi_2^2}{2\bar{\omega}_K}\exp\left(-\mathrm{i}\frac{2\pi}{3}\right),\\
Q_4 &= -\frac{\xi_1\xi_2}{2\bar{\omega}_K}\exp\left(\mathrm{i}\frac{2\pi}{3}\right)\{\exp[\mathrm{i}\mathbf{k}\cdot(\mathbf{a}_2-\mathbf{a}_1)] + \exp(\mathrm{i}\mathbf{k}\cdot\mathbf{a}_2)\}\\
&\qquad - \frac{\xi_2^2}{2\bar{\omega}_K}\exp\left(\mathrm{i}\frac{2\pi}{3}\right),\\
Q_5 &= -\frac{\xi_1\xi_2}{2\bar{\omega}_K}\exp\left(-\mathrm{i}\frac{2\pi}{3}\right)[\exp(\mathrm{i}\mathbf{k}\cdot\mathbf{a}_2) + \exp(\mathrm{i}\mathbf{k}\cdot\mathbf{a}_1)]\\
&\qquad - \frac{\xi_2^2}{2\bar{\omega}_K}\exp\left(-\mathrm{i}\frac{2\pi}{3}\right),\\
Q_6 &= -\frac{\xi_1\xi_2}{2\bar{\omega}_K}\exp\left(\mathrm{i}\frac{2\pi}{3}\right)[\exp(\mathrm{i}\mathbf{k}\cdot\mathbf{a}_2) + \exp(\mathrm{i}\mathbf{k}\cdot\mathbf{a}_1)]\\
&\qquad - \frac{\xi_2^2}{2\bar{\omega}_K}\exp\left(\mathrm{i}\frac{2\pi}{3}\right),\\
Q_7 &= -\frac{\xi_1\xi_2}{2\bar{\omega}_K}\exp\left(\mathrm{i}\frac{2\pi}{3}\right)\{\exp[\mathrm{i}\mathbf{k}\cdot(\mathbf{a}_1-\mathbf{a}_2)] + \exp(\mathrm{i}\mathbf{k}\cdot\mathbf{a}_1)\}\\
&\qquad - \frac{\xi_2^2}{2\bar{\omega}_K}\exp\left(\mathrm{i}\frac{2\pi}{3}\right).
\end{aligned}
\tag{7.50}
$$

It is well known that the Chern number (see chapter 2 for details) is a frequently used topological invariant to judge whether the system is in the FOTI phase. However, a different topological invariant must be considered if we want to discuss the HOTI phase. In addition to the bulk polarization, the Z_N Berry phase is also a powerful tool to describe the HOTI. In this model, the system has six-fold rotational (C_6) symmetry, thus the Z_6 Berry phase [45–48] is the proper topological invariant. The green line segment in figure 7.13(a) represents the integral path L_1: $G' \to \Gamma \to K'$. To avoid the difficulty of the gauge choice, we use the Wilson-loop method for calculating the Berry phase θ [49, 50]. Noticeably, because of the existence of C_6 symmetry, the six high-symmetry points G, K, G', K', G'', and K'' are

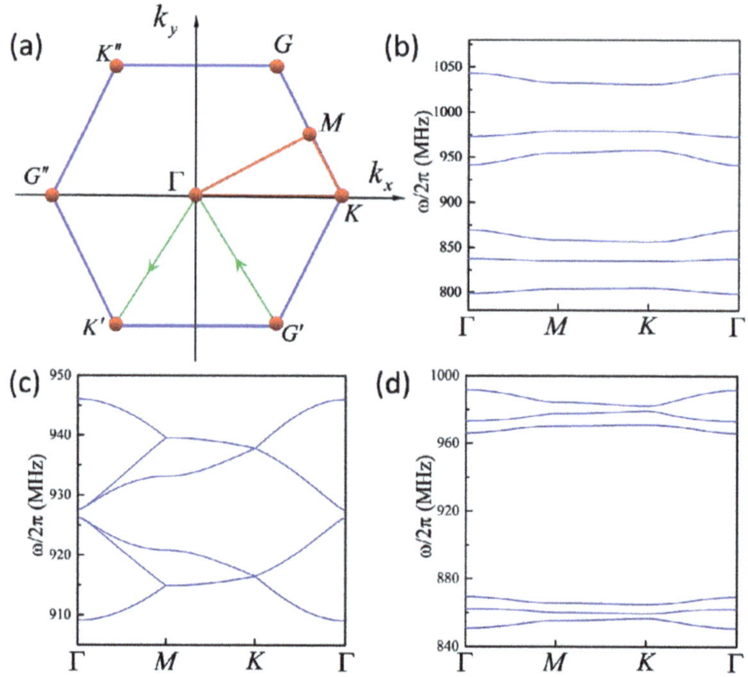

Figure 7.13. (a) The first Brillouin zone of the breathing honeycomb lattice. The dispersion relations of an infinite system along the route Γ–M–K–Γ for different geometric parameters: (b) $d_1 = 3.6$, $d_2 = 2.08$; (c) $d_1 = d_2 = 3.6$; and (d) $d_1 = 2.08$, $d_2 = 3.6$. Here $d_{1,2}$ has been normalized by r. (Reproduced with permission from [8]. Copyright 2020 The American Physical Society.)

isovalent. As a result, there exist other five equivalent integral paths (L_2: $K' \to \Gamma \to G''$, L_3: $G'' \to \Gamma \to K''$, L_4: $K'' \to \Gamma \to G$, L_5: $G \to \Gamma \to K$, and L_6: $K \to \Gamma \to G'$) which leads to the same θ. In addition, if we consider the integral path $L_1 + L_2 + L_3 + L_4 + L_5 + L_6$, then $\theta = 0$. Therefore, the Z_6 Berry phase $\theta = \frac{2n\pi}{6}$ ($n = 0, 1, 2, 3, 4, 5$) which shows a quantized value. Here different θ denotes different phases.

To obtain the spectra and the phase diagram of the system, we need to calculate the key parameters K, M, G_3, $I_\parallel$, and $I_\perp$ which have been defined in a previous section. Here we consider the parameters of Py [51, 52]: the exchange stiffness $A = 1.3 \times 10^{-11}$ J m^{-1} and the saturation magnetization $M_s = 0.86 \times 10^6$ A m^{-1}. In addition, we set the nanodisk thickness $w = 10$ nm and radius $r = 50$ nm. Then we have $G = -3.072\,5 \times 10^{-13}$ J s rad^{-1}m^{-2}. Then, the spring constant K, effective mass M, and non-Newtonian coefficient G_3 can be determined using the following identities [33, 53]: $\omega_0 = K/G$, $G_3\bar{\omega}^2 = G$, $G_3(\Delta\omega + \omega_0) = M$, $2\bar{\omega} = \omega_1 + \omega_2$, and $\Delta\omega = |\omega_2 - \omega_1|$, where ω_0 denotes the frequency of the lowest-order gyroscopic mode, while ω_1 and ω_2 represent the frequencies of the other two higher-order gyroscopic modes with opposite chiralities [33]. The excitation spectrum of a single vortex is plotted in figure 7.14(a), from which we can identify that the lowest mode is

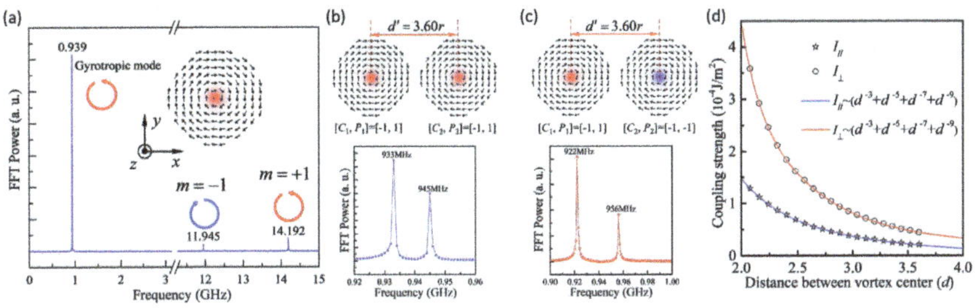

Figure 7.14. (a) The spectrum of a single vortex with the inset showing the micromagnetic structure. The colored arrows indicate the direction of the vortex gyrotropic motion for different modes. (b) and (c) The micromagnetic structures and corresponding spectra of a two-vortex system for different combinations of polarities with $d' = 3.60r$. Here C_i and P_i (i = 1 or 2) denote the chirality and polarity of the vortex, respectively. (d) The coupling parameters $I_\parallel$ and $I_\perp$ vary with d, where d is normalized by the disk radius r. Symbols and solid curves represent the results of micromagnetic simulation and analytical fitting, respectively. (Reproduced with permission from [7]. Copyright 2019 the Authors.)

an counterclockwise gyrotropic mode, while the other two modes with high-frequencies show opposite directions of gyrotropic motion [54, 55]. Therefore, we have $K = 1.812\,8 \times 10^{-3}$ J m^{-2}, $M = 9.122\,4 \times 10^{-25}$ kg, and $G_3 = -4.557\,1 \times 10^{-35}$ J s^3 rad^{-3} m^{-2}. In addition, it has been proved that the coupling parameters $I_\parallel$ and $I_\perp$ depend on the value of d [56–58], where $d = d'/r$ with d' is the distance between two vortices. For a coupled two-vortex system, the eigenfrequencies can be calculated as $\omega = \omega_0\sqrt{(1 \pm I_\parallel/K)(1 \mp P_1P_2I_\perp/K)}$ [56], where P_1 (or P_2) denotes the vortex polarity and is equal to +1 or −1. As a result, when the eigenfrequencies of a coupled two-vortex system with different polarity combinations ($P_1P_2 = 1$ and $P_1P_2 = -1$) are determined by micromagnetic simulations, we can obtain the values of $I_\parallel$ and $I_\perp$ by using the above expression. As an example, we set the parameter $d' = 3.60r$. The set-up and corresponding spectra are plotted in figures 7.14(b) and (c), respectively. By extracting the eigenfrequencies, we obtain $I_\parallel = 2.123\,7 \times 10^{-5}$ J m^{-2} and $I_\perp = 4.439\,9 \times 10^{-5}$ J m^{-2}. Similarly, we can systematically determine the parameters $I_\parallel$ and $I_\perp$ for different values of d, with the results being shown in figure 7.14(d) with symbols. Furthermore, it has been found that $I_\parallel$ and $I_\perp$ are functions of d^{-3}, d^{-5}, d^{-7}, and d^{-9} [56–58]. Finally, one can obtain the best fit of the numerical results

$$
\begin{aligned}
I_\parallel = \mu_0 M_s^2 r\bigg(&-1.720\,64 \times 10^{-4} + \frac{4.131\,66 \times 10^{-2}}{d^3} - \frac{0.246\,39}{d^5} + \frac{1.210\,66}{d^7} \\
&- \frac{1.818\,36}{d^9}\bigg),\ I_\perp = \mu_0 M_s^2 r\bigg(5.431\,58 \times 10^{-4} - \frac{4.346\,85 \times 10^{-2}}{d^3} \\
&+ \frac{1.237\,78}{d^5} - \frac{6.489\,07}{d^7} + \frac{13.642\,2}{d^9}\bigg),
\end{aligned} \tag{7.51}
$$

as shown by curves in figure 7.14(d). One can clearly see that the analytical formulas describe the simulation results well.

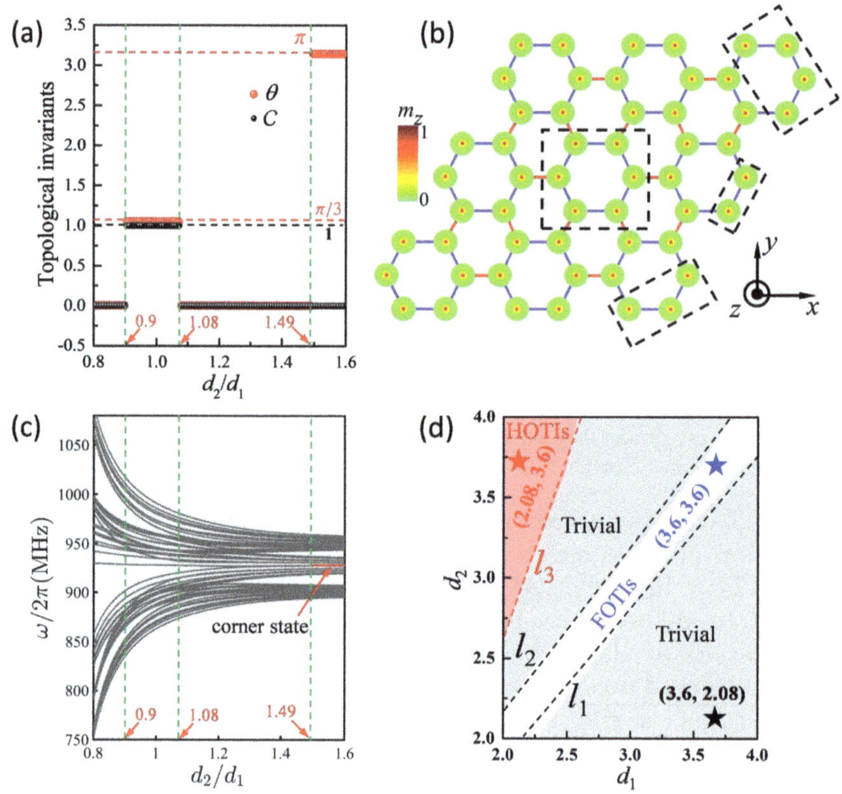

Figure 7.15. (a) The Chern number and Z_6 Berry phase as a function of d_2/d_1 with d_1 being fixed at 2.5. (b) A schematic diagram of a parallelogram-shaped vortex lattice with armchair edges. (c) The eigenfrequencies of the finite vortex lattice considered in (b) for different d_2/d_1. Here the red segment indicates the system supporting corner states. (d) Phase diagram of the vortex lattice. (Reproduced with permission from [8]. Copyright 2020 The American Physical Society.)

After these key parameters are identified, we can calculate the band structure of the vortex lattice, with the results shown in figures 7.13(b)–(d) with different geometric parameters. For $d_1 = d_2 = 3.6$ (see figure 7.13(c)), we can see that the highest three bands and the lowest three bands merged individually, leading to the emergence of a band gap. In this situation, the system supports the FOTI phase [2, 3]. However, if we consider $d_1 \neq d_2$, all bands are separated from each other and the system is in the insulating phase. By calculating the topological invariants Z_6 Berry phase and Chern number simultaneously, we can distinguish these insulating phases and obtain the phase diagram of the system.

The dependence of the Z_6 Berry phase (θ) and the Chern number (C) on the value of d_2/d_1 are plotted in figure 7.15(a), where we fix d_1 to 2.5. In addition, figure 7.15(c) shows the eigenfrequencies of a parallelogram-shaped vortex lattice (see figure 7.15(b)) for different ratios d_2/d_1. Combining figures 7.15(a) and (c), we conclude that when $d_2/d_1 < 0.9$ and $1.08 < d_2/d_1 < 1.49$, the system is in the trivial insulating phase, while the FOTI phase and HOTI phase will appear for parameters $0.9 < d_2/d_1 < 1.08$ and $d_2/d_1 > 1.49$, respectively. By systematically changing the parameters d_1 and d_2, we obtain the phase diagram of the system, as shown in figure 7.15(d). We can see that the boundary between the trivial and FOTI phases only determined by the value of d_2/d_1, as marked with dashed black lines (l_1: $d_2/d_1 = 0.94$ and l_2: $d_2/d_1 = 1.05$). In addition, the boundary between the trivial and HOTI phases exhibits a linear function, as marked by the dashed red line (l_3: $d_2 = 2.24d_1 - 1.88$). Noticeably, from equation (7.48), one can see clearly that the higher-order topological properties of the vortex lattice are not affected by the topological charge, because the sign of the topological charge only determines the direction of gyrotropic motion (clockwise or counterclockwise). However, it indeed has an influence on the chiral edge modes (first-order topology): when the sign of the topological charge switches, the chirality of the edge mode also reverses.

It is well known that the topological insulating phase can support symmetry-protected edge states. To show this exotic feature, we consider the strip structure with armchair edges, as plotted in figure 7.16(a). Figures 7.16(b)–(d) show the corresponding dispersion relation for different values of d_1 and d_2. We can see that for $d_1 = 3.6$ and $d_2 = 2.08$, the system is in the trivial insulating phase and no edge modes are supported (see figure 7.16(b)). When $d_1 = d_2 = 3.6$, the vortex lattice can be regarded as the magnetic soliton version of graphene. Different from the gapless dispersion relation for the perfect graphene nanostrips, the imaginary next-nearest hopping term in the dynamic equation of the vortex lattice opens a gap and a first-order topological chiral edge state will emerge [2, 3], as shown in figure 7.16(c). For $d_1 = 2.08$ and $d_2 = 3.6$, we can see that in addition to the bulk bands, two evident edge bands also emerge, as plotted in figure 7.16(d). By calculating the group velocity $\mathrm{d}\omega/\mathrm{d}k_x$ of the edge bands, we can easily identify that these localized edge modes are non-chiral and therefore not topologically protected. Interestingly, in this case, the higher-order topological edge states (corner states) will appear near these edge bands when we consider a reduced system which is finite in both dimensions.

Then we consider a parallelogram-shaped vortex lattice to show the second-order topological corner states. We choose $d_1 = 2.08$ and $d_2 = 3.6$. The spectrum is plotted in figure 7.17(a), from which we can see that there exist several degenerate states in the band gap. Furthermore, we plot the oscillation distribution of the vortex lattice for different modes, as shown in figures 7.17(b)–(f). Three different corner states are observed with the oscillations highly confined at obtuse-angled or acute-angled corners, as plotted in figures 7.17(c), (d), and (f). In addition, we also identify the bulk and edge states, with the results shown in figures 7.17(b) and (e), respectively.

It has been proved that topologically protected edge states and corner states can be immune from moderate disorder and defects. By analyzing the spectrum of vortex oscillation when the defects and disorder are introduced, we can judge whether these localized modes emerging in figure 7.17 are topological. Numerical results are shown in figures 7.18(a) and (b). Here, we assume that the coupling parameters ζ and ξ have a shift, i.e. $\zeta \to 1.5\zeta$, $\xi \to 1.5\xi$, which can act as a type of defect, while the disorder

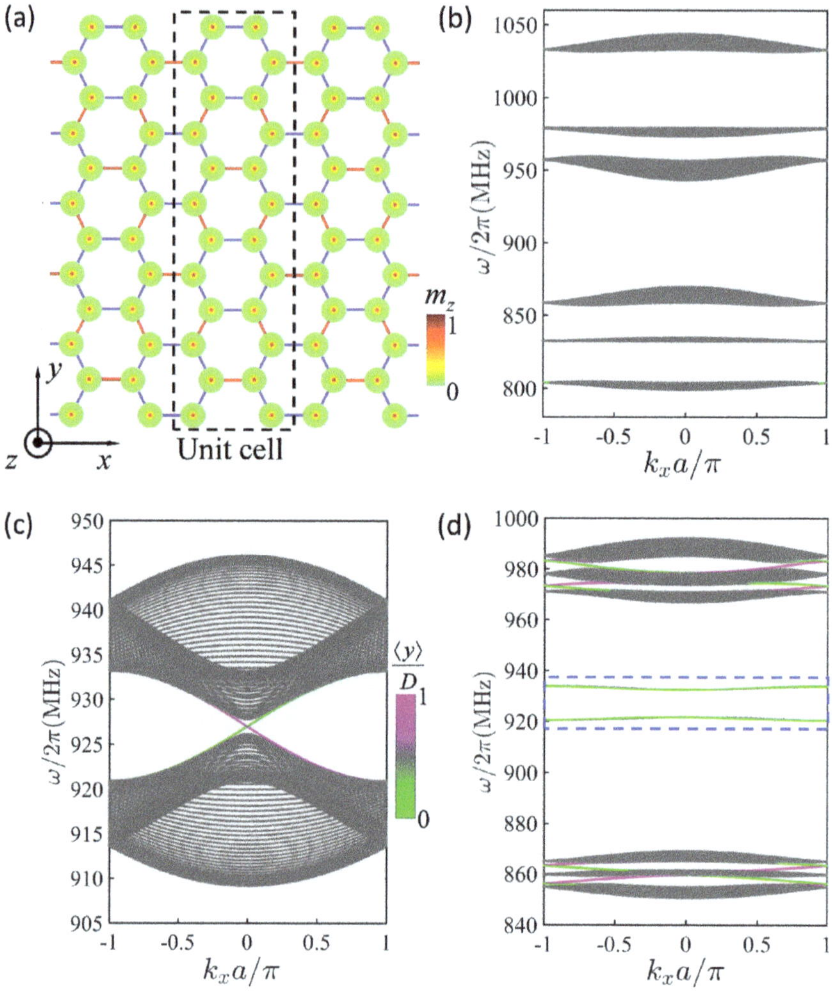

Figure 7.16. (a) A sketch map of a nanostrip with armchair edges based on the vortex lattice. The band structures for different parameters (b) $d_1 = 3.6$, $d_2 = 2.08$, (c) $d_1 = d_2 = 3.6$, and (d) $d_1 = 2.08$, $d_2 = 3.6$. The dashed blue box in (d) denotes the bands supporting non-chiral edge modes. D represents the width of the nanostrip. (Reproduced with permission from [8]. Copyright 2020 The American Physical Society.)

can be introduced by using the same method as those in previous sections discussing the DW chain. From figure 7.18(a) we can see that the frequency of the corner state 3 is robust against defects. However, the other edge and corner states have an obvious frequency shift compared to the clean system (black spheres). Furthermore, with an increase of disorder strength, the frequencies of the edge state, corner states 1 and 2, have significant modification and even move to the bulk band when the strength of the disorder is sufficiently large, while the frequency of corner state 3 is relatively robust, although the values change to some degree. Therefore, we conclude that the 'zero-energy' mode (corner state 3) protected by the generalized chiral symmetry is

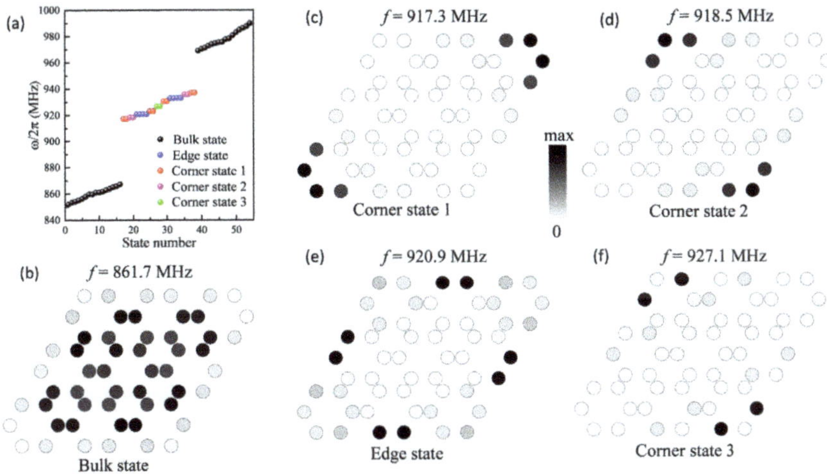

Figure 7.17. (a) The eigenfrequencies of the parallelogram-shaped vortex lattice with parameters $d_1 = 2.08$ and $d_2 = 3.6$. The oscillation distribution of the vortex lattice for the (b) bulk, (c), (d), (f) corner, and (e) edge modes with different frequencies. (Reproduced with permission from [8]. Copyright 2020 The American Physical Society.)

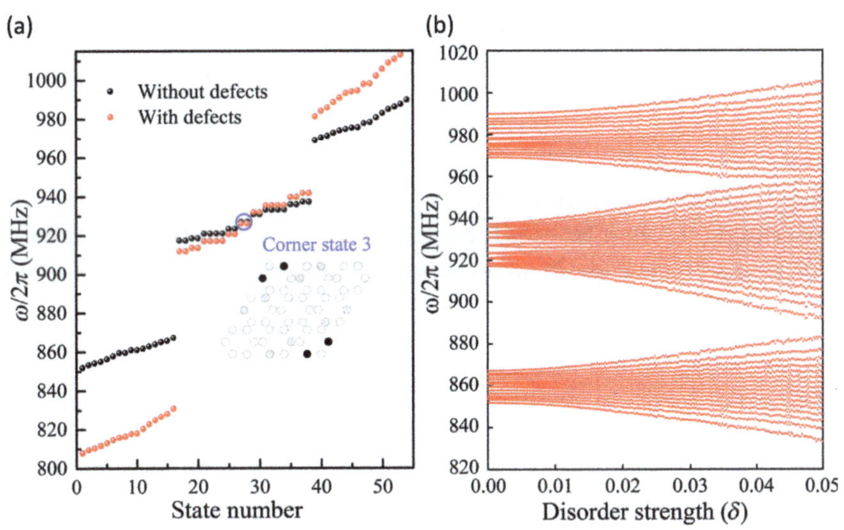

Figure 7.18. (a) Eigenfrequencies of a parallelogram-shaped vortex lattice with armchair edges in the absence (black dots) and presence (red dots) of defects. The blue circle indicates the topological robust corner state 3. (b) Eigenfrequencies of a parallelogram-shaped vortex lattice for different disorder strengths. (Reproduced with permission from [8]. Copyright 2020 The American Physical Society.)

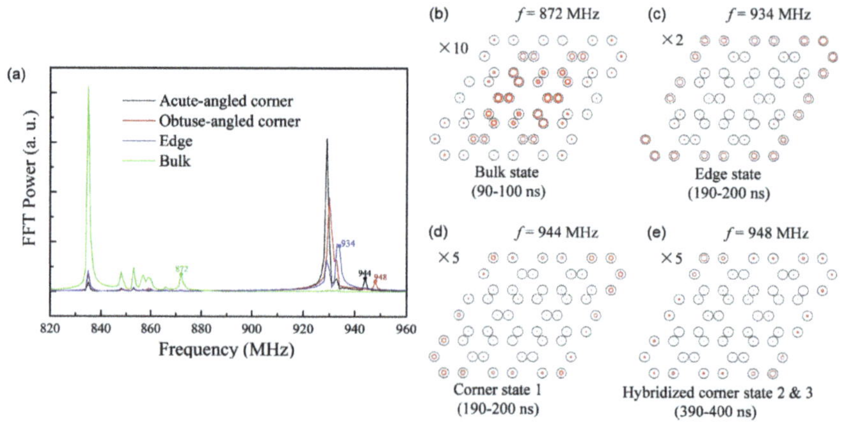

Figure 7.19. (a) The spectra of vortex oscillations at different positions. The 10 ns trajectories of all vortices under exciting fields with different frequencies, (b) 872 MHz, (c) 934 MHz, (d) 944 MHz, and (e) 948 MHz. (Reproduced with permission from [8]. Copyright 2020 The American Physical Society.)

much more robust than the other localized modes (corner states 1 and 2) protected by the crystalline symmetry. While the trivial edge state (see figure 7.17(e)) is described by the so-called Tamm–Shockley mechanism [39, 40].

One can perform micromagnetic simulations to confirm the theoretical predictions. Here we consider a magnetic vortex based parallelogram-shaped breathing honeycomb lattice (see figure 7.15(b)), where the parameters used are the same as those in figure 7.17. The spectra for the oscillation of vortices located at different positions are plotted in figure 7.19(a). One can clearly see that near the frequency of 944 MHz (948 MHz), there exists an obvious peak for an acute-angled corner (obtuse-angled corner) spectrum, while this does not happen for the edge and bulk bands. We can therefore infer that these two peaks represent two corner states with the oscillations confined at the acute-angled or obtuse-angled corners. Similarly, we can also identify the frequency range supporting bulk and edge states. In addition, we use four representative frequencies (as marked in figure 7.19(a)) to further show the difference of oscillation distribution between these modes. Then we excite their collective dynamics with a sinusoidal magnetic field applied to the whole lattice. When the vortex dynamics in the steady-state region, we record the 10 ns trajectories of gyrotropic motion for all vortices, as shown in figures 7.19(b)–(e). One can easily see the typical characteristics of oscillation distribution for bulk and edge states, as shown in figures 7.19(b) and (c), respectively. In addition, we identify corner state 1 with the oscillation confined to the acute-angled corner (see figure 7.19(d)). Interestingly, we observe the hybridization between the corner state 2 and corner state 3 (see figure 7.19(e)) for the reasons that their frequencies are very close and the corresponding wave functions have a great overlap (see figures 7.17(a) and 7.19(a)).

It has been proved that the corner states are protected by the symmetry of the Hamiltonian. Below, we show that the topological zero modes emerging in this model are indeed protected by the generalized chiral symmetry. First, because $(\xi_1^2 + 2\xi_2^2)/2\omega_K \ll \omega_0$, we can treat the diagonal element of H as a constant, i.e. $Q_0 = \omega_0$, which is also called 'zero-energy' for the original Hamiltonian. In addition, the terms $Q_{1,2,3,4,5,6}$ denote the next-nearest hopping. At first sight, the system has no chiral symmetry to protect the 'zero-energy' state for the reason that the breathing honeycomb lattice is not a bipartite lattice.

Here, we can generalize the chiral symmetry with unit cell containing six positions by defining

$$\begin{aligned} \Gamma_6^{-1}H_1\Gamma_6 &= H_2, \\ \Gamma_6^{-1}H_2\Gamma_6 &= H_3, \\ \Gamma_6^{-1}H_3\Gamma_6 &= H_4, \\ \Gamma_6^{-1}H_4\Gamma_6 &= H_5, \\ \Gamma_6^{-1}H_5\Gamma_6 &= H_6, \\ H_1 + H_2 + H_3 + H_4 + H_5 + H_6 &= 0, \end{aligned} \tag{7.52}$$

where Γ_6 is a diagonal chiral operator matrix, and $H_1 = H - Q_0 I$. To demonstrate that the system indeed has generalized chiral symmetry, we can divide the system into six subgroups where the components of the Hamiltonian are nonzero only between different subgroups. This property is essential for chiral symmetry which implies that there is no interaction within sublattices. Combining the six equations in (7.52), we obtain $\Gamma_6^{-1}H_6\Gamma_6 = H_1$, indicating that $[H_1, \Gamma_6^6] = 0$; therefore, $\Gamma_6^6 = I$, which is similar to the SSH model [4]. In addition, each Hamiltonian $H_{1,2,3,4,5,6}$ has the same eigenvalues $\lambda_{1,2,3,4,5,6}$. It is straightforward to know that Γ_6 has six eigenvalues as 1, $\exp(2\pi i/6)$, $\exp(4\pi i/6)$, $\exp(\pi i)$, $\exp(8\pi i/6)$, and $\exp(10\pi i/6)$. We thus can write

$$\Gamma_6 = \begin{pmatrix} 1 & 0 & 0 & 0 & 0 & 0 \\ 0 & e^{\frac{2\pi i}{6}} & 0 & 0 & 0 & 0 \\ 0 & 0 & e^{\frac{4\pi i}{6}} & 0 & 0 & 0 \\ 0 & 0 & 0 & e^{\pi i} & 0 & 0 \\ 0 & 0 & 0 & 0 & e^{\frac{8\pi i}{6}} & 0 \\ 0 & 0 & 0 & 0 & 0 & e^{\frac{10\pi i}{6}} \end{pmatrix}, \tag{7.53}$$

in the same basis as that for expressing Hamiltonian (7.49). By taking the trace of the sixth line from equations (7.52), we have $\sum_{i=1}^{6}\mathrm{Tr}(H_i) = 6\mathrm{Tr}(H_1) = 0$, which implies that the sum of the six eigenvalues is zero, i.e. $\sum_{i=1}^{6}\lambda_i = 0$. For an eigenstate ϕ_j that is only supported in sublattice j, it will satisfy the relations $H_1\phi_j = \lambda\phi_j$ and $\Gamma_6\phi_j = \exp[2\pi i(j-1)/6]\phi_j$ with $j = 1, 2, 3, 4, 5, 6$. By utilizing these formulas, we have $\sum_{i=1}^{6}H_i\phi_j = \sum_{i=1}^{6}\Gamma_6^{-(i-1)}H_1\Gamma_6^{i-1}\phi_j = 6\lambda\phi_j = 0$. Finally, we conclude that $\lambda = 0$ for any state that has support in only one sublattice, i.e. the zero-energy corner state.

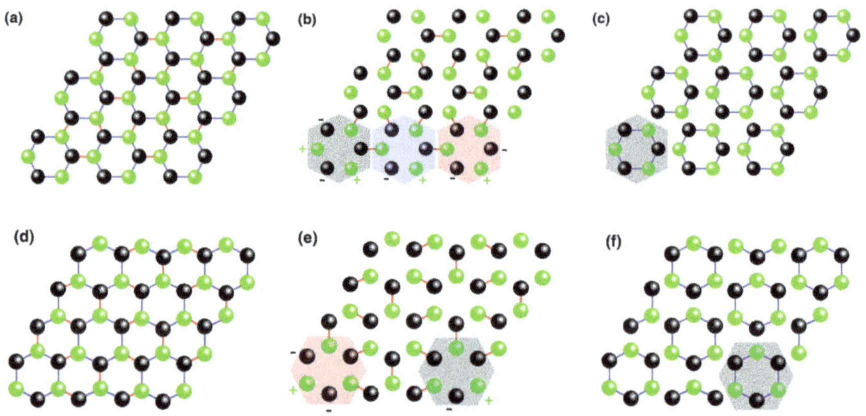

Figure 7.20. The configurations of breathing honeycomb lattices based on magnetic vortices with (a) armchair and (d) zigzag edges. Panels (b) and (e) ((c) and (f)) show the configurations in the zero-correlation-length limit $d_2 \to \infty$ ($d_1 \to \infty$) considered in (a) and (d), respectively. The green and black spheres represent the eigenvalues $+1$ and -1 of the chiral-symmetry operator, respectively. The shaded areas denote the different unit cells. (Reproduced with permission from [8]. Copyright 2020 The American Physical Society.)

Interestingly, if we consider the magnetic vortex based breathing honeycomb lattice with zigzag edges, the topologically stable corner states will appear in the acute angle instead of the obtuse one. The position of the topological (chiral-symmetry-protected) corner states emerging in a finite system for different edges can be explained well by the topological index $N = N_+ - N_-$ which describes the interplay between the topology of the system and the defects [59]. Here, the parameter N counts the number of topologically protected modes localized to corners, and N_+ and N_- are the integers counting the numbers of eigenstates of the chiral-symmetry operator $\widehat{\Pi}$ with eigenvalues $+1$ and -1, respectively. Figures 7.20(a) and (d) plot the parallelogram-shaped vortex lattices with armchair and zigzag edges, respectively. In addition, we also plot the configuration under the zero-correlation-length limit, i.e. the bond length $d_2 \to \infty$ ($d_1 \to \infty$) with the coupling within (between) unit cells close to zero, as shown in figures 7.20(b) and (e) (figures 7.20(c) and (f)), respectively. From the phase diagram as shown in figure 7.15(d), we know that the configurations in figures 7.20(b) and (e) (figures 7.20(c) and (f)) belong to the HOTI (trivial) phase. For figure 7.20(b), one can see that there exist four modes in the acute-angled corner unit cell (marked by the area shaded gray), two of them have eigenvalues $+1$ and the other two have eigenvalues -1; therefore, $N = |N_+ - N_-| = |2 - 2| = 0$. Similarly, we can obtain $N = 0$ for the edge unit cell and $N = 1$ for the obtuse-angled-corner unit cell. As a result, we conclude that the topological protected corner states appear only at obtuse-angled corners for armchair edges. However, if we consider zigzag edges (see figure 7.20(e)), we can obtain $N = 1$ for the acute-angled corner while $N = 0$ for the obtuse-angled corner, implying that the topological protected corner states emerge at acute-angled corners instead of obtuse-angled corners. On the other side, for the limit of $d_1 \to \infty$ (see figures 7.20(c) and (f)), we see that there are no uncoupled vortices. The six corners of the isolated hexagon are equivalent and no special edge or corner states appear.

In various breathing lattices, the topological corner states are protected by both the generalized chiral symmetry and the crystalline symmetry. To identify this conclusion experimentally, we can check whether the frequencies of corner states are stable after the generalized chiral symmetry or crystalline symmetry is broken. Generally speaking, by adding the NNN hopping terms to a specific position of the lattice, one can realize symmetry breaking. However, for the magnetic and condensed matter systems, it is very difficult to introduce NNN hopping with a desired manner. Fortunately, it has been shown recently that the topolectrical circuits can easily solve this issue for the reason that the coupling between any two unit cells can be conveniently modified by introducing extra circuit elements (for example capacitor and/or inductor). Yang *et al* [60] observed the symmetry-protected corner states (zero modes) in breathing kagome lattice based on topolectrical circuit. They proved that the frequencies of corner states have a visible variation when the NNN hopping term is added through connecting a capacitor C_N between the A sites at the corner. Figure 7.21(a) shows the corresponding illustration and experimental set-up. The experimental measurement results of X with and without C_N are plotted in figure 7.21(b), where X is the impedance between the corner and bulk nodes. Furthermore, they also confirmed that the frequency of corner states does not shift when the NNN hopping is located at the edge or in the bulk. However, the generalized chiral symmetry and C_3 symmetry are broken simultaneously in figure 7.21, thus the experimental results can only support that the corner states are protected by either C_3 symmetry or both C_3 and the generalized chiral symmetries. Noticeably, the generalized chiral symmetry is built upon crystalline symmetry (here it is C_3 symmetry). If the C_3 symmetry is broken, the generalized chiral symmetry is also broken. The corner states are therefore impossible to only protect by generalized chiral symmetry. Very recently, Song *et al* [61] investigated the topological stability of corner states based on a hybrid

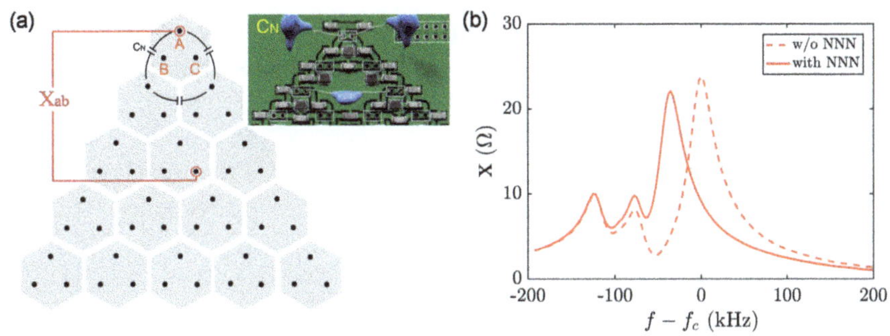

Figure 7.21. (a) A schematic diagram of a breathing kagome LC circuit, where three capacitors C_N are connected between sublattices A at the top corner. The experiment set-up is also shown in inset. (b) Experimental measurements of the system considered in (a). The solid and dashed curves denote the results with and without C_N, respectively. (Reproduced with permission from [60]. Copyright 2020 The American Physical Society.)

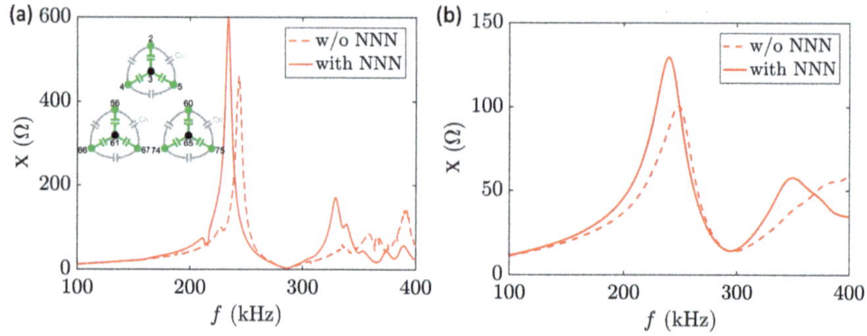

Figure 7.22. (a) The theoretical calculated impedance as a function of the driving frequency in a hybrid honeycomb–kagome LC circuit. Here NNN hopping terms are introduced to all three corners. The inset shows the configurations of the NNN coupling. (b) The experimental measurement for the configuration considered in (a). (Reproduced with permission from [61]. Copyright 2020 The American Physical Society.)

honeycomb–kagome circuit, where the C_3 symmetry is respected while the generalized chiral symmetry is broken. The inset of figure 7.22(a) plots the experiment set-up, where the NNN hoppings are introduced to all three corners. Both the theoretical calculation and experimental measurement prove that the 'zero-energy' corner state disappears and the corresponding frequency experiences a blueshift, as shown in figures 7.22(a) and (b), respectively. We thus conclude that the corner states are protected by both the crystalline symmetry (C_3) and generalized chiral symmetry instead of crystalline symmetry (C_3) alone.

In addition, it is worth noting that the second-order corner states can also emerge in other crystal structures, such as breathing kagome [7] and square [9] lattice of vortices.

7.1.3 Third-order topological insulating phases

In the above section, we have introduced second-order topological insulators based on magnetic vortices. In this section, we discuss third-order topological insulators in three-dimensional vortex lattice.

We consider a three-dimensional breathing cuboid lattice composed of magnetic vortices, as shown in figure 7.23. As discussed in previous sections, we can use the generalized Thiele's equation to describe the collective dynamics of vortex lattice. It is worth noting that the interaction energy must be modified as $U_{jk} = I_{\parallel}U_j^{\parallel}U_k^{\parallel} - I_{\perp}U_j^{\perp}U_k^{\perp} + \mu\mathbf{U}_j \cdot \mathbf{U}_k$, where the last term is the interlayer interaction with μ being the coupling parameter. Similarly, by defining $\mathbf{U}_j = (u_j, v_j)$, $\psi_j = u_j + iv_j$, the Thiele's equation takes the following form

$$\begin{aligned} -\mathrm{i}\dot{\psi}_j = & \left(\omega_0 - \frac{\xi_1^2 + \xi_2^2}{\omega_0}\right)\psi_j + \sum_{k\in\langle j\rangle,l} \zeta_l\psi_k + \sum_{k\in\langle j'\rangle,n} \eta_n\psi_k - \frac{\xi_1\xi_2}{2\omega_0}\sum_{s\in\langle\langle j_1\rangle\rangle} \mathrm{e}^{\mathrm{i}2\bar{\theta}_{js}}\psi_s \\ & - \frac{\xi_2^2}{2\omega_0}\sum_{s\in\langle\langle j_2\rangle\rangle} \mathrm{e}^{\mathrm{i}2\bar{\theta}_{js}}\psi_s - \frac{\xi_1^2}{2\omega_0}\sum_{s\in\langle\langle j_3\rangle\rangle} \mathrm{e}^{\mathrm{i}2\bar{\theta}_{js}}\psi_s, \end{aligned} \tag{7.54}$$

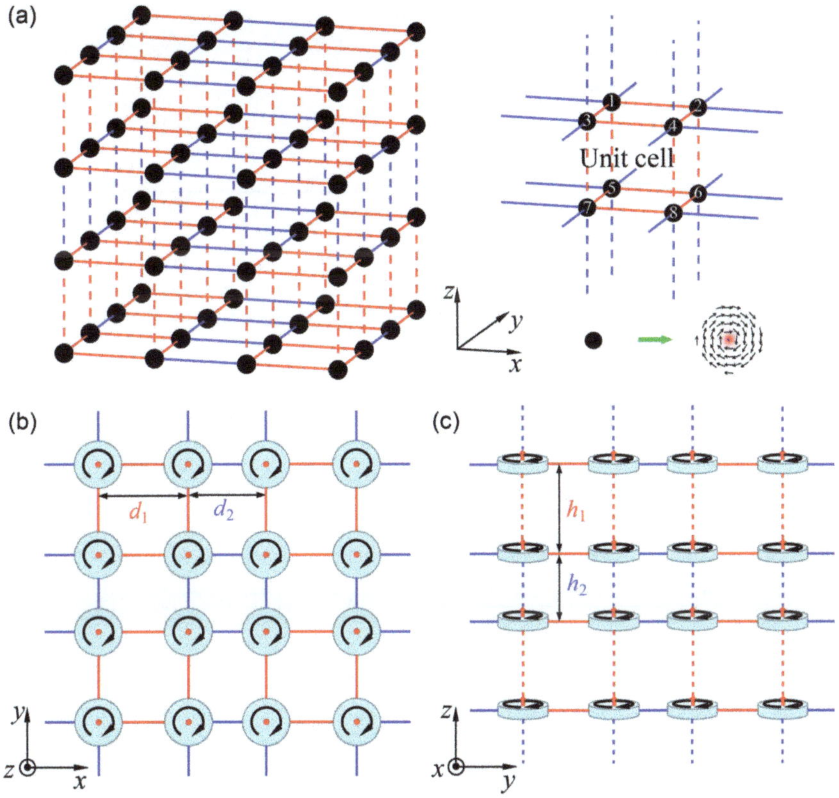

Figure 7.23. (a) A schematic diagram of a three-dimensional vortex lattice, where the black spheres represent nanodisks with the vortex state. The solid (dashed) red and blue segments denote the alternate distances of intralayer (interlayer) intracellular and intercellular bonds, respectively. The (b) top and (c) side views of the vortex lattice with d_1, d_2, h_1, and h_2 denoting the different bond lengths. (Reproduced with permission from [10]. Copyright 2021 The American Physical Society.)

where $\eta_n = \mu_n/|G|$, $n = 1$ and 2 represents the interlayer intracellular and intercellular bond, respectively; $\langle j \rangle$ ($\langle j' \rangle$) denotes the intralayer (interlayer) nearest neighbor of j; $\langle\langle j_1 \rangle\rangle$, $\langle\langle j_2 \rangle\rangle$, and $\langle\langle j_3 \rangle\rangle$ are the next-nearest neighbors of j for intralayer coupling.

The parameters K, $I_{\parallel}$, $I_{\perp}$, and μ must be calculated first if we want to solve equation (7.54). In previous sections, we have presented the method to determine parameters K, $I_{\parallel}$, and $I_{\perp}$. The parameter μ can be similarly determined from micromagnetic simulations. We consider a stacked two-vortex system (see figure 7.24(a)). Imposing $\psi_j = u_j + iv_j$ and $\psi_j = \psi_j e^{i\omega t}$, the equations describing the two-vortex system can be simplified as

$$\begin{aligned} \omega\psi_1 &= \frac{K}{|G|}\psi_1 + \frac{\mu}{|G|}\psi_2, \\ \omega\psi_2 &= \frac{K}{|G|}\psi_2 + \frac{\mu}{|G|}\psi_1. \end{aligned} \tag{7.55}$$

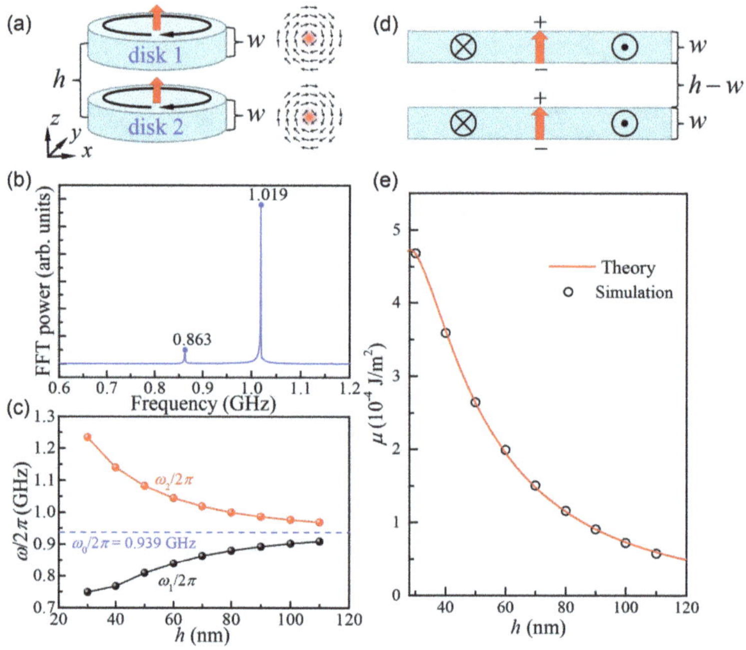

Figure 7.24. (a) A sketch map of a stacked two-vortex system. (b) The spectrum of the system considered in (a) with parameter $h = 70$ nm. (c) The eigenfrequencies of the two-vortex system as a function of h. (d) An illustration of the interaction between two vortices. (e) The dependence of the parameter μ on h. The black circles and red line represent the simulation results and analytical formula, respectively. (Reproduced with permission from [10]. Copyright 2021 The American Physical Society.)

After solving equation (7.55), we get two eigenfrequencies: $\omega_1 = (K - \mu)/|G|$ and $\omega_2 = (K + \mu)/|G|$. Then we obtain the expression of parameter $\mu = (\omega_2 - \omega_1)|G|/2$. The Fourier spectrum for vortex oscillation is plotted in figure 7.24(b), where we chose $h = 70$ nm. The two eigenfrequencies are $\omega_1/2\pi = 0.863$ GHz and $\omega_2/2\pi = 1.019$ GHz. Therefore, we obtain $\mu = 1.505\,8 \times 10^{-4}$ J m^{-2}. Similarly, for other value of h, we can also get the frequencies ω_1 and ω_2, with the results shown in figure 7.24(c). One can clearly see that when h increases gradually, ω_1 increases while ω_2 decreases. We naturally infer that when $h \to \infty$, $\omega_1 = \omega_2 = \omega_0$. It has been showed that the magnetostatic interaction between vortices can be regarded as the interaction between vortex cores [34, 62], with the illustration shown in figure 7.24(d). Therefore, the interaction between two vortices can be simplified as three parts: (i) the attractive interaction between negative magnetic charge of disk 1 and positive magnetic charge of disk 2 [$\propto 1/(h - w)$]; (ii) the repulsive interaction of the same magnetic charge between disk 1 and disk 2 ($\propto 1/h$); (iii) the attractive interaction between positive magnetic charge of disk 1 and negative magnetic charge of disk 2 [$\propto 1/(h + w)$]. To obtain the expression of parameter μ on h, we use the formula

$$\mu = c_1/(h - w) + c_2/h + c_3/(h + w) \tag{7.56}$$

to fit the micromagnetic simulation data. Figure 7.24(e) plots the numerical data (black circles) and fitting curve with $c_1 = -5.747\,8 \times 10^{-11}$ J m^{-1}, $c_2 = 2.369\,5 \times 10^{-10}$ J m^{-1}, and $c_3 = -1.822\,9 \times 10^{-10}$ J m^{-1}, which shows a high degree of consistency.

When these key parameters are determined, one can calculate the dispersion relations of the system. For an infinite system, we use the cube plotted in figure 7.23(a) as the unit cell. Then the basis vectors are $\mathbf{a}_1 = (d_1 + d_2)\hat{x}$, $\mathbf{a}_2 = (d_1 + d_2)\hat{y}$, and $\mathbf{a}_3 = (h_1 + h_2)\hat{z}$. By using the plane-wave expansion to ψ_j, we obtain the Hamiltonian which can be expressed as a matrix form

$$H = \begin{pmatrix} Q_0 & Q_1 & Q_2 & Q_3 & Q_4 & 0 & 0 & 0 \\ Q_1^* & Q_0 & Q_5 & Q_2 & 0 & Q_4 & 0 & 0 \\ Q_2^* & Q_5^* & Q_0 & Q_1 & 0 & 0 & Q_4 & 0 \\ Q_3^* & Q_2^* & Q_1^* & Q_0 & 0 & 0 & 0 & Q_4 \\ Q_4^* & 0 & 0 & 0 & Q_0 & Q_1 & Q_2 & Q_3 \\ 0 & Q_4^* & 0 & 0 & Q_1^* & Q_0 & Q_5 & Q_2 \\ 0 & 0 & Q_4^* & 0 & Q_2^* & Q_5^* & Q_0 & Q_1 \\ 0 & 0 & 0 & Q_4^* & Q_3^* & Q_2^* & Q_1^* & Q_0 \end{pmatrix}, \tag{7.57}$$

where the matrix elements have the accurate expression as

$$\begin{aligned} Q_0 &= \omega_0 - \frac{\xi_1^2 + \xi_2^2}{\omega_0} - \frac{\xi_1\xi_2}{\omega_0}[\cos(\mathbf{k} \cdot \mathbf{a}_1) + \cos(\mathbf{k} \cdot \mathbf{a}_2)], \\ Q_1 &= \zeta_2 + \zeta_1 \exp(-i\mathbf{k} \cdot \mathbf{a}_1), \\ Q_2 &= \zeta_2 + \zeta_1 \exp(i\mathbf{k} \cdot \mathbf{a}_2), \\ Q_3 &= \frac{\xi_1\xi_2}{\omega_0}[\exp(i\mathbf{k} \cdot \mathbf{a}_2) + \exp(-i\mathbf{k} \cdot \mathbf{a}_1)] + \frac{\xi_1^2}{\omega_0} \exp[i\mathbf{k} \cdot (\mathbf{a}_2 - \mathbf{a}_1)] + \frac{\xi_2^2}{\omega_0}, \\ Q_4 &= \eta_1 + \eta_2 \exp(i\mathbf{k} \cdot \mathbf{a}_3), \\ Q_5 &= \frac{\xi_1\xi_2}{\omega_0}[\exp(i\mathbf{k} \cdot \mathbf{a}_2) + \exp(i\mathbf{k} \cdot \mathbf{a}_1)] + \frac{\xi_1^2}{\omega_0} \exp[i\mathbf{k} \cdot (\mathbf{a}_2 + \mathbf{a}_1)] + \frac{\xi_2^2}{\omega_0}, \end{aligned} \tag{7.58}$$

where $\mathbf{k}$ denotes the wave vector. By diagonalizing Hamiltonian (7.57), we can obtain the spectrum of the system which determines the phase diagram because of the bulk–boundary correspondence. We will provide detailed calculations and discussion below.

First, we determine the band structures along the path Γ–K–M–N–Γ (see figure 7.25(a)) with different geometric parameters. The results are shown in figures 7.25(c)–(k), where d_2 and h_2 are fixed to 150 nm and 70 nm, respectively. One can clearly see that the spectra support eight bands because of the interlayer coupling, in contrast to four bands for the two-dimensional breathing square lattice [9]. Normally speaking, energy degeneracy is accompanied by some symmetries, and

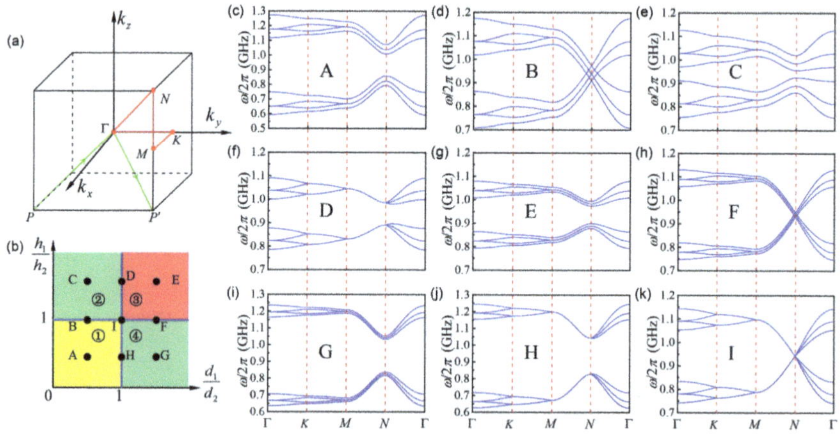

Figure 7.25. (a) The first Brillouin zone of the three-dimensional vortex lattice, where the high-symmetry points Γ, K, M, and N are located at $(k_x, k_y, k_z) = (0, 0, 0)$, $\left(0, \frac{\pi}{d_1+d_2}, 0\right)$, $\left(\frac{\pi}{d_1+d_2}, \frac{\pi}{d_1+d_2}, 0\right)$, and $\left(\frac{\pi}{d_1+d_2}, \frac{\pi}{d_1+d_2}, \frac{\pi}{h_1+h_2}\right)$, respectively. (b) The phase diagram of the system. (c)–(k) The band structures of the infinite system for different parameters as marked with black dots in (b). (Reproduced with permission from [10]. Copyright 2021 The American Physical Society.)

a gap opening is caused by the symmetry breaking. For (d_1, h_1)= (150 nm, 70 nm) marked by letter 'I', one can see that all the bands merge together with a gapless structure (see figure 7.25(k)). Here the system holds the inversion symmetry. On the one side, if we keep $d_1/d_2 = 1$ while letting $h_1/h_2 \neq 1$, as a result, the inversion symmetry along the z-axis is broken and the gap will open at point N, as shown in figures 7.25(f) and (j), where the letters 'D' and 'H' denote (d_1, h_1)= (150 nm, 110 nm) and (150 nm, 40 nm), respectively. On the other side, if we keep $h_1/h_2 = 1$ and let $d_1/d_2 \neq 1$, the inversion symmetries along x-axis and y-axis are broken simultaneously, then the gaps emerge at points K and M. The results are shown in figures 7.25(d) and (h), where the letters 'B' and 'F' represent (d_1, h_1)= (120 nm, 70 nm) and (180 nm, 70 nm), respectively. Finally, if the conditions $d_1/d_2 \neq 1$ and $h_1/h_2 \neq 1$ are satisfied simultaneously, the inversion symmetries along all directions (x-axis, y-axis, and z-axis) are broken. The band gaps appear at all high-symmetry points (K, M, and N), with the corresponding band structures being plotted in figures 7.25(c) (d_1, h_1)= (120 nm, 40 nm), 7.25(e) (d_1, h_1)= (120 nm, 110 nm), 7.25(g) (d_1, h_1)= (180 nm, 110 nm), and 7.25(i) (d_1, h_1)= (180 nm, 40 nm). Therefore, we conclude that the removal of energy degeneracy comes from the breaking of inversion symmetry. To further judge whether these insulating phases are topological, we can analyze the distribution of wave functions of eigen-modes and their robustness in finite systems.

We consider a finite system containing 512 (8 × 8 × 8) nanodisks with four different geometric parameters. In region ①, we set (d_1, d_2, h_1, h_2)= (104 nm, 180 nm, 40 nm, 110 nm), and plot the eigenfrequencies in figure 7.26(a). After analyzing the wave function, we find that the system can only support bulk states

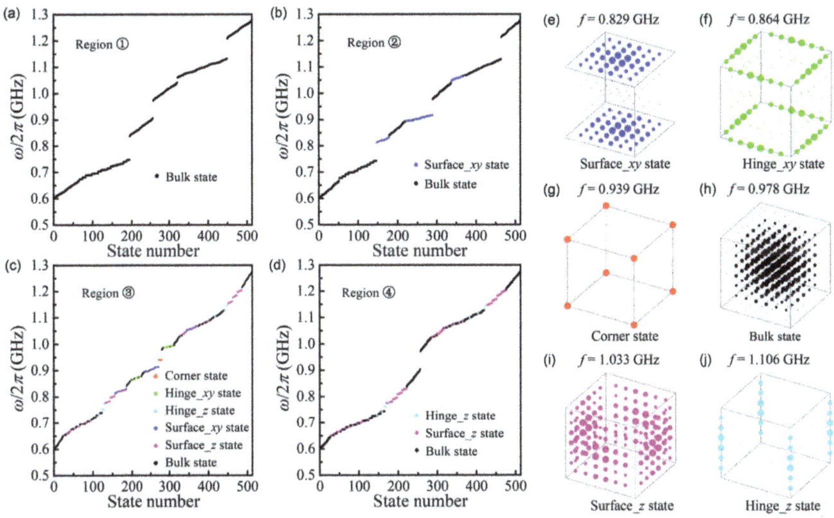

Figure 7.26. The eigenfrequencies of the $8 \times 8 \times 8$ vortex lattice for (a) parameter region ①, (b) parameter region ②, (c) parameter region ③, and (d) parameter region ④, as marked in figure 7.25(b). (e)–(j) The oscillation distribution in the vortex lattice for different states with the size of the colored spheres indicating the amplitude of the wave function. (Reproduced with permission from [10]. Copyright 2021 The American Physical Society.)

(see figure 7.26(h)). Then we consider region ② and set parameters (d_1, d_2, h_1, h_2) = (104 nm, 180 nm, 110 nm, 40 nm), and plot the eigenfrequencies in figure 7.26(b). We can see that the finite system can support both the surface xy states (see figure 7.26(e)) and bulk states. For region ③, we set (d_1, d_2, h_1, h_2)= (180 nm, 104 nm, 110 nm, 40 nm), with the spectrum shown in figure 7.26(c). We can observe five different edge states: zero-dimensional corner states (see figure 7.26(g)), two one-dimensional hinge states (see figures 7.26(f) and (j)), and two two-dimensional surface states (see figures 7.26(e) and (i)). Finally, for region ④, we set parameters (d_1, d_2, h_1, h_2)= (180 nm, 104 nm, 40 nm, 110 nm). In this case, we identify the hinge z (figures 7.26(j)), surface z (figures 7.26(i)), and bulk states.

To judge whether the edge states emerging in figure 7.26(c) are topologically robust, we study the influence of disorder and defects on the spectra of the finite systems. The results are shown in figures 7.27(a) and (b). We clearly see that only the corner states are robust against disorder and defects, while other edge states, including hinge and surface modes, are sensitive to the disturbances. Therefore, we have realized a third-order TI based on vortex lattices. These results can be understood from the perspective of the SSH model. On the one side, it has been proved that the breathing square lattice based on vortices can support the second-order TI when $d_1/d_2 > 1$ [9]. On the other side, the stacked structure can be regarded as an analogy to the SSH model. In addition, the number of layers considered here is even (eight). Therefore, the topological edge states emerge only when $h_1/h_2 > 1$ [5]. Naturally, we conclude that the third-order TI can appear if and only if the geometric conditions $d_1/d_2 > 1$ and $h_1/h_2 > 1$ are satisfied simultaneously.

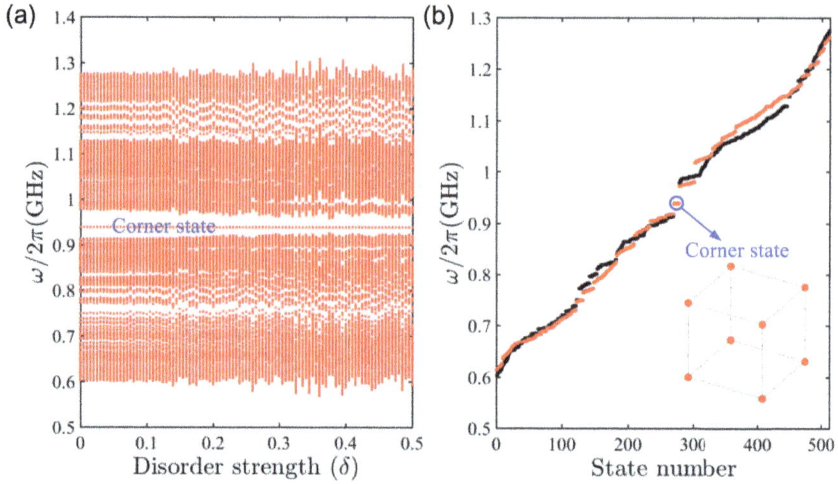

Figure 7.27. (a) Eigenfrequencies of an $8 \times 8 \times 8$ vortex lattice with different disorder strengths. (b) Eigenfrequencies of an $8 \times 8 \times 8$ vortex lattice in the absence (black dots) or in the presence (red dots) of defects. (Reproduced with permission from [10]. Copyright 2021 The American Physical Society.)

Noticeably, if we consider odd layers, the conditions for the appearance of corner states are $d_1/d_2 > 1$ and $h_1 \neq h_2$. Interestingly, in such a case, the corner states will only emerge the lower (upper) surface when $h_1/h_2 < 1$ ($h_1/h_2 > 1$).

Furthermore, the Z_4 Berry phase can also be used to identify whether the system is in the HOTI phase. After calculations, we obtain that the Z_4 Berry phase is quantized to π and 0 for region ③ and region ①, respectively, which indicates that these two regions belong to different topological phases.

Full micromagnetic simulations can be used to verify the theoretical calculations. The parameters adopted in the simulations are the same as those in figure 7.26(c). Here we consider a $6 \times 6 \times 6$ nanodisks system. Figure 7.28(a) plots the spectra of vortex oscillations with the red, green, cyan, blue, magenta, and black curves representing the positions of the corner, hinge xy, hinge z, surface xy, surface z, and bulk bands, respectively, from which we can easily identify the frequency range for different modes. To show the difference between various modes, we plot the spatial distribution of FFT intensity for six representative frequencies marked in figure 7.28(a), with the results being shown in figures 7.28(b)–(g). Here, the diameters of the colorful dots are proportional to the strength of FFT. We therefore identify the following different states: (i) surface xy state with oscillation confined to top and bottom surfaces (figure 7.28(b)); (ii) hinge xy state with oscillation confined to eight edges of top and bottom surfaces (figure 7.28(c)); (iii) corner state with oscillation confined to eight corners (figure 7.28(d)); (iv) bulk state with oscillation spreading over the system except the surfaces (figure 7.28(e)); (v) surface z state with oscillation confined to four side surfaces (figure 7.28(f)); (vi) hinge z state with oscillation confined to four side edges (figure 7.28(g)). These results compare well with theoretical predictions shown in figure 7.26.

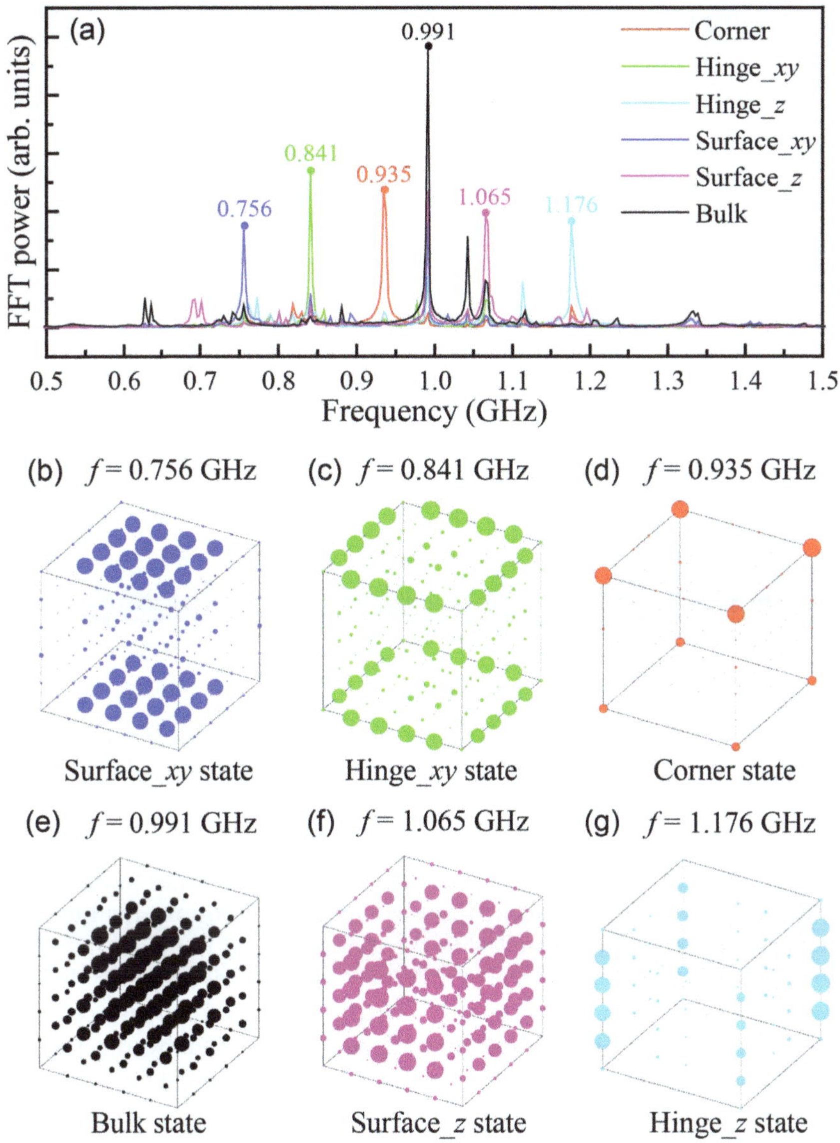

Figure 7.28. (a) The spectra of vortex oscillations at different positions with the exciting field applying to the whole system. (b)–(g) The oscillation distributions of the vortex lattice for different modes as marked in (a). Here the size of the colored spheres reflects the strength of the oscillation. (Reproduced with permission from [10]. Copyright 2021 The American Physical Society.)

In the above micromagnetic simulations, the external driving field (sinc function) is applied to the whole system, thus the information of the time-evolution of each state is missing. To observe this important information, we apply a sinusoidal driving field $\mathbf{h}(t) = h_0 \sin(2\pi f t)\hat{x}$ with $h_0 = 0.01$ mT and $t = 60$ ns confined to eight vortices at one corner, as represented by red cubes (see figure 7.29). Then we plot the

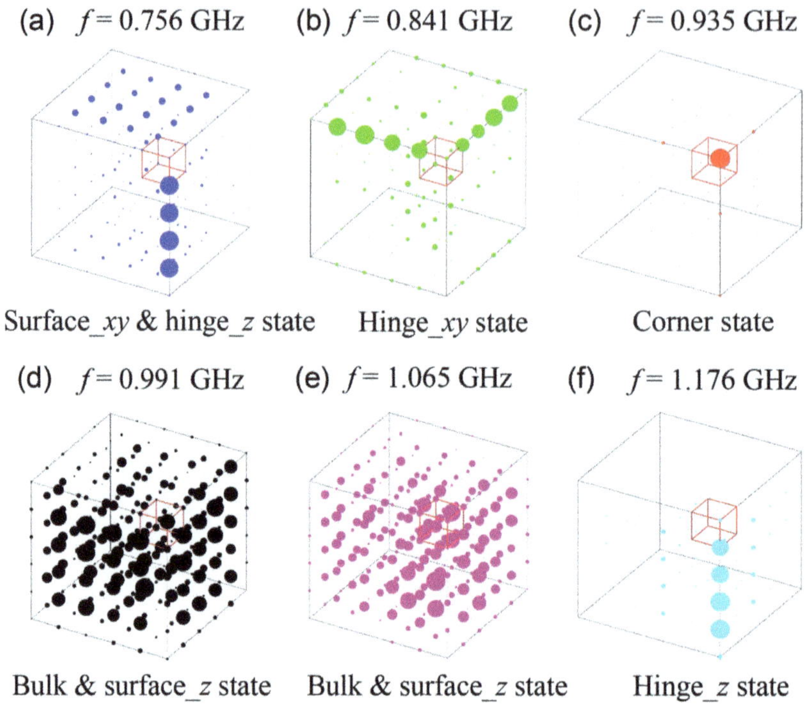

Figure 7.29. The oscillation distribution of a vortex lattice when local exciting fields with different frequencies are applied. (Reproduced with permission from [10]. Copyright 2021 The American Physical Society.)

spatial distribution of vortex oscillations with different frequencies, the results are shown in figures 7.29(a)–(f), where the size of the spheres denotes the strength of the oscillations. We can clearly see that the oscillation distributions for hinge z, hinge xy, and corner states compare well with the modes shown in figure 7.28. However, the other three modes do not. Interestingly, we observe several hybridized modes: surface xy and hinge z state (see figure 7.29(a)), and bulk and surface z state (see figures 7.29(d) and (e)). The mode hybridization is caused by the fact that the frequencies of these modes are close and the corresponding wave functions have great overlap (see figure 7.28(a)). In addition, we can identify that the hinge xy states are not chiral, as shown in figure 7.29(b). However, we cannot judge whether the hinge z states are chiral. To solve this issue, we apply an excitation field localized at a bottom corner with frequency $f = 1.176$ GHz and we observe that the hinge z states are also excited (not show here). Therefore, we conclude that all hinge states emerging in this system are non-chiral.

In this section, we realize the third-order TI based on 3D vortex lattices. For application about the emerging edge states, the 1D hinge state can be used to construct 3D magnonic waveguide, in contrast to ordinary 2D magnonic waveguides constructed with magnetic nanowires [63–67] or domain walls [68–70]. In addition, the 0D corner state can act as topologically protected nano-oscillators. Noticeably, in this model, we ignore both the non-Newtonian and inertial terms in

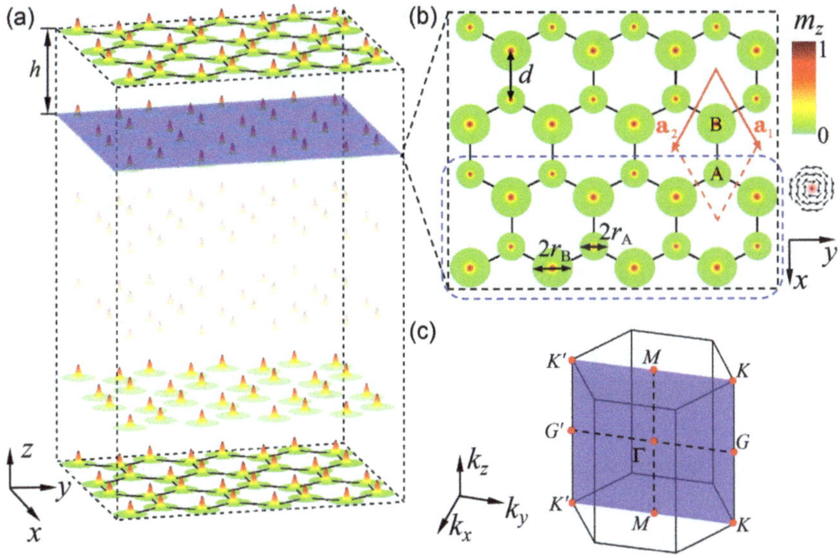

Figure 7.30. (a) A schematic diagram of a 3D stacked honeycomb lattice based on magnetic nanodisks with vortex states. (b) Zoomed in details of one layer. (c) The first Brillouin zone of the lattice. (Reproduced with permission from [80]. Copyright 2022 The American Physical Society.)

the generalized Thiele's equation (7.39). Apparently, if we consider the higher-order terms, we can obtain three band structures around 0.939 GHz, 11.945 GHz, and 14.192 GHz. Then two extra corner states will appear with higher frequencies (~10 GHz).

7.2 Topological solitonic semimetals

Depending on different band structures, there exist two types of non-trivial topological phases, one is topological insulators [71–75] which have gapped band structures, the other is a topological semimetal [76–79] with gapless dispersion relations. In previous sections, we have discussed the magnetic soliton lattice based topological insulators. In this section, we introduce the topological Weyl semimetal in vortex arrays arranged as a 3D stacked honeycomb lattice.

The collective dynamics of magnetic vortices in a 3D stacked honeycomb lattice (see figure 7.30(a)) can be described by the Thiele equation. After some algebra, we obtain the simplified dynamic equation

$$-i\dot{\psi}_j = \omega_j\psi_j + \zeta\sum_{k\in\langle j\rangle}\psi_k + \xi\sum_{k\in\langle j\rangle}e^{i2\theta_{jk}}\psi_k^* + \sum_{k\in\langle j'\rangle}\eta_j\psi_k, \tag{7.59}$$

with $j = A, B$ for different sublattices.

It has been proved that only when the time-reversal symmetry or inversion symmetry (or both of them) is broken, can the system support the Weyl semimetal (WSM) phase [77]. For simplicity, in our model we break the sublattice symmetry to obtain the WSM phase. This can be realized by setting the contrast nanodisk radius at two sublattices, i.e. $r_A \neq r_B$ (see figure 7.30(b)). When $|r_A - r_B| \ll (r_A + r_B)/2$, we can safely assume that $\omega_A = \omega_0 - \delta\omega$ and $\omega_B = \omega_0 + \delta\omega$, where ω_A and ω_B are the gyroscopic frequencies of vortices at A and B sites, respectively, and $\delta\omega$ is much smaller than ω_0. By adopting the method of rotating wave approximation (just the same as previous sections), we can map equation (7.59) to the familiar two-band Weyl model which gives a quite accurate band structures of vortex oscillations. It is worth noting that the excitation spectrum of vortices can also be obtained by performing the Bogoliubov transformation to ψ_j and ψ_j^*. Equation (7.59) then can be recast as

$$-\mathrm{i}\dot{\psi}_j = \left(\omega_j - \frac{3\xi^2}{2\omega_0}\right)\psi_j + \zeta \sum_{k\in\langle j\rangle} \psi_k - \frac{\xi^2}{2\omega_0} \sum_{s\in\langle\langle j\rangle\rangle} \mathrm{e}^{\mathrm{i}2\bar{\theta}_{js}}\psi_s + \sum_{k\in\langle j'\rangle} \eta_j\psi_k. \tag{7.60}$$

For an infinite lattice, we can use the rhomboid consisting of $\mathbf{a}_1$ and $\mathbf{a}_2$ as the unit cell for monolayer structure, as shown in figure 7.30(b). Then we obtain the three basis vectors: $\mathbf{a}_1 = \frac{3}{2}\mathrm{d}\hat{x} + \frac{\sqrt{3}}{2}\mathrm{d}\hat{y}$, $\mathbf{a}_2 = \frac{3}{2}\mathrm{d}\hat{x} - \frac{\sqrt{3}}{2}\mathrm{d}\hat{y}$, and $\mathbf{a}_3 = h\hat{z}$. Here the parameters d and h represent the distance between two nearest neighbor vortex cores for intralayer and interlayer coupling, respectively. By using the plane-wave expansion to ψ_j, we obtain the Hamiltonian of system

$$H = h_0\sigma_0 + h_x\sigma_x + h_y\sigma_y + h_z\sigma_z, \tag{7.61}$$

where the parameters h_i ($i = 0, x, y, z$) have explicitly expressions

$$\begin{aligned}
h_0 &= \frac{\omega_A + \omega_B}{2} - \frac{3\xi^2}{2\omega_0} + \frac{\xi^2}{2\omega_0}[2\cos(3k_xd/2)\cos(\sqrt{3}k_yd/2) + \cos(\sqrt{3}k_yd)] \\
&\quad + (\eta_A + \eta_B)\cos(k_zh), \\
h_x &= \zeta[1 + 2\cos(3k_xd/2)\cos(\sqrt{3}k_yd/2)], \\
h_y &= -2\zeta\sin(3k_xd/2)\cos(\sqrt{3}k_yd/2), \\
h_z &= \frac{\omega_A - \omega_B}{2} - \frac{\sqrt{3}\xi^2}{2\omega_0}[-2\cos(3k_xd/2)\sin(\sqrt{3}k_yd/2) + \sin(\sqrt{3}k_yd)] \\
&\quad + (\eta_A - \eta_B)\cos(k_zh).
\end{aligned} \tag{7.62}$$

We then obtain the spectrum of the system

$$\omega_\pm = h_0 \pm \sqrt{h_x^2 + h_y^2 + h_z^2}. \tag{7.63}$$

Equation (7.63) gives the dispersion relations of the vortex lattice. The band gap needs to be closed if we want to achieve the WSM phase, i.e. $h_x = h_y = h_z = 0$ for some specific $\mathbf{k}$. Then one can easily obtain four different points $\mathbf{k}_1^\pm = [0, 4\pi/3\sqrt{3}d, \pm\arccos(g_1)/h]$ and $\mathbf{k}_2^\pm = [0, -4\pi/3\sqrt{3}d, \pm\arccos(g_2)/h]$ in the $\mathbf{k}$

space where the band structures are gapless. Here $g_i = \frac{\omega_B - \omega_A}{2(\eta_A - \eta_B)} + (-1)^i \frac{9\xi^2}{4\omega_0(\eta_A - \eta_B)}$ with $i = 1, 2$. Using the Taylor expansion in the vicinity of $\mathbf{k}_1^\mp$ and $\mathbf{k}_2^\mp$, we arrive at the effective Hamiltonian

$$\begin{aligned}
\mathbf{k}_1^\mp: \mathcal{H}_1^\mp &= \left[\frac{\omega_A + \omega_B}{2} - \frac{9\xi^2}{4\omega_0} + (\eta_A + \eta_B)g_1 \mp (\eta_A + \eta_B)h\sqrt{1 - g_1^2}p_z\right]\sigma_0 - \frac{3}{2}\zeta d p_y \sigma_x \\
&\quad + \frac{3}{2}\zeta d p_x \sigma_y \mp (\eta_A - \eta_B)h\sqrt{1 - g_1^2}p_z\sigma_z, \\
\mathbf{k}_2^\mp: \mathcal{H}_2^\mp &= \left[\frac{\omega_A + \omega_B}{2} - \frac{9\xi^2}{4\omega_0} + (\eta_A + \eta_B)g_2 \mp (\eta_A + \eta_B)h\sqrt{1 - g_2^2}p_z\right]\sigma_0 + \frac{3}{2}\zeta d p_y \sigma_x \\
&\quad + \frac{3}{2}\zeta d p_x \sigma_y \mp (\eta_A - \eta_B)h\sqrt{1 - g_2^2}p_z\sigma_z,
\end{aligned} \tag{7.64}$$

with $\mathbf{p} = \mathbf{k} - \mathbf{k}_i^\mp$ ($i = 1, 2$). By analyzing equations (7.64), we conclude that the two bands of system are linearly touching at $\mathbf{k}_1$ and $\mathbf{k}_2$, which shows a typical characteristic for the band structure of WSMs. In addition, in our model, the Hamiltonian contains an extra tilting term, i.e. $(\eta_A + \eta_B)\cos(k_z h)\sigma_0$, which is absent for conventional Weyl Hamiltonian.

It has been showed that the type of WSM [81–83] is determined by the ratio of the tilting term to the z-term (here the z-term is $(\eta_A - \eta_B)\cos(k_z h)\sigma_z$). For simplicity, we use $b\cos(k_z)\sigma_0$ and $t\cos(k_z)\sigma_z$ as the tilting term and the z-term, respectively. Then the value of b/t dominates the type of WSMs: if $|b/t| < 1$ ($|b/t| > 1$), the WSM phase is type-I (type-II); if $|b/t| = 1$, the WSM phase belongs to type-III. In our model, we can clearly see that $|(\eta_A + \eta_B)/(\eta_A - \eta_B)| > 1$ is always satisfied for the reason that the parameters $\eta_A > 0$ and $\eta_B > 0$ (all vortices have the same polarity). Therefore, the type of WSM phase emerging in this model belongs to type-II. Noticeably, the other type (type-I for instance) WSM also may realize if we consider various different combinations for vortices polarities.

Apparently, the condition $|g_1| < 1$ (or $|g_2| < 1$) must be satisfied if we want to achieve WSM phase. To obtain the complete phase diagram, we calculate the topological invariant Chern number (C). As we know the Chern number has no definition in the 3D momentum space. Therefore, we compute the Chern number in $k_x - k_y$ space when k_z takes different value. If $C = 1$ ($C = 0$) for all k_z, the system is in topological (trivial) insulating phase. The phase diagram can be divided into five parts, with the results shown in figure 7.31(a). We can see that the system support three different WSM phases: (i) one pair of WNs located at $\mathbf{k}_1^\mp$ (blue region, WSM1); (ii) one pair of WNs located at $\mathbf{k}_2^\mp$ (green region, WSM1); and (iii) two pairs of WNs located at $\mathbf{k}_1^\mp$ and $\mathbf{k}_2^\mp$ simultaneously (red region, WSM2). In addition, the purple (gray) region denotes a topological (trivial) insulating phase with Chern number quantized to 1 (0) (see figure 7.31(f)).

The bulk band structure of the system can further show the difference between various phases. Figures 7.31(b)–(e) plot the spectra along the path K'–G'–K'–M–K–G–K in the momentum space (see figure 7.30(c)) for different phases. Here the material parameters for Py are used as follows: the thickness of the nanodisk

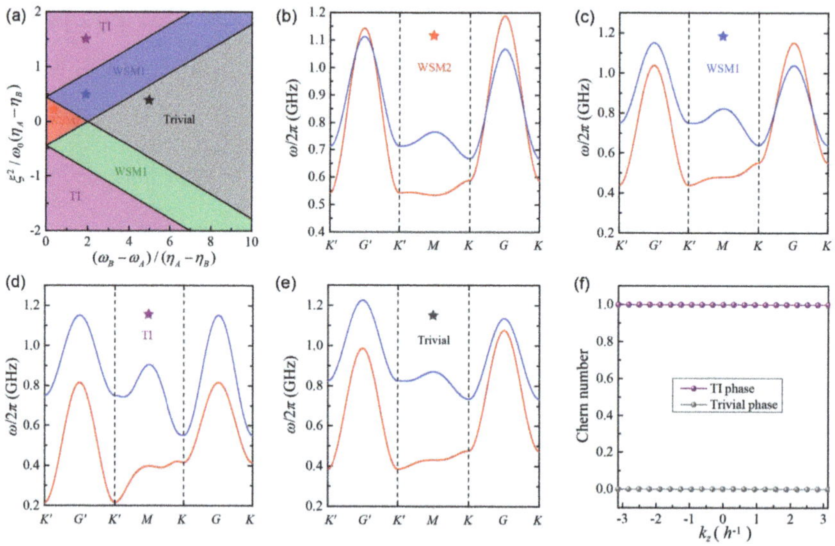

Figure 7.31. (a) The phase diagram of the system. The dispersion curves of the system for different parameters: (b) $(\omega_B - \omega_A)/(\eta_A - \eta_B) = 0.5$, $\xi^2/\omega_0(\eta_A - \eta_B) = 0.2$, (c) $(\omega_B - \omega_A)/(\eta_A - \eta_B) = 2$, $\xi^2/\omega_0(\eta_A - \eta_B) = 0.5$, (d) $(\omega_B - \omega_A)/(\eta_A - \eta_B) = 2$, $\xi^2/\omega_0(\eta_A - \eta_B) = 1.5$, and (e) $(\omega_B - \omega_A)/(\eta_A - \eta_B) = 5$, $\xi^2/\omega_0(\eta_A - \eta_B) = 0.4$. (f) The dependence of topological invariant Chern number on k_z for TI and trivial phases. (Reproduced with permission from [80]. Copyright 2022 The American Physical Society.)

$w = 10$ nm; the gyration frequency $\omega_0 = 0.9 \times 2\pi$ GHz; and the interlayer coupling coefficient $\eta_A = 0.15 \times 2\pi$ GHz, and $\eta_B = 0.1 \times 2\pi$ GHz. For the WSM1 phase (the blue star in figure 7.31(a)), we use $\omega_A = 0.85 \times 2\pi$ GHz, $\omega_B = 0.95 \times 2\pi$ GHz, $I_\parallel = 3.8 \times 10^{-4}$ J m^{-2}, and $I_\perp = 2 \times 10^{-4}$ J m^{-2}. For the WSM2 phase (the red star in figure 7.31(a)), we use $\omega_A = 0.887\,5 \times 2\pi$ GHz, $\omega_B = 0.912\,5 \times 2\pi$ GHz, $I_\parallel = 2.46 \times 10^{-4}$ J m^{-2}, and $I_\perp = 1.2 \times 10^{-4}$ J m^{-2}. For the TI phase (the purple star in figure 7.31(a)), we use $\omega_A = 0.85 \times 2\pi$ GHz, $\omega_B = 0.95 \times 2\pi$ GHz, $I_\parallel = 6.5 \times 10^{-4}$ J m^{-2}, and $I_\perp = 3.5 \times 10^{-4}$ J m^{-2}. For the trivial phase (the gray star in figure 7.31(a)), we use $\omega_A = 0.775 \times 2\pi$ GHz, $\omega_B = 1.025 \times 2\pi$ GHz, $I_\parallel = 3.45 \times 10^{-4}$ J m^{-2}, and $I_\perp = 1.73 \times 10^{-4}$ J m^{-2}. One can clearly see that for the WSM2 phase, the band crossings appear simultaneously at $\mathbf{k}_1^\pm$ and $\mathbf{k}_2^\pm$, while the band crossing only emerges at $\mathbf{k}_1^\pm$ ($\mathbf{k}_2^\pm$) for the blue (green) WSM1 phase. However, the band structures are gapped and no crossings are supported for the TI and trivial phases.

The Berry curvature has pairs of monopoles locating at the WNs is one of the important features for the WSM. To show this characteristic in our model, we plot the distribution of Berry curvature defined as [84–87]

$$\Omega_n^r(\mathbf{k}) = \mathrm{iTr}\left[P_n \epsilon_{rst}\left(\frac{\partial P_n}{\partial k_s}\frac{\partial P_n}{\partial k_t} - \frac{\partial P_n}{\partial k_t}\frac{\partial P_n}{\partial k_s}\right)\right], \tag{7.65}$$

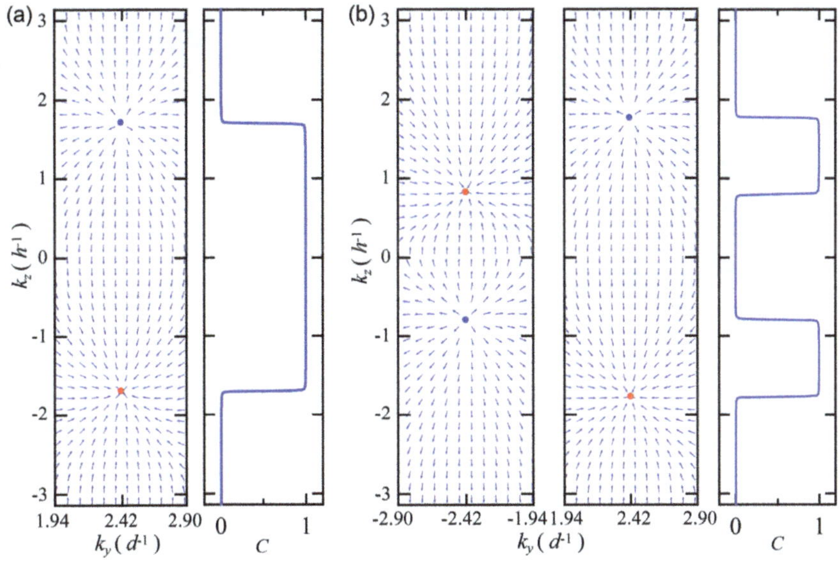

Figure 7.32. The spatial distribution of Berry curvature of the lowest band for the (a) WSM1 and (b) WSM2 phases. The corresponding Chern number varying with k_z is shown in the right panels. (Reproduced with permission from [80]. Copyright 2022 The American Physical Society.)

where ϵ_{rst} is the third-order Levi-Civita symbol and P_n is the projection matrix $P_n(\mathbf{k}) = \phi_n(\mathbf{k})\phi_n(\mathbf{k})^\dagger$ with $\phi_n(\mathbf{k})$ being the normalized eigenstate of (7.61) for nth band. For clarity, we can project the Berry curvature into the k_y–k_z-plane (k_x is fixed to 0). The results are plotted on the left sides of figures 7.32(a) and (b), where we only focus the Berry curvature near the WNs. The blue arrows denote the direction of Berry curvature. One can see that the Berry curvature indeed shows the feature of monopoles which serve as sources or sinks. Moreover, the dependence of the Chern number on k_z is plotted on the right sides of figures 7.32(a) and (b). We see that the Chern number is quantized to 1 when k_z is varied between the pair of WNs and to 0 otherwise, which indicates that the topological insulating phase appears only between the pair of WNs.

Similar to normal metals, WSMs are conducting in the bulk region. However, WSMs can support topologically robust surface states between the pair of WNs. To show this feature, we compute the spectra for a finite system, by adopting periodic boundary conditions along the x- and z-axes with an open boundary condition along the y-axis (armchair surface). We consider a unit cell which is denoted by a blue dashed box and contains 400 nanodisks (see figure 7.30(b)). The corresponding band structures for the WSM1 and WSM2 phases are plotted in figures 7.33(a) and (c), respectively. We also highlight the average position of the wave function along the y-direction $\langle y \rangle \equiv \sum_j R_{j,y}^0 |\mathbf{U}_j|^2 / \sum_j |\mathbf{U}_j|^2$: a color closer to red indicates that the mode is more localized to the surface. Here $R_{j,y}^0$ represents the equilibrium position of

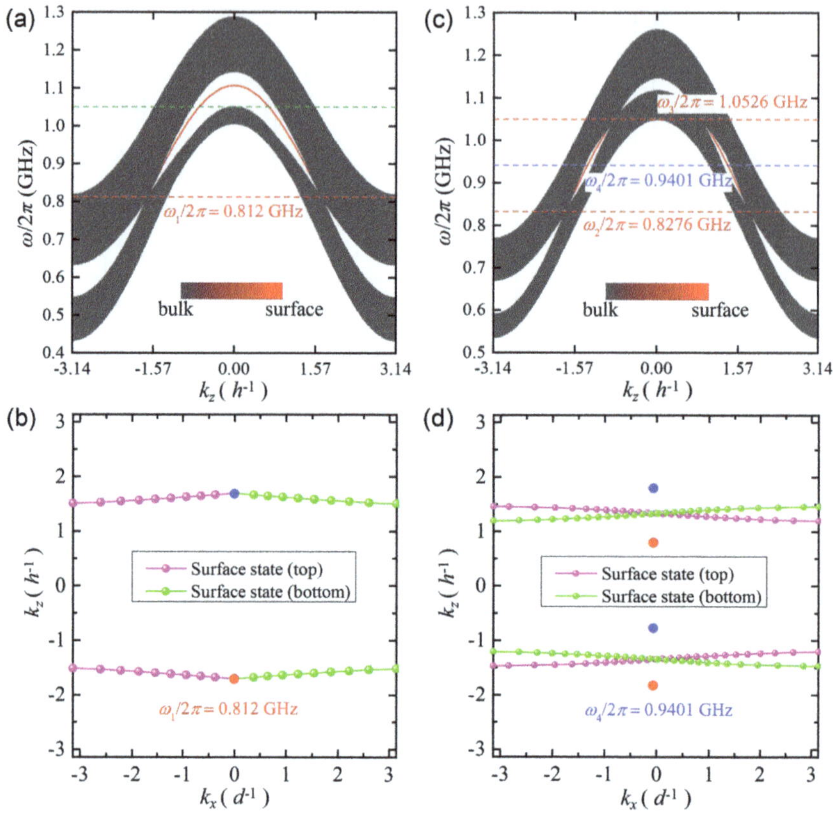

Figure 7.33. The dispersion relations of the semi-infinite system with periodic boundary conditions along the x- and z-axes and the armchair termination along the y-axis for the (a) WSM1 and (c) WSM2 phases. ω_1, ω_2, and ω_3 are frequencies corresponding to the WNs and $\omega_4 = (\omega_2 + \omega_3)/2$. (b) and (d) The isofrequency vortex arcs with frequency $\omega = \omega_1$ and ω_4, as marked in (a) and (c), respectively. (Reproduced with permission from [80]. Copyright 2022 The American Physical Society.)

vortices projected onto the y-axis. Actually, the in-gap modes emerging in figures 7.33(a) and (c) are arc surface states. We will verify this point in what follows.

To visualize the vortex arc, we calculate the isofrequency curves in the k_x–k_z-plane with $\omega = \omega_1$ (for figure 7.33(a)) and $\omega = (\omega_2 + \omega_3)/2$ (for figure 7.33(c)) in figures 7.33(b) and (d), respectively. Here the magenta and green lines denote the surface state localized at the top and bottom of the finite vortex lattice, respectively. Interestingly, the surface states form vortex arcs in a dual-segment style due to the tilted band structure of the type-II WSMs. Here the blue and red spheres represent the WNs with different chiralities. Remarkably, for the WSM1 phase, if the frequency we chose is not located at the WNs, the two vortex arcs will gradually move close to each other with the increase of frequency. When the frequency exceeds a critical value (the green dotted line in figure 7.33(a)), the two arcs will touch each other. While for the WSM2 phase, no matter which frequency we choose between the WNs, there are always bulk states existing between the surface states, the two arcs therefore can never touch each other.

Although both the type-I and type-II WSMs can support specific arcs, the corresponding shapes are markedly different. For type-I WSMs, since there are no bulk states existing between the WNs, the topologically protected surface states can connect the pairs of WNs and form complete arcs. While for type-II WSMs, because the Weyl cone is strongly tilted and Lorentz-violating, bulk states therefore emerge inevitably between the WNs. As a result, the surface states cannot connect the pairs of WNs and form partitioned arcs in a dual-segment fashion (see figure 7.33). For a conventional WSM, i.e. type-I WSM, the system keeps the Lorentz invariance and the Weyl cone has linear dispersion along all **k**-directions. However, the band crossings in crystals are maintained by the symmetries of space groups instead of the Lorentz invariance. Therefore, it is possible to find other types of band crossings beyond the type-I WSM [81]. For example, when we consider the Hamiltonian containing a tilting term, the Weyl cone becomes tilted. If the degree of tilting is large enough, the Fermi velocity will have same sign for different bands along k_z-directions. Then the WSM phase evolves to type-II. For the vortex lattice, because the interlayer coupling has the form of $\mu_j \mathbf{U}_j \cdot \mathbf{U}_k$ it leads to a tilting term $(\eta_A + \eta_B)\cos(k_z h)\sigma_0$. Meanwhile, one can see that the tilting term is always greater than the z-term $(\eta_A - \eta_B)\cos(k_z h)\sigma_z$; the WSM phase discussed in this section thus belongs to type-II.

For the vortex based WSM, we can use the following processes to experimentally detect WNs. First, by utilizing electron-beam lithography [88, 89] or x-ray illumination [90], we can fabricate the magnetic vortices in nanodisks. Then we arrange the nanodisks in a 3D stacked honeycomb lattice. Next, we place a microstrip antenna onto a nanodisk to stimulate the oscillations. After that, an excitation signal is applied to this antenna (a sinusoidal wave sweeps from 0.2 to 2 GHz). Then we record the positions of vortices for different times. It is noted that the nanometer-scale vortex orbits can be tracked by using a biased conductive scanning nanoscale tip [91]. Finally, the dispersion relation can be obtained from the Fourier analysis of the electrical signals. When we focus on the specific points, i.e. $\mathbf{k}_1^{\pm}$ or $\mathbf{k}_2^{\pm}$, linear crossings can be detected.

Because of the topological robustness of the surface arc states, the Weyl excitation in vortex lattices can be used to steadily transmit information. In addition, we can predict a 'wormhole' tunneling by utilizing the Weyl nodes which should pave the route for guiding the wave propagation without a real waveguide.

References

[1] Haldane F D M 1988 *Phys. Rev. Lett.* **61** 2015
[2] Kim S K and Tserkovnyak Y 2017 *Phys. Rev. Lett.* **119** 077204
[3] Li Z X *et al* 2018 *Phys. Rev.* B **98** 180407(R)
[4] Su W P, Schrieffer J R and Heeger A J 1979 *Phys. Rev. Lett.* **42** 1698
[5] Li Z X *et al* 2021 *Phys. Rev.* B **103** 054438
[6] Go G *et al* 2020 *Phys. Rev.* B **101** 134423

[7] Li Z X *et al* 2019 *npj Comput. Mater.* **5** 107
[8] Li Z X *et al* 2020 *Phys. Rev. Appl.* **13** 064058
[9] Li Z X *et al* 2020 *Phys. Rev.* B **101** 184404
[10] Li Z X *et al* 2021 *Phys. Rev.* B **103** 214442
[11] Zhang S and Li Z 2004 *Phys. Rev. Lett.* **93** 127204
[12] Thiaville A *et al* 2005 *Europhys. Lett.* **69** 990
[13] Schryer N L and Walker L R 1974 *J. Appl. Phys.* **45** 5406
[14] Li Z and Zhang S 2004 *Phys. Rev.* B **70** 024417
[15] Pivano A and Dolocan V O 2020 *Phys. Rev.* B **101** 014438
[16] Aharoni A 1998 *J. Appl. Phys.* **83** 3432
[17] Schindler F 2020 *J. Appl. Phys.* **128** 221102
[18] Zak J 1989 *Phys. Rev. Lett.* **62** 2747
[19] Meier G *et al* 2007 *Phys. Rev. Lett.* **98** 187202
[20] Moriya R *et al* 2008 *Nat. Phys.* **4** 368–72
[21] Thomas L *et al* 2006 *Nature* **443** 197–200
[22] Heyne L *et al* 2010 *Phys. Rev. Lett.* **105** 187203
[23] Jiang W *et al* 2013 *Phys. Rev.* B **87** 014427
[24] Badea R and Berezovsky J 2016 *Phys. Rev. Appl.* **5** 064003
[25] Shibata J, Shigeto K and Otani Y 2003 *Phys. Rev.* B **67** 224404
[26] Shibata J and Otani Y 2004 *Phys. Rev.* B **70** 012404
[27] Kane C L and Mele E J 2005 *Phys. Rev. Lett.* **95** 226801
[28] Makhfudz I, Krüger B and Tchernyshyov O 2012 *Phys. Rev. Lett.* **109** 217201
[29] Schütte C *et al* 2014 *Phys. Rev.* B **90** 174434
[30] Büttner F *et al* 2015 *Nat. Phys.* **11** 225–8
[31] Yang W R *et al* 2018 *Opt. Express* **26** 8778–90
[32] Mertens F G, Schnitzer H J and Bishop A R 1997 *Phys. Rev.* B **56** 2510
[33] Ivanov B A *et al* 2010 *JETP Lett.* **91** 178–82
[34] Cherepov S S *et al* 2012 *Phys. Rev. Lett.* **109** 097204
[35] Tomasello R *et al* 2014 *Sci. Rep.* **4** 6784
[36] Yu X Z *et al* 2010 *Nature* **465** 901–4
[37] Seki S *et al* 2012 *Science* **336** 198–201
[38] Nagaosa N and Tokura Y 2013 *Nat. Nanotechnol.* **8** 899–911
[39] Tamm I 1932 *Z. Phys.* **76** 849–50
[40] Shockley W 1939 *Phys. Rev.* **56** 317
[41] Neto A H C *et al* 2009 *Rev. Mod. Phys.* **81** 109
[42] Yao Y *et al* 2007 *Phys. Rev.* B **75** 041401(R)
[43] Noh J *et al* 2018 *Nat. Photon.* **12** 408–15
[44] Fan H *et al* 2019 *Phys. Rev. Lett.* **122** 204301
[45] Kariyado T, Morimoto T and Hatsugai Y 2018 *Phys. Rev. Lett.* **120** 247202
[46] Hatsugai Y and Maruyama I 2011 *Europhys. Lett.* **95** 20003
[47] Wakao H *et al* 2020 *Phys. Rev.* B **101** 094107
[48] Araki H, Mizoguchi T and Hatsugai Y 2020 *Phys. Rev. Res.* **2** 012009(R)
[49] Benalcazar W A, Bernevig B A and Hughes T L 2017 *Science* **357** 61–6
[50] Benalcazar W A, Bernevig B A and Hughes T L 2017 *Phys. Rev.* B **96** 245115
[51] Yoo M W, Lee J and Kim S K 2012 *Appl. Phys. Lett.* **100** 172413
[52] Velten S *et al* 2017 *Appl. Phys. Lett.* **110** 262406

[53] Ivanov B A *et al* 1998 *Phys. Rev.* B **58** 8464
[54] Yoo M W and Kim S K 2015 *J. Appl. Phys.* **117** 023904
[55] Kammerer M *et al* 2011 *Nat. Commun.* **2** 279
[56] Lee K S *et al* 2011 *J. Appl. Phys.* **110** 113903
[57] Sukhostavets O V, González J and Guslienko K Y 2013 *Phys. Rev.* B **87** 094402
[58] Sinnecker J P *et al* 2014 *J. Appl. Phys.* **115** 203902
[59] Teo J C Y and Kane C L 2010 *Phys. Rev.* B **82** 115120
[60] Yang H H *et al* 2020 *Phys. Rev. Res.* **2** 022028(R)
[61] Song L L *et al* 2020 *Nano Lett.* **20** 7566–71
[62] Bondarenko A V *et al* 2019 *Phys. Rev.* B **99** 054402
[63] Chumak A V, Serga A A and Hillebrands B 2014 *Nat. Commun.* **5** 4700
[64] Kruglyak V V, Demokritov S O and Grundler D 2010 *J. Phys. D: Appl. Phys.* **43** 264001
[65] Chumak A V *et al* 2015 *Nat. Phys.* **11** 453–61
[66] Lenk B *et al* 2011 *Phys. Rep.* **507** 107–36
[67] Serga A A, Chumak A V and Hillebrands B 2010 *J. Phys. D: Appl. Phys.* **43** 264002
[68] Lan J *et al* 2015 *Phys. Rev.* X **5** 041049
[69] Garcia-Sanchez F *et al* 2015 *Phys. Rev. Lett.* **114** 247206
[70] Wagner K *et al* 2016 *Nat. Nanotechnol.* **11** 432–6
[71] Hasan M Z and Kane C L 2010 *Rev. Mod. Phys.* **82** 3045
[72] Qi X L and Zhang S C 2011 *Rev. Mod. Phys.* **83** 1057
[73] Schindler F *et al* 2018 *Sci. Adv.* **4** eaat0346
[74] Xie B *et al* 2021 *Nat. Rev. Phys.* **3** 520–32
[75] Li Z X, Cao Y S and Yan P 2021 *Phys. Rep.* **915** 1–64
[76] Burkov A A 2016 *Nat. Mater.* **15** 1145–8
[77] Armitage N P, Mele E J and Vishwanath A 2018 *Rev. Mod. Phys.* **90** 015001
[78] Ghorashi S A A, Li T and Hughes T L 2020 *Phys. Rev. Lett.* **125** 266804
[79] Wang H X *et al* 2020 *Phys. Rev. Lett.* **125** 146401
[80] Li Z X *et al* 2022 *Phys. Rev. Appl.* **17** 024054
[81] Soluyanov A A *et al* 2015 *Nature* **527** 495–8
[82] Lv B Q, Qian T and Ding H 2021 *Rev. Mod. Phys.* **93** 025002
[83] Wu Y *et al* 2017 *Phys. Rev.* B **96** 024201
[84] Shindou R *et al* 2013 *Phys. Rev.* B **87** 174402
[85] Shindou R *et al* 2013 *Phys. Rev.* B **87** 174427
[86] Wang X S, Brataas A and Troncoso R E 2020 *Phys. Rev. Lett.* **125** 217202
[87] Wang X S and Wang X R 2021 *J. Appl. Phys.* **129** 151101
[88] Behncke C *et al* 2015 *Phys. Rev.* B **91** 224417
[89] Sun L *et al* 2013 *Phys. Rev. Lett.* **110** 167201
[90] Guang Y *et al* 2020 *Nat. Commun.* **11** 949
[91] Yu D *et al* 2021 *Phys. Rev. Appl.* **16** 034032

Chapter 8

Conclusion and outlook

In this book, we systematically developed the theory and introduced the research status of topological insulators (TIs) and topological semimetals (TSMs) in classical magnetic systems. The contents presented in this book should greatly deepen our understanding of topological phases and phase transitions in magnetic systems and provide important references for designing robust topological spintronic devices in the future.

In fermionic systems (electrons for instance), one can easily identify the topological property of the fermions through linear transport measurements [1, 2]. However, in a magnetic system, magnons belong to bosons which have a simple condensate or vacuum ground state when they are in topological phases [3]. Therefore, describing the topological property of magnons generally becomes difficult. Fortunately, when the system is in the excited state, the band structures of magnons can carry signatures (for example thermal Hall conductivity) for judging whether the magnetic system is in topological phases.

Generally speaking, topological magnon insulators emerge in gapped systems and the robust edge modes are localized at the boundaries or corners, while the topological magnon semimetals appear in gapless systems and the corresponding magnon arcs are composed of massive topologically protected surface states. The excitation and detection of magnon topological edge states can be realized by utilizing the same techniques adopted in ordinary magnonic devices. For instance, microwave antennae and Brillouin light-scattering spectroscopy can be used to excite and to detect the topological edge magnons [4–6], respectively. It is noted that Bonetti *et al* [7] showed that with the help of high-sensitivity time-resolved magnetic x-ray microscopy, one can obtain a real-space magnon propagation image, which can be conveniently utilized to probe topological magnons. From the perspective of application, using topological magnons, one can design various topological spintronic devices, such as magnon beam splitters, magnon diodes, magnon interferometers, etc.

doi:10.1088/978-0-7503-5082-2ch8

In many cases, the topological properties of magnons and electrons are similar. However, magnons also exhibit a lot of distinctive features. First, the magnon number is not a conserved quantity, thus the robustness of magnon transport is more important than that in electronic systems. Second, magnons are bosons, and the corresponding spectra may be described by a non-Hermitian Hamiltonian [8, 9], which is totally different from the Hermitian Hamiltonian for electrons. For example, to completely characterize the topological properties of Bogoliubov–de Gennes magnons, different invariants are required. Finally, magnons obey the Bose–Einstein statistics [10, 11], therefore their propagation characteristic is quite different from electrons which follow the Fermi–Dirac statistics. The novel superfluidity state may appear in magnon systems with appropriate conditions. Thus one may envision topological magnon superfluidity.

The emerging second-order topological edge states (corner states) in magnetic soliton lattices can be utilized to design different topological spintronic devices. For applications, it is still a challenge to realize a stable display function in various materials. The main obstacle originates from the difficulty of precisely controlling both the frequency and the position of local oscillations. Fortunately, the TIs provide a new route to solve this issue. The chiral edge modes emerging in TIs can perfectly confine the oscillation to the boundaries of the system. If the boundaries are designed to a desired shape, one can easily obtain the display function when the topological edge modes are excited by external sources. However, this method has some disadvantages. On the one side, to present a clear image, one needs to consider a very large system. On the other side, the target image must be continuous and it is difficult to display discrete ones. Remarkably, the emerging HOTIs can thoroughly solve these issues. Zhang *et al* [12] showed a programmable imaging device by using the acoustic corner states emerging in second-order TIs, as plotted in figures 8.1(a)–(e). The imaging device is composed of two digital elements of '0' and '1', which present the trivial and second-order topological phases, respectively. Figure 8.1(a) shows a schematic diagram of a heart-like imaging device with red dots representing the positions of the corners. When the corner states are excited, both the simulated and experimental results identify that most oscillations are localized at the corners, an acoustic imaging device is therefore realized, as shown in figures 8.1(b) and (c). Similarly, the acoustic imaging of characters can also be obtained, as shown in figures 8.1(d) and (e). For magnetic soliton crystals, Li *et al* [13] proposed an imaging device based on a two-dimensional breathing vortex lattice, as plotted in figures 8.1(f) and (g). It is worth noting that these imaging devices are topologically protected and robust against external disturbances. We therefore expect that the magnetic HOTIs have great potential for designing topologically protected spintronic imaging devices. In recent years, the emerging nano-oscillators in magnetic systems have attracted growing attention [14–17]. In particular, because magnetic solitons have the outstanding features of small size, easy manipulation, and high tunability, oscillators based on magnetic solitons [18–23] can be used to design advanced microwave sources. However, the working frequency of these nano-oscillators is very sensitive to external disturbances, which hinders their application. If the topological corner states are considered, the magnetic soliton-based nano-oscillators should have extraordinary

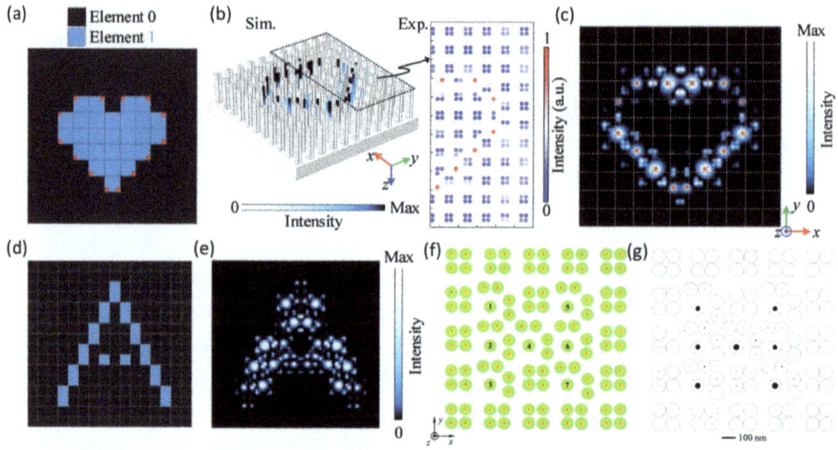

Figure 8.1. (a) A programmable imaging device based on an acoustic second-order TI. Here the heart-shaped region is in nontrivial phase (elements 1) while the surrounding regions are in the trivial phases (elements 0). (b) The simulated (left panel) and experimental (right panel) results of the sound intensity field distributions considered in (a). (c) The distribution of the sound intensity in the x–y-plane at a distance of 0.1 mm above the holey concentric square crystal surface. The red dots indicate the corners as marked in (a). (d) The coding lattices and (e) corresponding distributions of sound intensity for the image 'A'. (f) A schematic diagram of the 'H' display based on a magnetic-vortex lattice. (g) The oscillation distribution of the vortex lattice when the corner state is excited. ((a)-(e) Reproduced with permission from [12]. Copyright 2019 the Authors. (f), (g) Reproduced with permission from [13]. Copyright 2020 The American Physical Society.)

stability immune from defects and disorder. As a result, one can realize the so-called topological microwave sources. The multiband feature of the corner states (ranging from 1 GHz to dozens of gigahertz) is helpful for designing broadband topological spintronic devices.

For experimental realization of magnetic soliton-based higher-order TIs, we have observed that both fabricating the magnetic soliton lattice in a desired manner and detecting the localized corner modes are already within the reach of current technology. On the one side, by utilizing electron-beam lithography [24–26] or x-ray illumination [27], one can obtain various artificial magnetic soliton lattices. On the other side, by using the ultrafast Lorentz microscopy technique [28], the nanometer-scale vortex orbits can be tracked precisely. The topological corner states emerging in magnetic soliton lattices can therefore be observed directly. Figure 8.2 plots a schematic diagram of time-resolved Lorentz microscopy and the corresponding experimental measurements for tracking the vortex core. In addition, the detection of edge states in magnetic soliton crystals can be realized through the interaction between edge and bulk states, which is analogous to the nonlinear three-magnon processes [29]. Figure 8.3(a) plots a sketch map of nonlinear three-magnon processes in the magnetic nanostrip with DM interaction, where a propagating magnon (ω_i, $\mathbf{k}_i$) interacts with the localized magnon (ω_b, $\mathbf{k}_b$) confined in the nanostrip. Generally speaking, there exist two different three-magnon processes, i.e. confluence and splitting, as shown in figure 8.3(a). By analyzing the difference

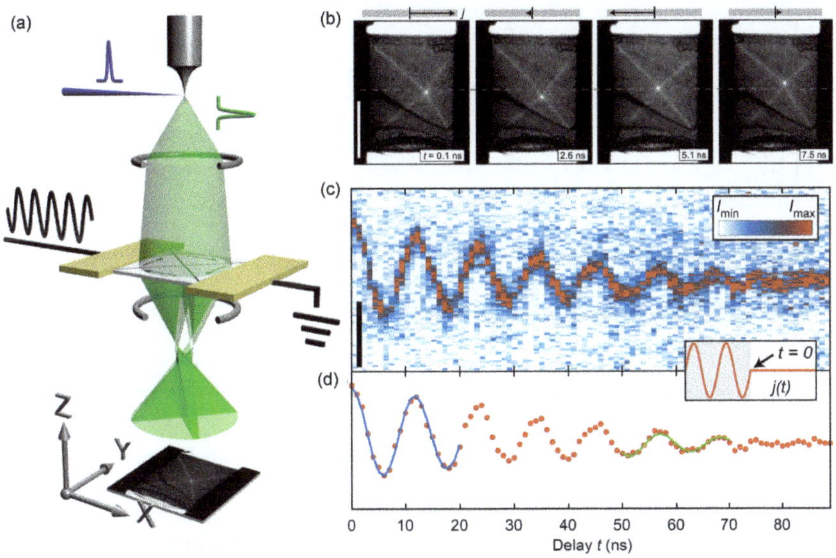

Figure 8.2. (a) A schematic diagram of time-resolved Lorentz microscopy combined with synchronous radiofrequency sample excitation. The electron pulse (green curve) is created from an optical laser pulse (blue curve) through the photoemission process. The magnetic sample (light gray) is stroboscopically illuminated by using a nearparallel electron beam, then the corresponding image is obtained with the help of a transmission electron microscope camera. The dynamics of the vortex are excited *in situ* by radiofrequency currents (black curve) phase-locked to multiples of the laser repetition rate, then a static picture is created. (b) Time-resolved Lorentz micrographs of samples with time steps of 2.5 ns at a defocus of 520 μm. (c) The measured positions of the vortex core along the x-axis varying with time (scale bar: 200 nm). (d) The dependence of the tracked x-position of the vortex core on time. The red dots denote the experimental data, while the blue and green curves represent the formula fitting to the vortex trajectory. (Reproduced with permission from [28]. Copyright 2020 the Authors.)

between $\mathbf{k}_i$ and $\mathbf{k}$ (or $\mathbf{k}_2$), we can obtain information on the bounded state magnon. Following this idea, if we simultaneously excite a bulk state (ω_2, $\mathbf{k}_2$) and an edge state (ω_1, $\mathbf{k}_1$) in the magnetic soliton lattice, by detecting the reflected bulk state (ω, $\mathbf{k}$), one can obtain information on the edge state by using the energy–momentum conservation law $\omega = \omega_2 + \omega_1$ and $(\mathbf{k}_2 + \mathbf{k}_1 - \mathbf{k}) \cdot \hat{x} = 0$, and $\omega = \omega_2 - \omega_1$ and $(\mathbf{k}_2 - \mathbf{k}_1 - \mathbf{k}) \cdot \hat{x} = 0$ for the three-wave process of confluence and splitting, respectively, as shown in figure 8.3(b). In a word, because of the experimental feasibility, the magnetic soliton lattice has provided a unique playground to study the fundamental topological phases, including FOTIs, HOTIs, TSMs, etc. Figure 8.3(c) plots schematically the typical characteristics of various topological phases in magnetic solitonic system.

Finally, we point out that the research on the topological properties in magnon and magnetic soliton systems is quite active, and a lot of open questions and novel phenomena are waiting to be solved and discovered:

1. In this book, for the calculations of HOTIs based on magnetic solitons, we assume that the nanodisks are identical. However, if the translational

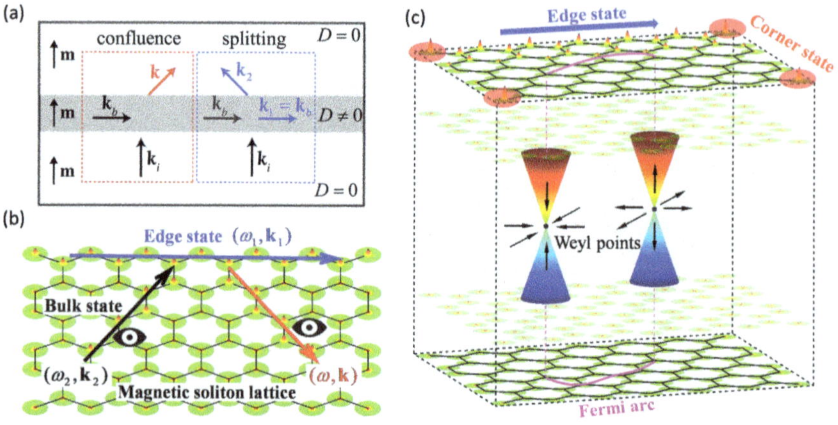

Figure 8.3. (a) A schematic diagram of a nonlinear three-magnon processes in the magnetic nanostrip where DM interaction is considered. The dashed red and blue squares plot the three-magnon confluence ($\mathbf{k}_i$ and $\mathbf{k}_b$ into $\mathbf{k}$) and splitting ($\mathbf{k}_i$ into $\mathbf{k}_1$ and $\mathbf{k}_2$) processes, respectively. (b) A sketch map for probing the topological edge state in a magnetic soliton array. (c) Illustration of typical characteristics of different topological phases: the edge state in FOTIs, the corner state in HOTIs, and the Fermi arc in Weyl SMs. ((a) Reproduced with permission from [29]. Copyright 2018 The American Physical Society.)

symmetry of the magnetic soliton lattice is broken (for example by introducing Kekulé distortions to the disk sizes), we may realize topologically protected Majorana-like zero modes confined at the geometric center of the lattice [30]. In addition, the influence of the chirality and polarity of magnetic solitons on topological phases is also an interesting problem.

2. The twisted bilayer graphene structure (moiré superlattice) has attracted growing attention owing to its novel physical properties [31, 32]. Very recently, a topological magnon band structure based on a moiré superlattice has been reported [33]. It is naturally expected that the twisted bilayer of magnetic soliton lattices may show exotic topological properties, which is an appealing direction for future study.
3. The practical applications of topological phases (including FOTIs and HOTIs) in magnetic systems are still lacking. Designing various topological spintronic devices based on magnons and magnetic solitons is worth further study.
4. It is well known that the LLG equation is intrinsically nonlinear. When the oscillation strength of the magnetization or magnetic soliton is large enough, the emerging nonlinear effect should be considered. The influence of the nonlinear effect on the topological properties of magnons and magnetic solitons is an interesting topic to investigate.
5. Although the topological phases of conventional magnons have been researched extensively, reports of topologically twisted magnons are still rare, which is an appealing topic for study.
6. The essential physics in magnetic soliton insulators and semimetals can be tested by topolectric circuit experiments, which is quite an active research area at present.

We encourage interested readers and young researchers to enter the realm of topological magnetism and to explore the fascinating physics and applications in this field.

References

[1] Hasan M Z and Kane C L 2010 *Rev. Mod. Phys.* **82** 3045
[2] Qi X L and Zhang S C 2011 *Rev. Mod. Phys.* **83** 1057
[3] Vishwanath A and Senthil T 2013 *Phys. Rev.* X **3** 011016
[4] Kruglyak V V, Demokritov S O and Grundler D 2010 *J. Phys. D: Appl. Phys.* **43** 264001
[5] Serga A A, Chumak A V and Hillebrands B 2010 *J. Phys. D: Appl. Phys.* **43** 264002
[6] Demokritov S O, Hillebrands B and Slavin A N 2001 *Phys. Rep.* **348** 441–89
[7] Bonetti S *et al* 2015 *Nat. Commun.* **6** 8889
[8] Akagi Y 2020 *J. Phys. Soc. Jpn.* **89** 123601
[9] Kondo H, Akagi Y and Katsura H 2020 *Prog. Theor. Exp. Phys.* **2020** 12A104
[10] Demokritov S O *et al* 2006 *Nature* **443** 430–3
[11] Demidov V E *et al* 2008 *Phys. Rev. Lett.* **100** 047205
[12] Zhang Z *et al* 2019 *Adv. Mater.* **31** 1904682
[13] Li Z X *et al* 2020 *Phys. Rev.* B **101** 184404
[14] Kaka S *et al* 2005 *Nature* **437** 389–92
[15] Awad A A *et al* 2017 *Nat. Phys.* **13** 292–9
[16] Demidov V E, Urazhdin S and Demokritov S O 2010 *Nat. Mater.* **9** 984–8
[17] Zeng Z, Finocchio G and Jiang H 2013 *Nanoscale* **5** 2219–31
[18] Pribiag V S *et al* 2007 *Nat. Phys.* **3** 498–503
[19] Zhang S *et al* 2015 *New J. Phys.* **17** 023061
[20] Hrkac G *et al* 2015 *J. Phys. D: Appl. Phys.* **48** 453001
[21] Hamadeh A *et al* 2014 *Phys. Rev. Lett.* **112** 257201
[22] Dussaux A *et al* 2010 *Nat. Commun.* **1** 8
[23] Martinez E, Torres L and Lopez-Diaz L 2011 *Phys. Rev.* B **83** 174444
[24] Behncke C *et al* 2015 *Phys. Rev.* B **91** 224417
[25] Han D S *et al* 2013 *Sci. Rep.* **3** 2262
[26] Sun L *et al* 2013 *Phys. Rev. Lett.* **110** 167201
[27] Guang Y *et al* 2020 *Nat. Commun.* **11** 949
[28] Möller M *et al* 2020 *Commun. Phys.* **3** 36
[29] Wang Z Y *et al* 2018 *Phys. Rev. Appl.* **10** 054018
[30] Gao P *et al* 2019 *Phys. Rev. Lett.* **123** 196601
[31] Cao Y *et al* 2018 *Nature* **556** 43–50
[32] Cao Y *et al* 2018 *Nature* **556** 80–4
[33] Hua C B *et al* 2023 *Phys. Rev.* B **107** L020404

www.ingramcontent.com/pod-product-compliance
Ingram Content Group UK Ltd.
Pitfield, Milton Keynes, MK11 3LW, UK
UKHW051853100926
8556IPUK00005B/614